Glenda CHIDRAWI Stephanie HOLLIS

MW00911456

preliminary course

BIOLOGY
IN FOCUS

NELSON
CENGAGE Learning™

Australia • Brazil • Japan • Korea • Mexico • Singapore • Spain • United Kingdom • United States

Biology in Focus
Preliminary Course
1st Edition
Glenda Chidrawi
Stephanie Hollis

Acquisitions editor: Libby Houston
Managing editor: Jo Munnelly
Production editor: Kathryn Murphy
Editor: Kathryn Murphy
Rights & permissions manager: Jared Dunn
Editorial assistant: Vida Long
Text and cover design: Jenny Pace Walter
Illustrator: Alan Laver, Shelly Communications
Proofreader: Joy Window
Indexer: Glenda Browne
CD-rom design & preparation: Sunset Digital Pty Ltd, Brisbane
Reprint: Jess Lovell
Typeset in ITC Garamond, Helvetica & Eurostile by Sunset Digital Pty Ltd, Brisbane

Any URLs contained in this publication were checked for currency during the production process. Note, however, that the publisher cannot vouch for the ongoing currency of URLs.

First published in 2008 by McGraw Hill Australia.
Reprinted in 2009 by McGraw Hill Australia.
This edition published in 2010 by Cengage Learning Australia.

Acknowledgements
Additional owners of copyright are acknowledged on the acknowledgements page.

For product information and technology assistance,
in Australia call **1300 790 853**;
in New Zealand call **0800 449 725**

For permission to use material from this text or product, please email
aust.permissions@cengage.com

National Library of Australia Cataloguing-in-Publication Data
Chidrawi, Glenda.
Biology in focus : preliminary course / Glenda Chidrawi, Stephanie Hollis.

9780170226677 (pbk.)
Includes index.
For secondary students doing the NSW stage 6 biology syllabus.

Biology--Textbooks.
Hollis, Stephanie, 1968-

570.71

Cengage Learning Australia
Level 7, 80 Dorcas Street
South Melbourne, Victoria Australia 3205

Cengage Learning New Zealand
Unit 4B Rosedale Office Park
331 Rosedale Road, Albany, North Shore 0632, NZ

For learning solutions, visit **cengage.com.au**

Printed in China by China Translation & Printing Services.
8 9 10 11 12 13 14 20 19 18 17 16

Contents

A LOCAL ECOSYSTEM   1

Chapter 1   Characteristics of ecosystems are determined by biotic and abiotic factors   2

The distribution, diversity and numbers of plants and animals found in ecosystems are determined by biotic and abiotic factors

Chapter 2   Unique aquatic and terrestrial ecosystems   23

Each local aquatic or terrestrial ecosystem is unique

PATTERNS IN NATURE   63

Chapter 1   Cells and the cell theory   66

Organisms are made of cells that have similar structural characteristics

CONTENTS

LIFE ON EARTH — 195

EVOLUTION OF AUSTRALIAN BIOTA — 233

To the student

Biology in Focus: Preliminary Course is written specifically to address the rigorous content of the New South Wales Stage 6 Biology syllabus. This book follows the syllabus in a logical order to ensure that all dot points are covered completely. The first-hand and secondary source investigations from column 3 of the syllabus are dealt with at appropriate points in the text.

One of the features of modern biology is the rate at which new terminology is created. Knowledge of some of these terms is essential to understand the subject. Throughout the text, when important terms are introduced they are in **bold type**. These words are defined in the glossary at the back of the book.

A major feature of the New South Wales Biology syllabus is the use of keywords (see the list of definitions of verbs on page viii) in constructing examination questions. An understanding of how to use these key words is essential for success in the HSC. The revision questions in each chapter are designed to test your command of the keywords as well as your understanding of the content of the course.

There is an emphasis in the text on Prescribed Focus Areas (PFAs) and Biology Skills. For more information on how they are dealt with in the text please refer to pages ix–xii. The text is also supported by a NelsonNet student website that contains lists of relevant website references; student worksheets; extension and classroom activities; and sample answers to end of chapter revision questions. See www.nelsonnet.com.au

When preparing for exams, remember that the syllabus is the ultimate guide to what you need to know. But you can be reassured that *Biology In Focus: Preliminary Course* contains the information you need to complete the course work.

Please note: All resources listed throughout the book as available on the Student Resource CD-ROM can now be found on the NelsonNet student website.

List of Board of Studies verbs

Account	Account for: state reasons for, report on Give an account of: narrate a series of events or transactions
Analyse	Identify components and the relationship among them; draw out and relate implications
Apply	Use, utilise, employ to a particular situation
Appreciate	Make a judgement about the value of
Assess	Make a judgement of value, quality, outcomes, results or size
Calculate	Ascertain/determine from given facts, figures or information
Clarify	Make clear or plain
Classify	Arrange or include in classes/categories
Compare	Show how things are similar or different
Construct	Make; build, put together items or arguments
Contrast	Show how things are different or opposite
Critically (analyse/evaluate)	Add a degree or level of accuracy depth, knowledge and understanding, logic, questioning, reflection and quality to (analysis/evaluation)
Deduce	Draw conclusions
Define	State meaning and identify essential qualities
Demonstrate	Show by example
Describe	Provide characteristics and features
Discuss	Identify issues and provide points for and/or against
Distinguish	Recognise or note/indicate as being distinct or different from; to note differences between
Evaluate	Make a judgement based on criteria; determine the value of
Examine	Inquire into
Explain	Relate cause and effect; make the relationships between things evident; provide why and/or how
Extract	Choose relevant and/or appropriate details
Extrapolate	Infer what is known
Identify	Recognise and name
Interpret	Draw meaning from
Investigate	Plan, inquire into and draw conclusions about
Justify	Support an argument or conclusion
Outline	Sketch in general terms; indicate the main features of
Predict	Suggest what may happen based on available information
Propose	Put forward (for example a point of view, idea, argument, suggestion) for consideration and action
Recall	Present remembered ideas, facts or experiences
Recommend	Provide reasons in favour
Recount	Retell a series of events
Summarise	Express concisely the relevant details

Prescribed Focus Areas—an introduction

Many areas of the Preliminary Biology course lend themselves to the study of the *process of science* by focussing on five Prescribed Focus Areas (PFAs), as outlined in the table below. The application of these PFAs has become an important part of the New South Wales Board of Studies Biology Syllabus. PFAs are targeted for examination questions in both the Preliminary and HSC Biology courses.

Examples of how to apply each of the PFAs 1–5 have been provided in this textbook. Wherever an icon appears in the textbook, it signals that a PFA is being addressed and provides the opportunity for students to analyse course content in relation to a particular PFA and to become skilled at applying their area of learning to the particular PFA.

On the NelsonNet teacher website special reference has been given to 'unpacking' each PFA (breaking the 'dot point' down into its component parts) and to assisting teachers to facilitate students in developing the skills needed to address PFAs. Templates or 'scaffolds' have been provided that simplify the process of applying each PFA and these may be used by teachers and/or students in conjunction with any module of work.

In addition to this, the NelsonNet teacher website contains a table which links specific syllabus areas ('dot points') throughout the Preliminary course with each PFA. See www.nelsonnet.com.au.

Please note: All resources listed throughout the book as available on the Teacher Resource CD-ROM can now be found on the NelsonNet teacher website. Please note that complimentary access to NelsonNet and the NelsonNetBook is only available to teachers who use the accompanying student textbook as a core educational resource in their classroom. Contact your Education Consultant for information about access codes and conditions.

Linking syllabus
'dot points' to PFAs

Table of objectives and outcomes—Prescribed Focus Areas

Objectives	Preliminary Course outcomes
Students will develop knowledge and understanding of:	*A student:*
1 the history of biology	P1 outlines the historical development of major biological principles, concepts and ideas
2 the nature of biology	P2 applies the processes that are used to test and validate models, theories and laws of science, with particular emphasis on first-hand investigations in biology
3 applications and uses of biology	P3 assesses the impact of particular technological advances on understanding in biology
4 implications of biology for society and the environment	P4 describes applications of biology which affect society or the environment
5 current issues, research and developments in biology	P5 describes the scientific principles employed in particular areas of biological research

Biology Skills—an introduction

During the Preliminary Course, it is expected that students will further develop skills in planning and conducting investigations, communicating information and understanding, scientific thinking and problem-solving and working individually and in teams. Each module specifies content through which skill outcomes can be achieved. Teachers should develop activities based on that content to provide students with opportunities to develop the full range of skills.

Preliminary Course outcomes	Content
A student:	*Students will learn to:*
P11 identifies and implements improvements to investigation plans	**11.1 identify data sources to:** a) analyse complex problems to determine appropriate ways in which each aspect may be researched b) determine the type of data which needs to be collected and explain the qualitative or quantitative analysis that will be required for this data to be useful c) identify the orders of magnitude that will be appropriate and uncertainty that may be present in the measurement of data d) identify and use correct units for data that will be collected e) recommend the use of an appropriate technology or strategy for data collection or gathering information that will assist efficient future analysis
	11.2 plan first-hand investigations to: a) demonstrate the use of the terms 'dependent' and 'independent' to describe variables involved in the investigation b) identify variables that need to be kept constant, develop strategies to ensure that these variables are kept constant and demonstrate the use of a control c) design investigations that allow valid and reliable data and information to be collected d) design and trial procedures to undertake investigations and explain why a procedure, a sequence of procedures or repetition of procedures is appropriate e) predict possible issues that may arise during the course of an investigation and identify strategies to address these issues if necessary
	11.3 choose equipment or resources by: a) identifying and/or setting up the most appropriate equipment or combination of equipment needed to undertake the investigation b) carrying out a risk assessment of intended experimental procedures and identifying and addressing potential hazards c) identifying technology that could be used during investigating and determining its suitability and effectiveness for its potential role in the procedure or investigations d) recognising the difference between destructive and non-destructive testing of material and analysing the potentially different results of these two procedures

Preliminary Course outcomes	Content
A student:	*Students will learn to:*
P12 discusses the validity and reliability of data gathered from first-hand investigations and secondary sources	**12.1 perform first-hand investigations by:** a) carrying out the planned procedure, recognising where and when modifications are needed and analysing the effect of these adjustments b) efficiently undertaking the planned procedure to minimise hazards and wastage of resources c) disposing carefully and safely of any waste materials produced during the investigation d) identifying and using safe work practices during investigations
	12.2 gather first-hand information by: a) using appropriate data collection techniques, employing appropriate technologies, including data loggers and sensors b) measuring, observing and recording results in accessible and recognisable forms, carrying out repeat trials as appropriate
	12.3 gather information from secondary sources by: a) accessing information from a range of resources, including popular scientific journals, digital technologies and the Internet b) practising efficient data collection techniques to identify useful information in secondary sources c) extracting information from numerical data in graphs and tables as well as from written and spoken material in all its forms d) summarising and collating information from a range or resources e) identifying practising male and female Australian scientists, the areas in which they are currently working and information about their research
	12.4 process information by: a) assess the accuracy of any measurements and calculations and the relative importance of the data and information gathered b) apply mathematical formulae and concepts c) best illustrate trends and patterns by selecting and using appropriate methods, including computer-assisted analysis d) evaluate the relevance of first-hand and secondary information and data in relation to the area of investigation e) assess the reliability of first-hand and secondary information and data by considering information from various sources f) assess the accuracy of scientific information presented in mass media by comparison with similar information presented in scientific journals
P13 identifies appropriate terminology and reporting styles to communicate information and understanding in biology	**13.1 present information by:** a) selecting and using appropriate text types, or combinations thereof, for oral and written presentations b) selecting and using appropriate media to present data c) selecting and using appropriate formats to acknowledge sources of information d) using symbols and formulae to express relationships and using appropriate units for physical quantities e) using a variety of pictorial representations to show relationships and present information clearly and succinctly f) selecting and drawing appropriate graphs to convey information and relationships clearly and accurately g) identifying situations where use of a curve of best fit is appropriate to present graphical information

Preliminary Course outcomes	Content
A student:	*Students will learn to:*
P14 draws valid conclusions from gathered data and information	**14.1 analyse information to:** a) identify trends, patterns and relationships as well as contradictions in data and information b) justify inferences and conclusions c) identify and explain how data supports or refutes an hypothesis, a prediction or a proposed solution to a problem d) predict outcomes and generate plausible explanations related to the observations e) make and justify generalisations f) use models, including mathematical ones, to explain phenomena and/or make predictions g) use cause and effect relationships to explain phenomena h) identify examples of the interconnectedness of ideas or scientific principles
	14.2 solve problems by: a) identifying and explaining the nature of a problem b) describing and selecting from different strategies those which could be used to solve a problem c) using identified strategies to develop a range of possible solutions to a particular problem d) evaluating the appropriateness of different strategies for solving an identified problem
	14.3 use available evidence to: a) design and produce creative solutions to problems b) propose ideas that demonstrate coherence and logical progression and include correct use of scientific principles and ideas c) apply critical thinking in the consideration of predictions, hypotheses and the results of investigations d) formulate cause and effect relationships
P15 implements strategies to work effectively as an individual or as a member of a team	The Preliminary course further increases students' skills in working individually and in teams.

Acknowledgments and credits

Acknowledgements

Special thanks to our Acquisitions Editor **Libby Houston**, for her guidance, invaluable support and friendship. Libby's enthusiastic response to our sometimes unconventional ideas, as well as her clear, analytical thinking, helped turn our ideas into reality.

We would also like to thank our editor **Kathryn Murphy** for her dedication, effective suggestions and her amazing attention to detail.

I, Glenda, would like to acknowledge with gratitude my teacher, friend and mentor **Mrs Joyce Austoker-Smith**, who encouraged me so many years ago to embark on writing textbooks and mentored me through the early years.

I, Stephanie, would like to acknowledge and extend my thanks for the academic input voluntarily provided by **Dr Greg Hollis** towards aspects of biodiversity throughout the text.

We would both like to extend personal thanks to our families, friends and colleagues for their patience as well as their immeasurable support and encouragement while we were writing this book.

We also thank **Robert Farr** for his professional input and assistance with developing the Prescribed Focus Areas, **Jared Dunn** for his assistance with sourcing of photographs and the illustrator for their valuable artistic contribution. As a result of the combined effort of all, we believe that these resources will ease the workload of teachers and make Biology exciting and more meaningful to students.

Photographs

Cover image: Coloured SEM of diatoms, courtesy of Photolibrary

A Local Ecosystem

Corbis: p. 1; Digital Vision: Fig. 1.3 (left), Fig. 2.29(a, b, c); WH Dunn: Fig. 1.3 (right), Table 1.2 (desert); iStockphoto: Table 1.2 (grassland), Fig. 1.5, Fig. 2.7(a), Fig. 2.9, Fig. 2.20(a), Fig. 2.25; Michael Artup: Table 1.2 (shrubland); Jared Dunn: Table 1.2 (woodland, temperate forest, rainforest), Fig. 2.20(b), Fig. 2.29(d); Glenda Chidrawi: Fig. 1.8, Fig. 2.31; Courtesy of Brandon Bales, Department of Biology and Microbiology, South Dakota State University: Fig. 1.10(h); Newspix: Fig. 1.11; ANT Photo Library: Fig. 2.3, Fig. 2.20(d); David Albrecht: Fig. 2.4(a); Associate Professor Michael Keogh, University of Melbourne: Fig. 2.4(b), Fig. 2.23, Fig. 2.24; Dr Peter Harrison, Southern Cross University: Fig. 2.4(c), Fig. 2.8; Eiko Bron: Fig. 2.6; Big Stock Photo: Fig. 2.7(b), Fig. 2.22, Fig. 2.26, Fig. 2.27(b), Fig. 2.29(e); Photolibrary: Fig. 2.7(c, d); T Itoh and RM Brown Jr, *Planta Journal*, Vol. 160, pp. 372–81, Springer-Verlag, 1984: Fig. 2.10(a); Courtesy of the University of Chicago, model by Tyler Keillor: Fig. 2.19(b); Auscape: Fig. 2.20(c); Clare Barnes: Fig. 2.21; Gerry Marantelli: Fig. 2.27(a); Dr Greg Hollis, Senior Biodiversity Officer, Department of Sustainability and Environment: Fig. 2.30

Patterns in Nature

Corbis: p. 63; Digital Vision: Fig. 1.1 (left); Photodisc: Fig. 1.1 (right); Photolibrary: Fig. 1.3, Fig. 1.5, Fig. 1.8(a); Glenda Chidrawi: Fig. 1.7(a), pp. 98–99; Faculty of Biological Sciences, University of Leeds: Fig. 1.9 (right); Rob Farr: Fig. 1.6, Fig. 1.12(a, b, c), Fig. 1.15(a, b), Fig. 3.15(a), Fig. 3.16(b), Fig. 4.12 (bottom left), Fig. 4.14 (top right); Fig. 4.15 (top left), Fig. 5.6 (centre); Photoresearchers Inc: p. 95, Table 1.5 (centre top, centre bottom); Visuals Unlimited: p. 96, Table 1.5 (centre bottom); Professor Adrienne Hardham, Australian National University: p. 97, Table 1.5 (centre bottom); Artville: Fig. 3.6; Professor J Pickett-Heaps, University of Melbourne: Fig. 5.5.

Life on Earth

Corbis: p. 195; Photolibrary: Table 2.2, Fig. 2.2, Fig. 3.1(b), Fig. 3.2, Fig. 4.6; Photodisc: Fig. 2.4(a); Nature Focus: Fig. 2.4(b); Associate Professor Andrew Drinnon, University of Melbourne: Fig. 2.4(c, d); Associate Professor Christina Cheers, University of Melbourne: Fig. 3.4; Vivienne Cassie-Cooper: Fig. 3.5; Australian National Botanic Gardens: Fig. 4.1

Evolution of Australian Biota

Corbis: p. 233; Photolibrary: Fig. 1.4, Fig. 1.16, Fig. 2.1(a), Fig. 2.4, Table 2.4 (bottom left); Fig. 2.12 (top left), Fig. 2.13, Fig. 3.3, Fig. 3.10, Fig. 3.17, Fig. 3.25(a), Fig. 3.27, Fig. 3.29, Fig. 3.34; iStockPhoto: Fig. 1.6(a, d, e), Fig. 1.7(b), Fig. 2.2, Fig. 2.1(b), Fig. 2.8(a, b, c), Fig. 2.10 (b, left), Fig. 2.12 (bottom centre), Fig. 2.14, Fig. 3.4, Fig. 3.19, Fig. 4.6; ANT Photo Library: Fig. 1.6(b), Fig. 2.5(b), Fig. 2.12 (bottom left, bottom right), Fig. 3.6, Fig. 3.7, Fig. 3.18, Fig. 3.22, Fig. 3.30c, Fig. 4.5; Big Stock Photo: Fig. 1.6(c), Fig. 1.7(a), Fig. 2.3(a), Table 2.4 (top left), Fig. 2.5(a), Fig. 2.10(a), Fig. 2.12 (top centre, top right); © Commonwealth of Australia, Geoscience Australia 2007: Fig. 1.14; Dr Greg Kirby, Flinders University: Fig. 2.3(b); IWH Dunn: Fig. 2.8(d); L. Lumsden: Fig. 2.10 (b, right); marinethemes.com/Kelvin Atkinson: Fig. 3.5; Auscape: Fig. 3.8; EYEDEA: Fig. 3.9; Associate Professor Andrew Drinnon, University of Melbourne: Fig. 3.13; Leanne Poll: Fig. 3.14(a); Peter Taylor: Fig. 3.14(b, d); Pauline Ladiges: Fig. 3.14(c); Australian National Botanic Gardens: Fig. 3.15, Fig. 3.21, Fig. 3.32, Fig. 3.33; Greg Jordan, University of Tasmania: Fig. 3.16; Bab and Bert Wells: Fig. 3.20; A Flowers and L Newman: Fig. 3.25(b); Jared Dunn: Fig. 3.26; Dr Greg Hollis, Senior Biodiversity Officer, Department of Sustainability and Environment: Fig. 4.8

Tables and illustrations

A Local Ecosystem

Adapted from Charles J Krebs, *Ecology: The Experimental Analysis of Distribution and Abundance*, Fig. 13.9, p. 246 (Adison Wesley Longman): Fig. 2.2; Courtesy of Ecological Society of America, Washington DC: Fig. 2.5

Patterns in Nature

Thomas, *Biology: A Functional Approach*, third edition (Nelson & Sons, 1971): Fig. 4.5; Raven & Johnson, *Biology*, fourth edition (McGraw-Hill Irwin, 1995): Fig. 5.1, Fig. 5.3

Evolution of Australian Biota

White M, Australia's Prehistoric Plants, (Methuen Australia, 1984): Fig. 1.3; Raven & Johnson, *Biology*, fourth edition (McGraw-Hill Irwin, 1995): Fig. 3.23, Fig. 4.1, Fig. 4.2, Table 4.1, Table 4.2

A **LOCAL** ECOSYSTEM

Characteristics of ecosystems are determined by biotic and abiotic factors

The distribution, diversity and numbers of plants and animals found in ecosystems are determined by biotic and abiotic factors

Matching terms and definitions exercise

Ecology

Ecology is the study of the **distribution** and **abundance** of living organisms and how these properties are affected by interactions between the organisms and their environment.

There are a few important terms that need to be defined prior to understanding the study of ecology (see Table 1.1). These terms will be regularly used throughout this chapter so it is essential to know them well before proceeding.

Term	Definition
Abiotic	Non-living features—physical and chemical factors (e.g. temperature, rainfall, salinity)
Aquatic environment	An environment existing mainly in water: freshwater, saltwater or both
Biome	Large regional system characterised by major vegetation type (e.g. desert); region of earth with similar ecosystems grouped together
Biosphere	The part of the earth and atmosphere in which living organisms are found
Biotic	Living features—all living things (e.g. numbers, distribution, interactions)
Community	Groups of different populations in an area or habitat
Ecology	Study of the relationships living organisms have with each other and their environment
Ecosystem	A community together with its environment: any environment containing organisms interacting with each other and the non-living parts of the environment (e.g. rainforest, freshwater pond)
Environment	Both living (biotic) and non-living (abiotic) surroundings of an organism
Habitat	Place where an organism lives
Niche	Place of a species within a community involving relationships with other species
Organism	Living thing (e.g. plant, animal)
Population	Group of organisms of the same species living in the same area at a particular time
Species	Groups of similar individuals that can reproduce fertile offspring (e.g. kookaburra, snow gum)
Terrestrial environment	An environment existing mainly on land

Table 1.1 Key terms used in ecology and their meanings

Ecology is studied at different levels as illustrated in Figure 1.1.

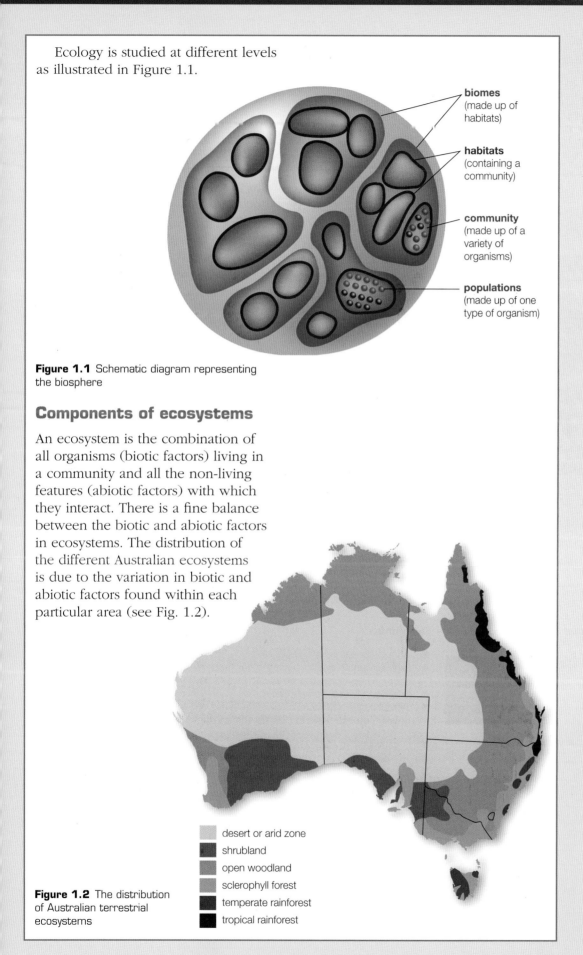

biomes
(made up of habitats)

habitats
(containing a community)

community
(made up of a variety of organisms)

populations
(made up of one type of organism)

Figure 1.1 Schematic diagram representing the biosphere

Components of ecosystems

An ecosystem is the combination of all organisms (biotic factors) living in a community and all the non-living features (abiotic factors) with which they interact. There is a fine balance between the biotic and abiotic factors in ecosystems. The distribution of the different Australian ecosystems is due to the variation in biotic and abiotic factors found within each particular area (see Fig. 1.2).

desert or arid zone
shrubland
open woodland
sclerophyll forest
temperate rainforest
tropical rainforest

Figure 1.2 The distribution of Australian terrestrial ecosystems

1.2

Terrestrial and aquatic environments

Terrestrial environments are those found on land, for example desert or rainforest ecosystems (see Table 1.2). There are two main types of aquatic (water) environments: saltwater or **marine environments** (e.g. coral reefs) and freshwater environments (e.g. lakes) (see Fig. 1.3).

However, some aquatic environments are exposed to both freshwater and saltwater, such as an **estuarine environment** affected by tidal changes. The major types of organisms found in aquatic environments are influenced by the level of water **salinity**.

Figure 1.3 Examples of aquatic environments

Examples of Australian terrestrial and aquatic environments

Table 1.2 Examples of common terrestrial ecosystems found in Australia (ordered from lowest to highest rainfall)

There are many diverse ecosystems found in Australia. Examples of aquatic ecosystems include wetlands and mangrove swamps; rock platforms—bare rock and littoral zones;

estuaries, rivers and lakes; oceans and coral reefs. Common terrestrial ecosystems found in Australia are described in Table 1.2.

Terrestrial ecosystem	Appearance	Description
Desert		■ Annual rainfall is low, <250 mm ■ High temperatures through the day (approx. 40°C) and cold temperatures through the night (approx. 0°C) ■ Often sandy soil, sometimes rocky ■ Typical organisms include sparse grasses and saltbushes; the spinifex hopping mouse; insects, lizards and snakes
Grassland		■ Annual rainfall 250–750 mm ■ Temperature can be hot or mild ■ Typical organisms include grasses (e.g. spinifex), kangaroos, rabbits and snakes
Shrubland		■ Annual rainfall 200–400 mm ■ Temperatures are hot ■ Typical organisms include mallee trees, mulga; kangaroos, rabbits and snakes

continued . . .

Terrestrial ecosystem	Appearance	Description
Woodland		■ Annual rainfall 400–750 mm ■ Temperature can be mild, and sometimes hot ■ Canopy cover 10–30% ■ Typical organisms include grasses, shrubs, eucalypt trees, mice, birds, insects, spiders and wallabies
Temperate forest		■ Annual rainfall >750 mm ■ Temperature is mild ■ Canopy cover 30–70% ■ Typical organisms include eucalypt trees of various types
Rainforest		■ Annual rainfall >1500 mm ■ Air is humid and temperature can be hot or mild ■ Canopy cover is dense (70–100%) and layers (strata) develop (i.e. canopy, understorey, forest floor) ■ Typical organisms include a diverse number of habitats and species (e.g. birds nest ferns, palms, lianas, bracken ferns, leaf litter organisms)

Abiotic factors in aquatic and terrestrial environments

Biotic and abiotic factors differ significantly between aquatic and terrestrial environments. Abiotic factors create various conditions which suit different types of organisms and hence affect biotic factors. First, we must look at the underlying abiotic factors of an environment in order to determine the possible effect that they may have on the living (biotic) component of that environment.

■ *compare the abiotic characteristics of aquatic and terrestrial environments*

Abiotic factors in aquatic and terrestrial environments are described comprehensively in Table 1.3. It must be noted that the abiotic factors of water environments differ depending on whether the water is saltwater or freshwater. For easy reference, variations in some of the abiotic characteristics described in Table 1.3 for a marine (saltwater) environment are illustrated in Figure 1.4. However, in order to approach this dot point, you must first understand the meaning of the verb used (refer to page viii for verb definitions). The verb 'compare' requires a discussion of *both* the similarities and differences in the abiotic characteristics of the two environments. Refer to the information provided above for a description of aquatic and terrestrial environments (or refer to Table 1.1 for definitions).

Table 1.3 Abiotic characteristics of aquatic and terrestrial environments

Abiotic characteristic	Aquatic environment	Terrestrial environment
Temperature	■ Only small temperature changes occur in aquatic environments ■ Temperature changes in water are more gradual ■ Large bodies of water such as the sea are more or less constant in temperature, whereas smaller bodies of water such as lakes are more variable ■ It is much easier for water organisms to adapt to a constant temperature environment than the varying temperatures on land. However, organisms in the water tend to lose heat more rapidly because water conducts heat better than air ■ Water organisms having body temperatures higher than the water must have adaptations to prevent heat loss by conduction	■ Large variations in temperature can occur in terrestrial environments over short periods of time. Daily temperatures may vary up to 20°C in one typical day ■ A desert environment may vary 40°C in one day (0°C at night and 40°C during the day) ■ Seasonal variations between summer and winter are quite significant ■ Different altitudes also demonstrate temperature variation. For every 1000 m above sea level, the average temperature drops by 1°C ■ Land organisms must therefore have adaptations to cope with such large temperature changes
Pressure	■ As the depth of water increases, pressure increases ■ Deep-sea organisms must be adapted to the crushing effects from the pressure of the water above them	■ Only small variations occur in pressure on land. Organisms in environments at sea level are under more pressure than those in environments on high mountains above sea level (high altitude) ■ Small daily fluctuations in pressure may occur due to weather changes
Light availability	■ Water surfaces reflect up to 55% of light reaching depths of 1 m or more. Only 1% of light reaches depths of 100 m or more ■ When the sun is high in the sky more light can be absorbed, whereas at sunset the light strikes the water surface at a more acute angle and less light is absorbed. The angle of the sun during times of the day will vary depending on the time of the year ■ Different coloured wavelengths of sunlight penetrate to different depths of water. Red wavelengths are absorbed first by water and do not penetrate as far as blue or violet wavelengths ■ Cloud cover will affect the intensity and the length of time the light strikes the water. Seasons will affect light availability as does latitude. Extreme latitudes receive 6 months of light and 6 months of darkness, whereas the equator receives roughly 12 hours of sunlight and darkness each day ■ The cloudiness (turbidity) of the water affects the light availability. This depends on the disturbance or movement in the body of water that may disturb the soil, organic material or other small particles or objects in the water ■ Aquatic organisms therefore must have adaptations to the amount of light available to them	■ Light availability is abundant on land ■ Cloud cover may only reduce a small proportion of light availability
Landscape position (slope/aspect)	■ The slope and aspect of the surrounding landscape may affect light availability and the temperature of aquatic environments ■ It may also affect exposure to currents, tides and waves	■ Slope (angle of the slope of the land) and aspect (direction facing, e.g. north) may affect temperature, water and light availability as well as impact on soil quality ■ Runoff and erosion can cause dramatic effects to an environment. However, drainage may be essential in some environments

continued . . .

Abiotic characteristic	Aquatic environment	Terrestrial environment
Gases (O_2 and CO_2)	■ Where air and water are in close contact, gases from the air, like oxygen and carbon dioxide, readily dissolve in the water. The more movement in the water, the more gases dissolve. For example, the rapids of a river would contain more oxygen and carbon dioxide gas than would a still pond ■ Temperature also affects the amount of gases in water. As temperature increases, dissolved gases decrease ■ Aquatic organisms must have adaptations to the variations in gas availability in the water	■ Oxygen and carbon dioxide are found in abundant quantities for terrestrial organisms TR A blank copy of Table 1.3
Rainfall and water availability	■ Water is abundant in aquatic environments but may not be readily available ■ In freshwater environments, water tends to move into organisms freely and this must be removed effectively ■ In marine (saltwater) environments, the opposite happens where water readily moves out of the organism and must therefore be replaced regularly by the organism (for more information see osmosis in 'Patterns in Nature') ■ Aquatic organisms must have adaptations to the type of water environment they exist in, in order to maintain water balance	■ Water is not freely abundant in land environments. It must be sourced from the soil or consumed ■ Organisms must have adaptations to the amount of water available in their environment and the process by which it needs to be sourced
Salinity and ion availability (dissolved salts)	■ In marine environments, most ions are available in abundance (e.g. Na⁺) ■ Ions from decomposing organisms are distributed by currents throughout the different water depths and therefore available to many organisms	■ Dissolved ions are available in soil water and the amount of ions in the soil is called the soil salinity ■ Different soils contain different salinity levels and plants must have adaptations to cope with the different levels
pH (acidity/alkalinity)	■ The pH may vary between different aquatic environments depending on things such as the organic material and dissolved gases available ■ Carbon dioxide can lower the pH of the water, making it more acidic	■ Soil pH can vary significantly between different terrestrial environments ■ Dissolved salts play a large role in determining the pH of the soil ■ Plants must have adaptations to the soil pH of their environment
Buoyancy (the amount of support provided by a medium)	■ Water provides much more buoyancy and support than air. Water may not only hold an organism up but may also maintain its shape (e.g. jellyfish)	■ Air provides a very small amount of buoyancy and support. Land organisms therefore need more supporting skeletal and muscular structures to assist with support
Viscosity (the amount of resistance provided by a medium)	■ Water is more viscous than air ■ Water provides more resistance and it is therefore more difficult for an organism to move through it than air ■ Organisms with more streamlined shapes (i.e. fish) find it easier to move through water than organisms of other shapes	■ Air is less viscous than water ■ Air provides less resistance and it is therefore easier for an organism to move through it than water ■ A human finds it easier to walk through air than to walk through water because the high viscosity of water provides greater resistance against any movement
Exposure to natural forces (wind, tide, waves)	■ Different aquatic environments are exposed to different natural forces ■ Marine environments may have to cope with varying strengths of tides, currents and waves which vary depending on the weather or season ■ Freshwater environments may have to cope with the varying strength of running water (e.g. rivers/waterfalls as opposed to still ponds with no water movement at all)	■ Different terrestrial environments are exposed to different strengths of wind and rain which vary depending upon weather patterns and seasonal changes ■ Exposure can range from the extremes of monsoons and cyclones, through to floods and drought

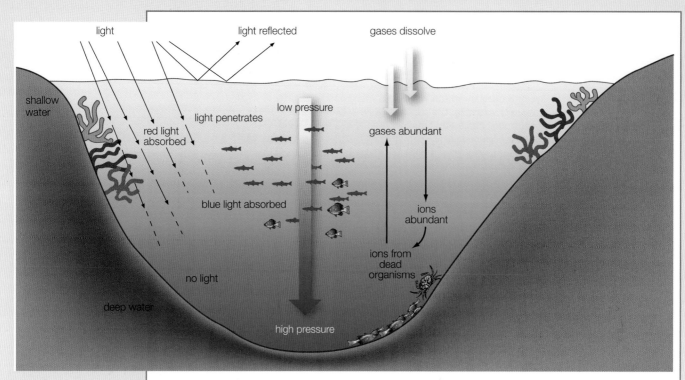

Figure 1.4 Comparing abiotic characteristics at different depths of an aquatic marine environment

Now that you know what the dot point requires, construct a table using Table 1.4 as a guide. Use the background information provided in Table 1.3 and Figure 1.4, and any other sources of information, to complete your table by listing the differences and similarities that exist between aquatic and terrestrial environments.

Table 1.4 Comparison of abiotic characteristics in aquatic and terrestrial environments

	Differences		
	Aquatic environments (e.g. freshwater or marine/saltwater)	**Terrestrial environments** (e.g. rainforest or woodland)	
Abiotic characteristic			**Similarities**
Temperature	Only small changes occur. Usually more gradual changes	Large variations can occur	Temperature is not stable and does change or vary
Pressure			
Light availability			
Landscape position (altitude/slope/aspect)			
Gases (O_2 and CO_2)			
Rainfall and water availability			
Salinity and ion availability (dissolved salts)			

continued . . .

| Abiotic characteristic | Differences | | Similarities |
	Aquatic environments (e.g. freshwater or marine/saltwater)	Terrestrial environments (e.g. rainforest or woodland)	
pH (acidity/alkalinity)			
Buoyancy			
Viscosity			
Space and shelter			
Exposure to natural forces (wind, tide, waves)			

 Blank copy of Table 1.4

 First-hand investigation— abiotic factors

The distribution and abundance of a species

1.3

■ *identify the factors determining the distribution and abundance of a species in each environment*

When an ecosystem is barren and unoccupied, new organisms colonising the environment rely on favourable environmental conditions in the area to allow them to successfully live and reproduce. These environmental conditions are abiotic factors. When a variety of species is present in the ecosystem, the actions of these species can affect the lives of other species in the area; these factors are biotic factors.

The distribution of a species describes where it is found and the abundance of a species determines how many members of that species live throughout the ecosystem. Abiotic and biotic factors affect the distribution and abundance of organisms in an ecosystem by causing fluctuations or changes in populations. Populations may occupy certain areas and not others due to the resources that are available.

Abiotic and biotic factors can affect the abundance and distribution of a species

Abiotic and biotic factors play a very important part in affecting how many species live in a particular area (abundance) and where they may be distributed. In terrestrial environments, abiotic factors such as temperature range, light and water availability most commonly affect a species' abundance and distribution. In aquatic environments, the importance of each abiotic factor differs between the two main types of environments: freshwater and saltwater (see Table 1.5). Of course, an estuarine environment (one exposed to both freshwater and saltwater) must deal with constant changes in the environment.

Table 1.5 The importance of abiotic factors differs between freshwater and marine environments in affecting the distribution and abundance of species

Freshwater environment	Marine (saltwater) environment
1. Temperature variation	1. Salinity
2. Dissolved gases (O_2 and CO_2)	2. Dissolved gases (O_2 and CO_2)
3. pH (acidity) of water	3. Tidal movements and wave action
4. Light availability	
5. Clarity of water	

Just as important as the abiotic factors are the biotic factors that may influence an organism's existence in an ecosystem. Even though there is a much greater variation in biotic factors between ecosystems (e.g. availability of worms as a food source for kookaburras in one ecosystem may be much higher than for kookaburras in another), a few key factors will affect organisms within an ecosystem. Examples of biotic factors that may determine the distribution and abundance of a species:

- availability and abundance of foods
- number of competitors
- number of mates
- number of predators
- number and variety of disease-causing organisms.

Generally, in Australia, rainfall, temperature and landform patterns significantly affect the abundance and distribution of vegetation and ecosystems. As can be seen in Figure 1.2, most of the rainforest ecosystems are distributed along the east coast of Australia, particularly in the northern regions. Abiotic factors such as high temperatures and high rainfall create a suitable **tropical** environment for this type of ecosystem. Desert ecosystems, however, are distributed among the central areas of Australia. Abiotic factors such as a high temperature range and low rainfall (**arid** conditions) create an environment suitable for desert ecosystems. Of course, the distribution

and abundance of organisms within these ecosystems may also vary due to biotic factors such as the availability of food, competition within and between species, the availability of mates for reproduction, exposure to predators, and exposure to disease.

Population ecologists, in order to understand and record environmental changes in plant and animal populations over time, must collect information on the distribution and abundance of organisms in each ecosystem. Ecologists usually wish to determine the size of a population (total number of organisms present); however, it is also useful to know the density of organisms (total number of organisms per unit area) as it is often a reflection of how many organisms a particular environment can support. Ecologists also need to determine the distribution of organisms in order to look at any patterns that are formed and the possible reasons for this. This information enables us to determine whether a population is increasing or decreasing in size and what particular aspects of the habitat are favoured over others. It comes down to two main questions:

- Why is a species only present in particular places?
- What determines the number of individuals (size) of a population in one particular place?

In other words, what determines the distribution and abundance of populations?

Measuring the distribution of a species in an ecosystem

1.4

Types of distribution

Where an organism can be found is the distribution of that organism. Distribution may differ from species to species. Some may randomly distribute themselves around an ecosystem or they may be found distributed in patterns, determined by the resources around them. Some organisms can be found in clumped patterns, others in uniform patterns (see Table 1.6). For example, barnacles clump together for shelter from the waves and to prevent water loss during low tides (see Fig. 1.5).

Table 1.6 Types of distribution of individuals or species in a community

Type of distribution	Example	Possible explanations of distribution
Regular distribution—very rare in nature		Territorial species
Random distribution (no clear pattern exists)		Random distribution of resources
Clumped distribution—possibly the most common pattern reflecting the fact that species need resources that are not distributed uniformly		Patchy distribution of resources

Transects

There are different techniques that may be used to measure the distribution of a species in an ecosystem. In large areas **transects** are commonly used to give an idea of the variation that may occur. A transect is a narrow strip that crosses the entire area being studied, from one side to the other. Transects provide an accurate and easy method of representing an area simply.

Two examples of transects are a **plan sketch** and a **profile sketch**.

A plan sketch is an aerial or surface view of a representative area within an ecosystem. It shows to scale the distribution of organisms in a measured and plotted view (see Fig. 1.6). A profile sketch is a side-on view of an area showing to scale the distribution of organisms along a line (see Fig. 1.7).

Figure 1.5 Clumping distribution pattern made by barnacles to maximise their resources of shelter and water

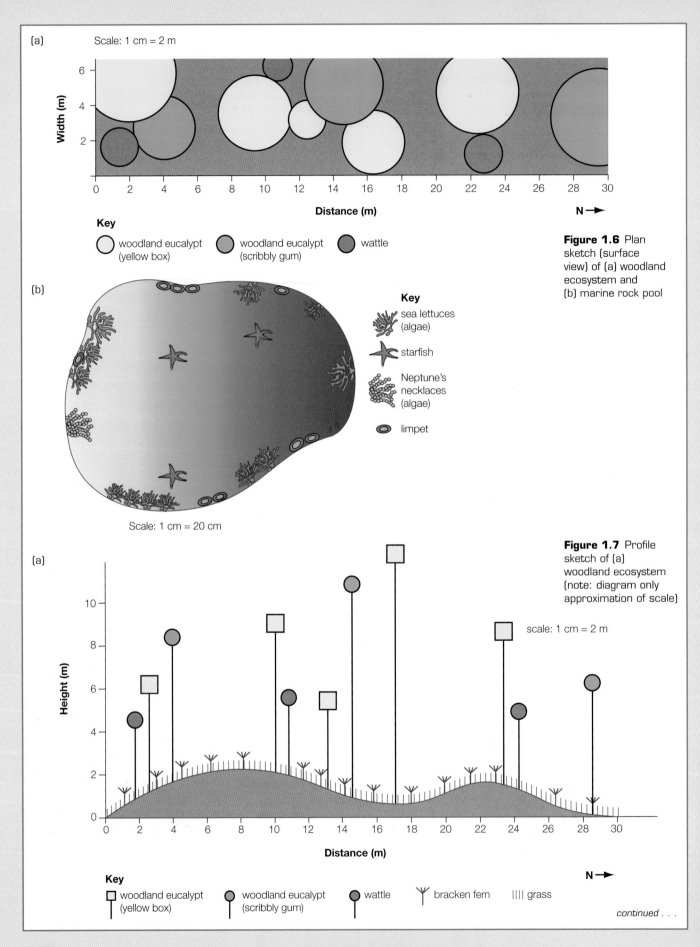

(a) Scale: 1 cm = 2 m

Key

○ woodland eucalypt (yellow box)　● woodland eucalypt (scribbly gum)　● wattle

N →

Figure 1.6 Plan sketch (surface view) of (a) woodland ecosystem and (b) marine rock pool

(b)

Key

🌿 sea lettuces (algae)

★ starfish

🌾 Neptune's necklaces (algae)

◎ limpet

Scale: 1 cm = 20 cm

Figure 1.7 Profile sketch of (a) woodland ecosystem (note: diagram only approximation of scale)

scale: 1 cm = 2 m

(a)

Key

□ woodland eucalypt (yellow box)　● woodland eucalypt (scribbly gum)　● wattle　🌿 bracken fern　||||| grass

N →

continued . . .

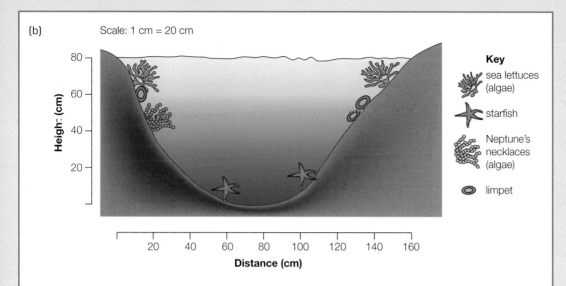

Figure 1.7 Profile sketch of (b) marine rock pool

It is more difficult to assess the distribution of animals due to their daily and seasonal movements. When assessing the animals in an ecosystem we can only observe evidence of their existence in the area. For example, personal sightings, hearing their call, observing their tracks or burrows and traces such as scats (animal faeces) or footprints. Transects can only be used to determine animal distribution for those that hardly move such as barnacles or snails.

Measuring abundance of a species in an ecosystem

1.5

One of the most important features of any population is its size. Population size has a direct bearing on the ability of a given population to survive. The number of individuals of a species found in an area is the abundance of that species.

Plants

It is much easier to calculate the abundance of plant species because they stay in the one place. However, calculating the number of an entire plant species in many cases would be an endless task, so ecologists commonly use sampling techniques to estimate plant species abundance. Usually one or more samples are taken randomly from a population and assumed to be representative of the total population.

There are a few different techniques used to estimate abundance in plants.

The one that is simple and easy to use in the field is the percentage cover method. This method uses **quadrats** (1 m × 1 m squares) to cover randomly-selected representative areas for estimating the percentage cover of an area (see Fig. 1.8). This method is beneficial when plant species are too high in number to count individually.

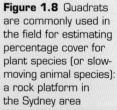

Figure 1.8 Quadrats are commonly used in the field for estimating percentage cover for plant species (or slow-moving animal species): a rock platform in the Sydney area

Percentage cover calculations require randomly plotting a number of quadrats (e.g. ten), estimating the percentage cover for each one and then finding an average percentage cover. If the area of the ecosystem is known or estimated, then the percentage cover can then be converted to area (see example below).

Estimating the abundance of grass

If the school gardener, Mr G, needs to purchase new turf for the football ground, he needs to know how much grass cover the football oval has. To find this out he uses the percentage cover method to estimate the grass cover. Ten 1 m × 1 m quadrats were randomly placed on the oval and the grass cover was drawn to scale and plotted for each one.

A sample of three of the quadrats is shown in Figure 1.9. Back in the office, estimates of percentage cover were made for each of the ten quadrat drawings (see results in Table 1.7), then the percentages were added up and averaged.

Total of the ten quadrats = 540
An average of the percentage of grass cover for the entire oval is calculated.

$$\text{Average \%} = \frac{\text{total\%}}{\text{no. of quadrats}}$$

$$= \frac{540}{10}$$

$$= 54\%$$

Therefore the oval is estimated as having 54% of grass cover.

If the area of the oval is measured at 250 m² then the estimated area of grass cover is 54% × 250 m² = 135 m². Mr G can now safely assume he needs to purchase 115 m² (250 m²–135 m²) of turf to fill the bare areas. He can then repeat the same process the following year to determine if the turf replacement has successfully changed the percentage grass cover of the football oval.

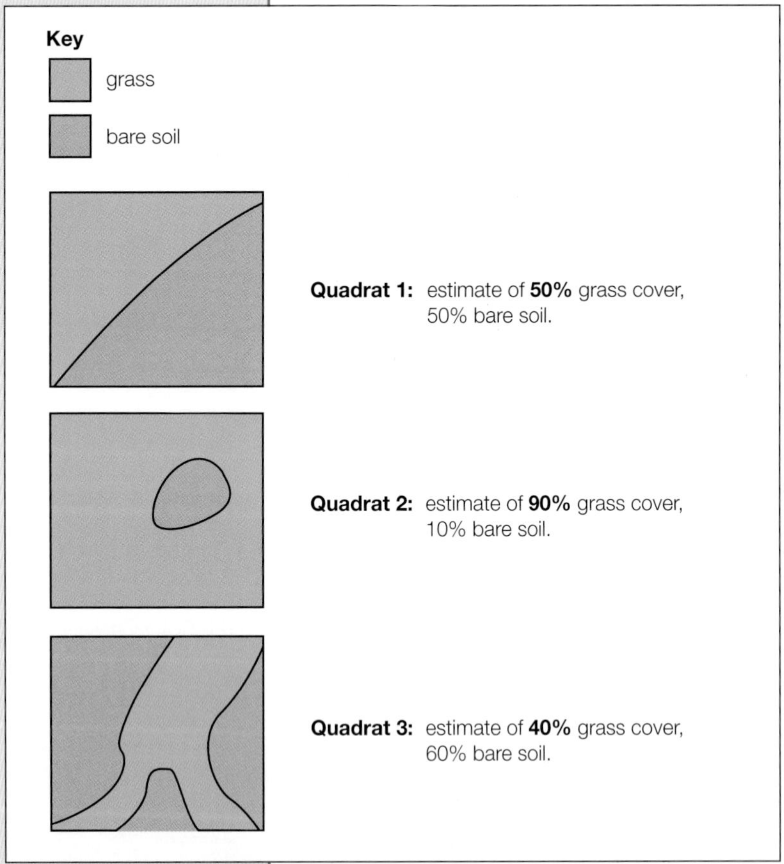

Animals

Obviously, it is a little more difficult to calculate the abundance of animals than plants, and attempting to count every animal species in an area is sometimes very difficult. For those that are slow moving counts can be made; however, for those that move around much more quickly estimates may need to be

Figure 1.9 Calculation of the percentage of grass coverage—sample of three quadrats from the ten recorded

Key
grass
bare soil

Quadrat 1: estimate of **50%** grass cover, 50% bare soil.

Quadrat 2: estimate of **90%** grass cover, 10% bare soil.

Quadrat 3: estimate of **40%** grass cover, 60% bare soil.

Table 1.7 Results of quadrat percentage cover

Quadrat	1	2	3	4	5	6	7	8	9	10	
Estimated percentage cover	50	90	40	60	20	70	90	80	10	30	**Total percentage cover** 540

taken. Hence, estimating abundance is a much easier way of finding out roughly how many animal species exist in an area. This is a little more difficult than the method used for plants as animals may constantly move around or hide. That is why ecologists use a **sampling technique** called the *mark–release–recapture technique*. Animals are captured, the sample animals are tagged then released, these animals are given time to mix again, recaptured and the number tagged in the sample are counted.

Estimating the abundance of an animal sample

The formula for calculating the estimated abundance of animals using the mark–release–recapture method is as follows:

Abundance =
$$\frac{\text{number captured} \times \text{number recaptured}}{\text{number marked in recapture}}$$

Example of the mark–release–recapture technique:
1. *Capture*—a random sample of animals from the population is selected
   —Twenty small birds (Superb Blue Wren) were captured, using bird nets
2. *Mark and release*—marked animals from the first capture are released back into the natural population and left for a period of time to mix with unmarked individuals
   —The 20 birds are tagged with leg bands and released back into their area and left for three weeks to mix with the population
3. *Recapture*—a sample is captured again to look at the proportion of animals marked from the previous sample
   —After three weeks a second sample of ten birds is captured to find four marked birds from the first capture

4. We now insert the following values into the formula:
   number captured = 20
   number recaptured = 10
   number marked in recapture = 4

Abundance =
$$\frac{\text{number captured} \times \text{number recaptured}}{\text{number marked in recapture}}$$

$$\text{Abundance} = \frac{20 \times 10}{4}$$
$$= 50$$

Therefore, we have estimated the total Superb Blue Wren population size in that area as being 50.

The mark–release–recapture technique is based on a number of assumptions for accurate estimates of the total population to be calculated:
1. There is no population change through migration, births or deaths between the sampling periods
2. All animals are equally able to be caught (individuals are not 'trap happy' or 'trap shy')
3. Marked animals are not hampered in their ability to move and mix freely with the rest of the population.

Capturing animals requires various trapping techniques, all designed so that animals are unhurt (e.g. traps, nets and small pits) (see Fig. 1.10). Some techniques avoid the need for recapturing the animals such as radio-tracking and the use of electronic detection devices. Conservationist Steve 'The Crocodile Hunter' Irwin, in his last television project *Ocean's Deadliest*, demonstrates the use of satellite-tracking devices for crocodiles. He is seen having to subdue a giant crocodile in order to attach the device; however, the crocodile remains unharmed.

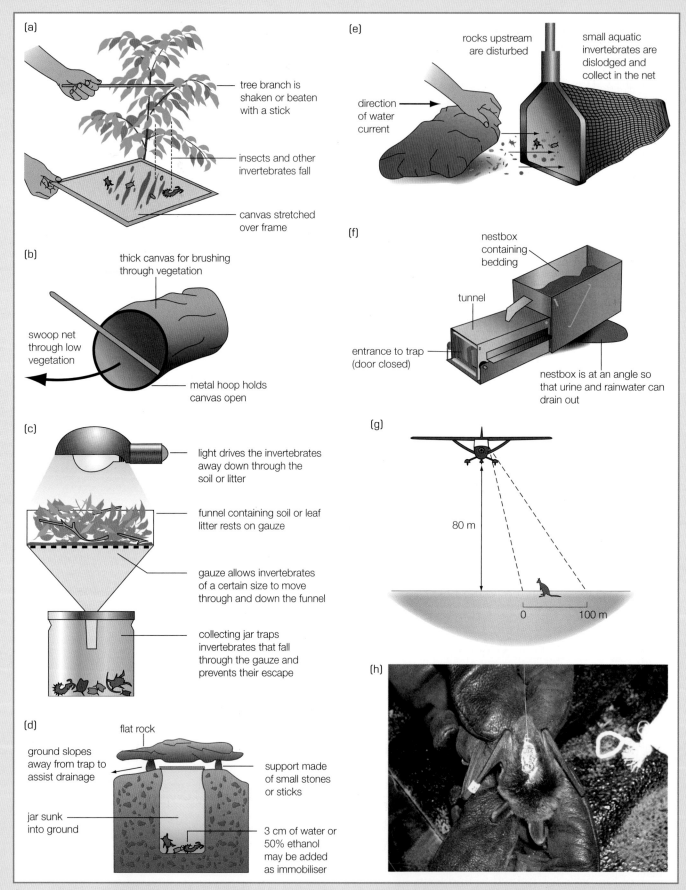

Figure 1.10 Examples of animal sampling techniques (a) beating tray; (b) sweep net; (c) Tullgren funnel; (d) pitfall trap; (e) kick sampling; (f) Longworth small mammal trap; (g) counting from a plane (h) *Myotis septentrionalis* with a Holohil LB-2 transmitter attached using skin-bond glue

The discussed sampling techniques, their uses, and their advantages and disadvantages can be best summarised in Table 1.8.

Table 1.8 Sampling techniques

Sampling technique	Equipment and method	Advantages	Disadvantages
Transects: plan sketches and profile sketches	■ Rope or measuring tape marks the line that is drawn to scale ■ The area is selected at random across the ecosystem ■ Species are plotted along the line, in surface view for a plan sketch or in side-on view for a profile sketch	■ Provides a quick, easy and inexpensive method for measuring species occurrence ■ Minimal disturbance to the environment	■ Only suitable for plants or slow-moving animals. ■ Species occurring in low numbers may be missed
Quadrat sampling	■ Measuring tape, metre rulers or quadrats are used to randomly place the 1 m × 1 m square areas ■ The occurrence of organisms in the quadrat is recorded and repeated a number of times ■ Individual species can be counted if in small numbers or percentage cover can be calculated for larger numbers by estimating the percentage cover for each quadrat and then finding an average of the quadrats taken	■ Easy and inexpensive method for measuring abundance in large populations ■ Minimal disturbance to the environment ■ Quadrats can also be used for determining the distribution of species along a transect	■ Only suited for plants and slow-moving animals
Mark–release–recapture	■ Animals are captured (e.g. traps, nets, pits), tagged or marked (e.g. limb bands) and then released ■ After a suitable time to mix with others, a sample is recaptured ■ The number of tagged or marked animals recaptured is counted ■ Numbers are then entered into the formula: Abundance = $\dfrac{\text{number captured} \times \text{number recaptured}}{\text{number marked in recapture}}$ *Note*: Technology is sometimes used in marking very mobile animals using tracking bands and tracing their movements by GPS systems and satellites	■ A simple method that provides an estimate of abundance for animals in large populations that are difficult to count	■ Only suitable for mobile animals ■ Can be time consuming depending on the type of species captured, method of tagging, and time suitable for waiting while tagged group mix with others ■ Can be disturbing to the environment

Sampling techniques for determining distribution are coloured red and techniques for determining abundance are coloured green.

Justification of different sampling techniques

■ *process and analyse information obtained from a variety of sampling studies to justify the use of different sampling techniques to make population estimates when total counts cannot be performed*

SECONDARY SOURCE INVESTIGATION

BIOLOGY SKILLS

P14

Background information

When ecologists attempt to analyse the distribution and abundance of a species in an ecosystem there are particular reasons why population estimates must be made when total counts cannot be performed. When observing very large populations of plant or animal species it is too difficult to count species individually. Mobile animal species are also very difficult to count individually. This is why population estimates are used instead. It is less time consuming, less expensive and less disturbing for the environment. Techniques such as quadrat sampling (percentage cover) for plants and mark–release–recapture methods for animals are commonly used.

Aim

1. To process and analyse information from a variety of sampling studies.
2. To justify the use of different sampling techniques to make population estimates when total counts cannot be performed.

Method A: Processing information obtained from a variety of sampling studies

Using the following information on the grey-headed flying-fox (see Fig. 1.11), and information from other sources, research sampling techniques used for estimating population size for three organisms that are either too high in number to count or are too mobile and quick to count (e.g. flying-foxes, crocodiles and sharks).

Information sources may include scientific journals, CD-ROMs, Internet sites, textbooks, library reference books, newspaper articles, microfiche and experts in the field. Using a variety of different sources increases the reliability of your information. Using recent sources, such as scientific journals, improves the accuracy and validity of your information.

The search for information usually begins with keywords. These are used when looking up the index of a textbook or using an Internet search engine such as google.com.au to find relevant websites. Once you have found a source of information you then need to analyse and sort through it to select what is important and relevant to your task. This information then needs to be collated for its specific use.

Carefully read the background information in this section and on pages 13–17 before beginning this task. Use this information to help you identify some keywords that you might use in your search for information from a variety of sampling studies.

Website for radio-tracking study of bats

Population estimates of the grey-headed flying-fox

Radio-tracking study of grey-headed flying-foxes (*Pteropus poliocephalus*) in Gordon, Sydney (Augee & Ford, 1999)

This study looked at the release and integration of hand-reared grey-headed flying-foxes in the northern suburbs of Sydney.

Radio-collared hand-reared bats (belonging to the Ku-ring-gai Bat Conservation Society) were found after release to move between Gordon and the Royal Botanical Gardens at Sydney and at Cabramatta Creek south of Sydney. Individual bats moved along the coast as far as 310 km north and 279 km south of Gordon.

When radio-tracking bats there is virtually no chance of recapturing the bat in order to remove the expired transmitter. Therefore, transmitters were designed in 1995 to be glued onto the back of the bat (see Fig. 1.10h) so it would fall off over a short period of time. The transmitters weighed 7 g and had 26 cm-long antennae that were detectable from up to 5 km; however, most transmitters fell off within 20 days (or were possibly removed by grooming). New transmitters weighing 16 g with shorter antennae were used in 1996. They were attached to collars made out of surgical tubing. Sunlight destroyed the rubber surgical tubing after approximately three to four months when the collars fell off. These collars were previously tested on captive flying-foxes and were tolerated very well by the bats. The position of individual flying-foxes was determined using directional antennae and magnetic compasses plotting bearings on a map. The bearings were taken from transmitters in a number of different locations in the area.

For the complete study, access www.sydneybats.org.au/cms/index.php?references and download 'Radio tracking study.pdf'

Figure 1.11 Grey-headed flying-fox

For recommended websites

Method B: Analysing information obtained from a variety of sampling studies to justify the use for making population estimates when total counts cannot be performed

Using the information you have processed and collated from Method A, analyse and search for information that justifies the use of such techniques. This information must support your argument or conclusion as to why these different sampling techniques are used to make population estimates when total counts cannot be performed. Use Table 1.9 as a guide for completing this task.

Table 1.9 Different sampling techniques to make population estimates when total counts cannot be performed

Example	Sampling technique	Method	Reasons why useful when total counts cannot be performed
Flying-foxes	Radio-tracking		
Crocodiles	Satellite-tracking		
Sharks	Tagging		

Analysis questions

1. **Explain** the reasons why ecologists may opt for population estimates rather than individual organism counts.
2. **Name** three examples of sampling techniques used for population estimates and briefly **describe** the methods involved.
3. **Justify** the use of the examples described in Question 2 for making population estimates when total counts cannot be performed.

Blank copy of Table 1.9

Energy use in ecosystems

1.6

■ *identify uses of energy by organisms*

Energy enters the ecosystem firstly from the sun. Sunlight is absorbed by producers (plants) and used in the process of **photosynthesis** to make glucose. Glucose is an energy source and a small amount of it is used by the plant for the production of organic molecules (e.g. proteins, carbohydrates); growth, repair and maintenance; fluid movement and transport; and for specialised cell functions.

Most of the glucose is used for plant growth or reproduction, which in turn may be consumed by animals. These consumers then receive the energy contained by the plant and use a small amount of this energy for cellular function, but lose most of it as heat. All organisms release energy as heat during the process of **respiration**. Most of the energy is used in respiration to fuel the many chemical reactions of cells in the body.

■ *describe the roles of photosynthesis and respiration in ecosystems*

Role of photosynthesis in ecosystems

Photosynthesis is the process in which plants use the energy from sunlight (absorbed by the green pigment **chlorophyll**) to convert carbon dioxide and water to glucose (a type of sugar) and oxygen (released back into the air). Plants take in carbon dioxide from the air through their leaves and absorb water in through their roots. A small amount of glucose is used by the plant for energy but most of it is used for plant growth or reproduction. Photosynthesis is carried out by green plants and even by photosynthetic bacteria (blue-green algae). The process of photosynthesis can be summarised by a word equation, expressed as an equation using chemical symbols or as a balanced chemical equation:

■ word equation:

$$\text{carbon dioxide + water} \xrightarrow[\text{chlorophyll}]{\text{sunlight}} \text{glucose + oxygen}$$

■ chemical equation:

$$CO_2 + H_2O \xrightarrow[\text{chlorophyll}]{\text{sunlight}} C_6H_{12}O_6 + O_2$$

■ balanced equation:

$$6CO_2 + 6H_2O \xrightarrow[\text{chlorophyll}]{\text{sunlight}} C_6H_{12}O_6 + 6O_2$$

Plants that manufacture their own food through the process of photosynthesis are sometimes referred to as **producers** as they produce or make their own food rather than eat or consume other organisms (plants or animals), which are referred to as **consumers**. Plants are responsible for harnessing the energy from sunlight for use in ecosystems. Their role as producers starts the food chain with high amounts of energy, ready for passing on to consumers. This involves the removal of carbon dioxide from the air, the return of oxygen to the air and the manufacture of food.

Role of respiration in ecosystems

Aerobic respiration (or mitochondrial respiration) takes place in the **mitochondria** of all living cells and results in the release of energy for organisms to use. Glucose is broken down in the presence of oxygen to produce carbon dioxide and water and in doing so energy is released. Energy in the form of ATP (adenosine triphosphate) is released as heat from this process and is used for cell functions such as growth, repair and maintenance. The role of respiration is to remove oxygen from the air, return carbon dioxide to the air and provide energy.

■ *identify the general equation for aerobic cellular respiration and outline this as a summary of a chain of biochemical reactions*

Aerobic respiration

Cellular respiration is the process in which glucose is converted into energy, usable for life processes. It allows organisms to use (release) the energy stored in glucose. The following general equation for aerobic cellular respiration is a summary of a chain of many biochemical reactions which occur in the cells of the organism and is generally used in order to understand the main changes through the respiration process.

General equation for aerobic cellular respiration expressed as a:

■ word equation:

$$\text{glucose + oxygen} \xrightarrow{\text{many chemical reactions}} \text{carbon dioxide + water + energy (ATP)}$$

■ chemical equation:

$$C_6H_{12}O_6 + O_2 \xrightarrow{\text{many chemical reactions}} CO_2 + H_2O + ATP$$

■ balanced equation:

$$C_6H_{12}O_6 + 6O_2 \xrightarrow{\text{many chemical reactions}} 6\,CO_2 + 6H_2O + ATP$$

The reaction appears to be the reverse of photosynthesis, but it is not. Respiration is a series of chemical reactions which have no similarity to the processes occurring during photosynthesis. In fact, the processes of respiration and photosynthesis actually work together as a cycle essential to plant life (see Fig. 1.12).

Anaerobic respiration

Cellular respiration may sometimes occur as **anaerobic respiration**, meaning without adequate oxygen. Anaerobic respiration does not produce the high quantity of ATP energy that aerobic respiration does. This is due to the lack of oxygen that can assist in releasing ATP when converted to carbon dioxide and water. Figure 1.13 illustrates the two reaction pathways that respiration can take depending upon the availability of oxygen. Aerobic respiration releases 36 ATP in total, while anaerobic respiration releases only 2 ATP in total. There is a large difference in the amount of energy released from each reaction.

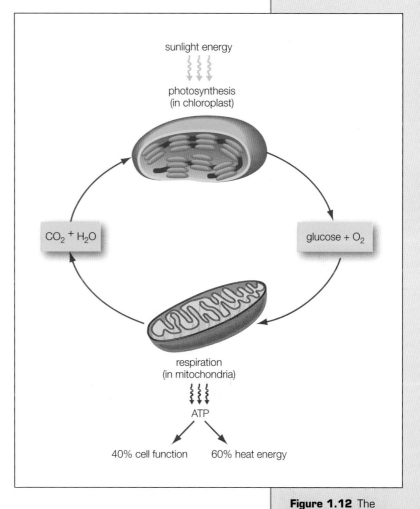

Figure 1.12 The processes of photosynthesis and respiration in cell function as a cycle essential to plant life

21

Figure 1.13 Aerobic and anaerobic respiration pathways occurring in a cell

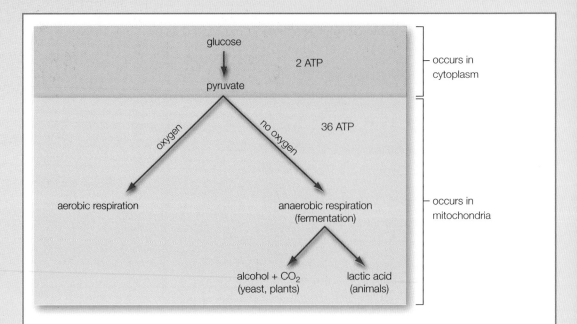

glucose

2 ATP

pyruvate

occurs in cytoplasm

oxygen

no oxygen

36 ATP

aerobic respiration

anaerobic respiration (fermentation)

occurs in mitochondria

alcohol + CO_2 (yeast, plants)

lactic acid (animals)

For recommended websites

An example of anaerobic respiration is sprinting. In a 100 m sprint our body is not providing adequate oxygen for the process of respiration and therefore is carrying out anaerobic respiration. This, of course, cannot be sustained for long periods of time as the amount of energy available is much less (2 ATP) than aerobic respiration (36 ATP). This is why we only have 100 m sprint races.

REVISION QUESTIONS

1. List three abiotic characteristics and briefly describe the main differences between aquatic and terrestrial environments.

2. Briefly describe how an ecologist would go about recording the distribution of plants in a rainforest ecosystem.

3. Describe an investigation you conducted using the quadrat method. List one advantage and one disadvantage of this method.

4. Describe a method you would use to estimate population numbers of kookaburras in an area.

5. Identify two factors that might determine the distribution and abundance of mangroves in their estuarine environment.

6. Describe the roles of photosynthesis and respiration in ecosystems.

7. Identify two uses of energy by organisms.

8. Identify the general word equation for aerobic cellular respiration.

Answers for revision questions

9. Justify the use of two different sampling techniques to make population estimates when total counts cannot be performed.

CHAPTER 2

Unique aquatic and terrestrial ecosystems

Each local aquatic or terrestrial ecosystem is unique

Trends in population estimates

2.1

- *examine trends in population estimates for some plant and animal species within an ecosystem*

Plant population trends

Mangrove species

When studying a tidal **estuary** you are most likely to look at the abundance of mangrove species along a transect line from the sea to inland. If we graphed the results we would see something like that shown in Figure 2.1(a). Mangrove species A is highest in abundance inland, species C is highest closest to the sea, and species B is most abundant in between A and C. Mangrove species C being most abundant closest to the sea appears to be the most tolerant of **saline** conditions, followed by species B, and then the least tolerant of saline conditions is species A.

Trends in population estimates can be seen easily when abundance values have been graphed. Examining trends can lead to inferences about the species and what abiotic or biotic characteristics they are most suited to.

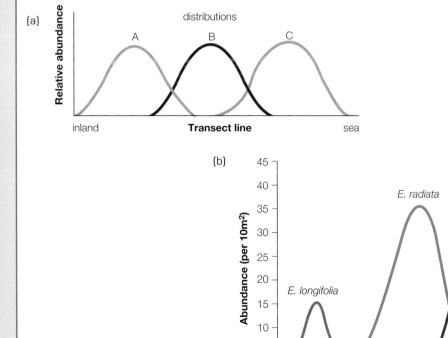

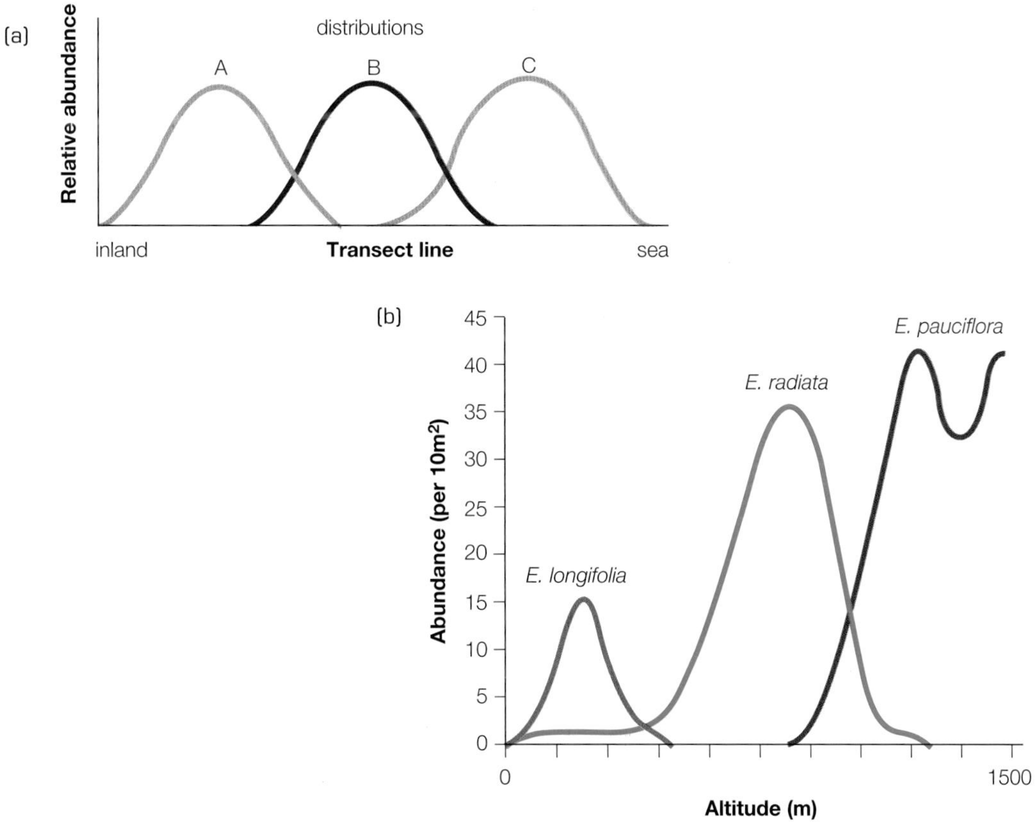

Figure 2.1 Trends in plant population estimates: (a) three mangrove species and distance from the sea; (b) three eucalypt species at different altitudes

23

Eucalyptus species

Figure 2.1(b) illustrates the abundance of different eucalyptus species at different altitudes in southeast New South Wales. Although the species tend to overlap as altitude increases, each different eucalypt species appears to be more abundant in, and therefore more suited to, a specific altitude range. *Eucalyptus pauciflora* numbers dominate the higher altitudes, while *Eucalyptus longifolia*, even though it has a smaller abundance, still dominates the lowest altitude ranges. *Eucalyptus radiata* appears to be suited to most altitudes; however, it is most abundant in the middle altitude range.

Animal population trends

Grain beetle species

The abundance of two grain beetle species (*Calandra* and *Rhizopertha*) in Figure 2.2(a) show different trends over the period of 180 weeks. Although both species rapidly increase over the first 20 weeks, and drop quickly over the second 20 weeks, *Rhizopertha* continues to drop to a certain abundance level and remains stable over the remaining time period. *Calandra*, however, goes through a series of fluctuations then drops a little during weeks 80 to 140, but then increases again, close to its highest-reached abundance level. This graph infers that *Calandra* appears to survive better than *Rhizopertha* when in competition with each other in this environment. Although we cannot predict what these two species were in competition for (e.g. food, space), we can assume that one species dominates the other in this particular situation through higher species abundance. This, of course, does not mean that the two species cannot successfully exist together in the same environment.

Figure 2.2 Trends in animal population estimates: (a) two grain beetle species in competition; (b) Australian marsupial *Antechinus stuartii* supplemented with food

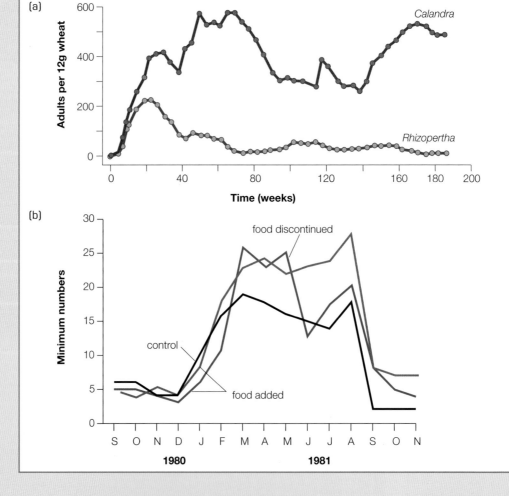

Australian marsupial species

Figure 2.2(b) illustrates the effect of increasing the food supply for a population of small marsupial carnivores, *Antechinus stuartii* (Fig. 2.3), over 12 months. Supplementary food was provided to two groups of *Antechinus stuartii*; however, the food supply was discontinued in one group along the way. Looking at the graph in Figure 2.2(b), all groups rapidly increase in numbers up to March 1981 and then stabilise for a few months. Each group reaches a peak in August 1981 before rapidly declining in numbers to a point similar to 1980 figures in November 1981. The control group (not supplemented with any food) have not reached as high numbers as the supplemented groups. When food was discontinued for one group a dramatic drop occurred in numbers, but it seemed to recover quickly and increase and return to the same pattern as the other groups. Food appears to play an important part in the abundance of *Antechinus stuartii* and indicates that the abundance of the populations studied was limited by food supply. Otherwise, we would not have seen any increase in species numbers for those supplemented with food.

Knowledge of the response of a species to different environmental factors can be used to make predictions about the potential distribution and abundance of species. You will be using your ability to examine trends in population estimates when analysing your own population estimate graphs collected on your field trip later in this chapter.

Figure 2.3 Australian marsupial *Antechinus stuartii*

Predator and prey populations

2.2

■ *outline factors that affect numbers in predator and prey populations in the area studied*

There are two types of interactions between organisms: detrimental and beneficial. **Detrimental interactions** exist when one or more organisms are harmed or disadvantaged from the relationship. **Beneficial interactions** exist when one or more organisms benefit from the relationship. One example of a detrimental interaction is the **predator–prey relationship**.

Predation (predator–prey relationship)

A predator–prey relationship is a feeding relationship where the **predator** (consumer) obtains its food by killing an animal (**prey**), for example spiders eating flies or eagles eating bush rats. Not only are land animals predators, some plants and marine organisms also play this role (see Fig. 2.4).

Predators affect the abundance of their prey. Providing the prey species reproduces as fast as it is predated upon its population will stay at a constant size. For example, if the rabbits in a grassland ecosystem reproduce faster than the foxes that predate them then the rabbit population will increase. In natural communities, the abundance of a predator and its prey can fluctuate through time, with the predator numbers copying those of the prey. When there are large numbers of prey available, the predator population increases in size. As prey are consumed, their numbers decline, leading to a shortage of food for the predators, whose numbers also decline. This pattern is illustrated in Figure 2.5.

Figure 2.4 Plants and marine animals can be predators: (a) the pitcher plant traps and digests insects in highly modified leaves; (b) a predatory snail feeds on soft coral; (c) the Australian 'blue bottle' with its large blue float captures its prey using long stinging tentacles

(a)

(b)

(c)

(a)

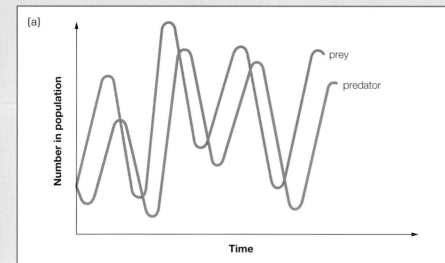

(b)

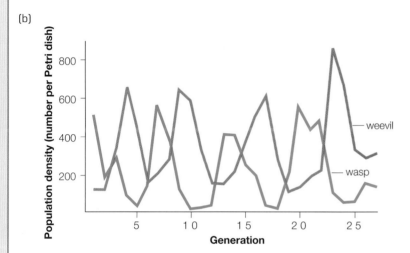

Figure 2.5 (a) The typical pattern found in predator–prey relationships; (b) The number of parasitoid wasps fluctuates in relation to its host the bean weevil

Factors affecting numbers of predator and prey populations

There are a number of different factors that may affect the numbers of predator and prey populations:

- number of predators competing for same prey
- availability of prey's food
- birth rate (depending on the age of reproductive maturity and the number of reproductive episodes per lifetime)
- death rate (increased by exposure to disease)
- number of males and females
- size of ecosystem for supporting the predator and prey numbers
- movement between ecosystems
- number of shelter sites available.

Factors relevant to your chosen study area may not include all of those listed above. For example, in a mangrove ecosystem you may find that the factors affecting numbers in predator and prey populations are directly related to human impact. Use the above examples as a reference when you are conducting the field study of your area and looking for possible factors affecting the numbers of predator and prey populations.

2.3 Allelopathy, parasitism, mutualism and commensalism

- *identify examples of allelopathy, parasitism, mutualism and commensalism in an ecosystem and the role of organisms in each type of relationship*

Allelopathy

Allelopathy is the production of specific biomolecules by one plant that can be beneficial or detrimental to another plant. This concept suggests that biomolecules (**allelochemicals**) produced by a plant escape into the environment and subsequently influence the growth and development of other surrounding plants. Not all plants have **allelopathic** tendencies, but most plants that do use it to compete with other plants and therefore negatively influence the existence of neighbouring plants. Basically, it is mainly used by plants to keep other plants out of its space. Space is crucial to the survival of plants. The fewer plants around, the more water to absorb from the soil, the more soil to support the roots for plant stability, and the more sunlight available to absorb.

Types of allelopathy

There are a number of different types of allelopathy. In one type, the plant that is protecting its space releases growth-compounds from its roots into the ground. New plants trying to grow near the allelopathic plant absorb those chemicals from the soil inhibiting root/shoot growth or seed germination. Another type of allelopathy involves the release of chemicals that slow or stop the process of respiration or photosynthesis, some may just inhibit nutrient uptake. Plants may also release chemicals that can change the amount of chlorophyll in another plant. The plant cannot then make food with the changed chlorophyll levels and dies.

Allelopathic chemicals can be present in any part of the plant. They can be found in roots, stems, flowers, fruits and leaves.

Examples of allelopathy

- The black walnut plant releases a chemical that inhibits respiration. The chemical is found in all parts of the plant but it is concentrated in the buds and roots. Plants exposed to this chemical exhibit symptoms such as wilting, yellowing of foliage and eventually death.
- Sorghum species (cereal grass) release a chemical in the root exudates that disrupts mitochondrial functions and inhibits photosynthesis. It is currently being researched extensively as a weed suppressant.
- Eucalyptus leaf litter and root exudates are allelopathic for certain soil microbes and plant species (see Fig. 2.6). Some pine trees are also allelopathic. When their needles fall to the ground, they begin to decompose and release acid into the soil. This acid in the soil keeps unwanted plants from growing near the pine tree.

The more that is learnt about allelopathy the more we can find out about healthier alternatives to herbicides. That is, we could prevent unwanted plants or weeds from growing in an area by selecting plants that specifically produce chemicals against them.

Figure 2.6 Eucalyptus leaf litter is allelopathic and doesn't allow other plants to grow near it

Beneficial interactions

Symbiosis is the term used for interactions in which two organisms live together in a close relationship that is beneficial to at least one of them. Symbiosis usually involves providing protection, food, cleaning or transportation.

There are three types of beneficial or **symbiotic interactions**:

- **parasitism**—one species benefits and the other is harmed
- **mutualism**—both species in the relationship benefit from the association
- **commensalism**—one species benefits and the other is unaffected.

Parasitism

Relationships where one species benefits and the other is harmed are **parasitic**.

A **parasite** obtains food and shelter from the host organism. They feed upon the tissues or fluids of the host organism, but do not usually kill it, as this would destroy the parasite's food supply. The parasite is often smaller than their host and they may live on the surface of their host (**ectoparasites**, e.g. ticks, fleas and tinea) or internally (**endoparasites**, e.g. tapeworms) (see Fig. 2.7).

(a) (b)
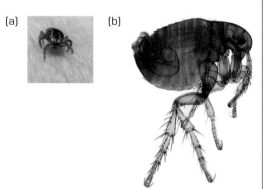

Figure 2.7 Parasites can be ectoparasites: (a) ticks, (b) fleas, (c) tinea; or endoparasites, (d) tapeworm

(c)
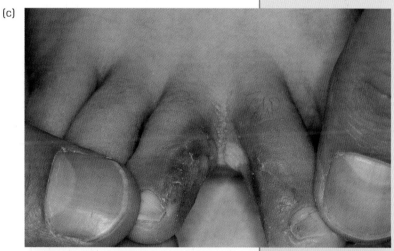

(d)

Mutualism

Relationships where both species benefit from the association are **mutualistic**.

Reef-building corals have symbiotic algae within their tissues which provide the yellow-brown pigments that give the coral its colour (see Fig. 2.8). The algae live, reproduce and photosynthesise in the host and use the waste products of the host. In turn, the coral uses oxygen and food produced by the algae during photosynthesis to grow, reproduce and form its hard skeleton, which is the basis of the reef. The formation of the Great Barrier Reef depends on this mutualistic relationship. When corals are stressed (i.e. when disturbance turns water murky, or sea temperatures increase) they expel the algae, which in turn causes the corals to starve, leaving white skeletons.

The relationship between the sea anemone and the anemone fish (or 'clown fish') was once thought to only benefit the anemone fish; however, recent studies have suggested that, in fact, both organisms benefit. The anemone fish is neither stung nor eaten by the anemone. The anemone fish repeatedly brushes against the anemone's tentacles until its own mucous coating inhibits the anemone's sting. The anemone fish is therefore protected from predators by hiding in the anemone's tentacles unharmed. It feeds on the anemone's food scraps. The anemone benefits as the anemone fish cleans its host and lures other animals into the anemone's tentacles (see Fig. 2.9).

Figure 2.9 The mutualistic relationship between the sea anemone and the anemone fish ('clown fish')

The 'ant plant' has a mutualistic relationship with a species of ant. The plant has a swollen base in which there are specialised chambers. The ants form large colonies within these chambers and carry their prey corpses and excreta to parts of the chambers (cemeteries) where the plant is able to absorb the waste nutrients.

Figure 2.8
Reef-building corals have a mutualistic relationship with algal cells

Commensalism

Relationships where one species benefits and the other is unaffected are **commensal**.

Epiphytes such as mosses, small ferns and orchids can be seen on tree trunks in moist forests. They appear to benefit from living on the trunk of the host tree by catching rainwater to dissolve nutrients and by being closer to sunlight. Epiphytes do not appear to affect the host tree negatively. The strangler fig commences its life as an epiphyte. The seed germinates from bird droppings on the host tree and the young fig starts to grow.

The fig benefits and the host at this stage is not affected. However, the fig grows and extends its roots down into the soil below. It envelops its host and prevents trunk growth (see Fig. 2.10). The relationship changes from commensalism to competition for space.

The barnacle is a crustacean that normally adheres to a fixed surface; however, some barnacles adhere to the surface of whales and turtles. This does not affect the whales or turtles, but benefits the barnacles as they are transported to diverse areas rich in food (plankton).

Extension activity—
allelopathy, parasitism, mutualism and commensalism

(a)

(b)

Figure 2.10 (a) Young fig starts to grow and extend its roots; (b) Fig envelops its host

2.4

Decomposers in ecosystems

■ *describe the role of decomposers in ecosystems*

Decomposers use the organic nutrients of dead organisms for energy and break these dead bodies down to inorganic nutrients which can be recycled for use by palnts. Bacteria and fungi in the soil are very important because they return nutrients to the soil when they decompose dead animals and plants. The highly important cycle operating in this process is the nitrogen cycle (see Fig. 2.11). Nitrogen is essential to all living things.

The nitrogen cycle

Atmospheric nitrogen becomes part of living organisms in two ways. Firstly, bacteria in the soil form nitrates from the nitrogen in the air. Secondly, during electrical storms (lightning), nitrogen is combined with oxygen and water to produce an acid that falls to the earth in rainfall and deposits nitrates in the soil. Plants take up the nitrates and convert them to proteins that then travel up the food chain.

Figure 2.11
The nitrogen cycle

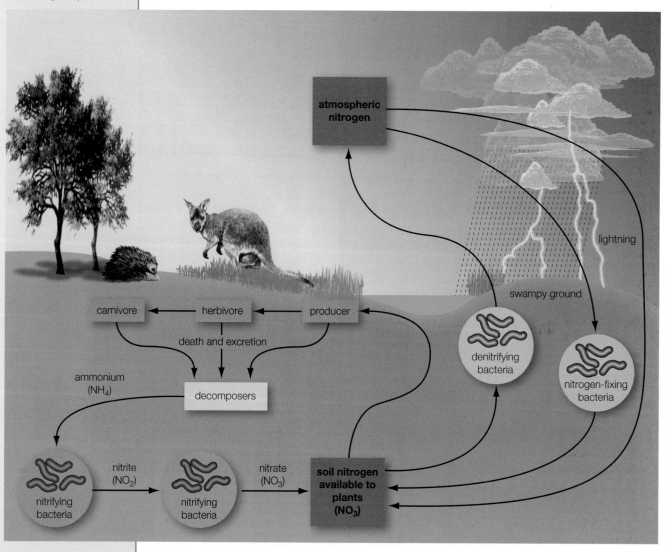

When organisms excrete wastes the nitrogen is released back in to the environment. When organisms die and decompose the nitrogen is broken down and converted to ammonia. Plants absorb some of this ammonia, the remainder stays in the soil where bacteria convert it back to nitrates. The nitrates may be stored in humus or are leached from the soil and carried into lakes and streams. Nitrates may also be converted to gaseous nitrogen through a process called denitrification and returned to the atmosphere, continuing the cycle.

Optional website activity. Visit www.biology.ualberta.ca/facilities/ multimedia and click on 'Ecology', then animate the nitrogen cycle.

Website activity

Trophic interactions between organisms

2.5

■ *explain trophic interactions between organisms in an ecosystem using food chains, food webs and pyramids of biomass and energy*

Autotrophs and heterotrophs

Ecological interactions are the exchanges and flows of energy and matter, and these interactions are determined by the ways in which organisms obtain their food. Ecosystems are often described in terms of their **trophic** or feeding relationships. **Autotrophs** or *producers* are organisms that make their own food by converting inorganic molecules to organic compounds (see page 20). The majority of autotrophs are green plants and algae that use sunlight for photosynthesis. A small group of autotrophs that do not carry out photosynthesis are chemosynthetic bacteria.

All other organisms are **heterotrophs** or *consumers*. They must consume other organisms in order to gain the organic molecules they need for life. There are several types of consumer organisms:

■ **primary consumers —herbivores**: organisms that eat plants only (e.g. koala)
■ **secondary or tertiary consumers —carnivores**: organisms that eat animals only (e.g. crocodile)
 —omnivores: organisms that eat both plants and animals (e.g. ants).

Primary consumers are eaten by secondary consumers, and secondary consumers are eaten by tertiary consumers.

Among the heterotrophs there are also organisms that feed on dead organisms and organic waste from different **trophic (feeding) levels**. These are called **degraders**, which include:

■ **scavengers**—animals that eat dead organisms
■ **detritivores**—animals that ingest organic litter or detritus (and then digest it)
■ decomposers—fungi and bacteria that cause chemical decay of organic matter and absorb the broken-down material.

Of course, some of the consumers may overlap more than one trophic level. Some organisms may even change trophic levels during their life cycles. For example, some insects are **carnivorous** as larvae but become **herbivorous** as adults (e.g. stoneflies). Figure 2.12 illustrates the general characteristics of an ecosystem in a simplified diagram of the flow of energy and materials.

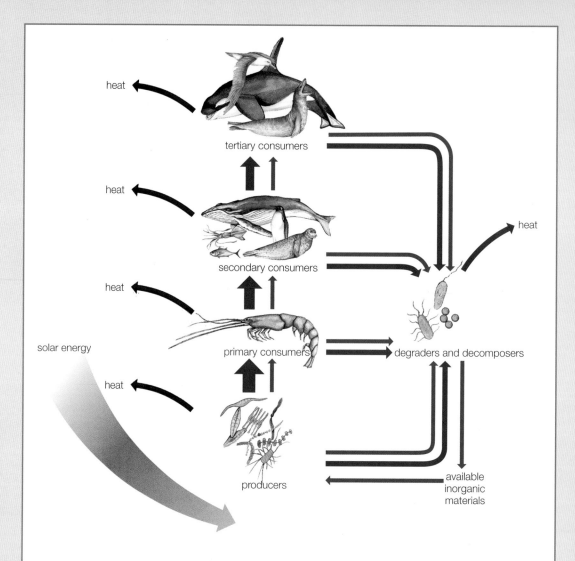

Figure 2.12
A simplified diagram of the flow of energy (red) and materials (blue) through an ecosystem. Although there is a net loss of energy, materials may be recycled

Worksheet on
food chains

Worksheet on
a food web

Food chains and food webs

Food chains show the energy movement from one living thing to another. They describe the feeding order in which plants or animals eat or are eaten by other animals. A simple food chain is comprised of:

producer ⟶ herbivore ⟶ carnivore

The diagram in Figure 2.13 represents examples of food chains where the energy flows up each trophic level. The trophic level of an organism is its position in a food chain, the sequence of feeding and energy transfer through the environment. Arrows indicate the direction of the flow of energy from one organism to the other.

However, food chains are not isolated in ecosystems; they are more realistically shown as a **food web**. Food webs show complex food interactions in an ecosystem. They are made up of two or more food chains (see Fig. 2.14). Even food webs may not be isolated in ecosystems. Trophic interactions usually take place within a particular habitat however these sometimes overlap.

Ecological pyramids

A patch of eucalypt woodland can support *thousands* of leaf-eating insects but only *hundreds* of insect-eating birds and only *one or two* bird-eating foxes or cats. Food webs such as that shown in Figure 2.14 represent a **pyramid of**

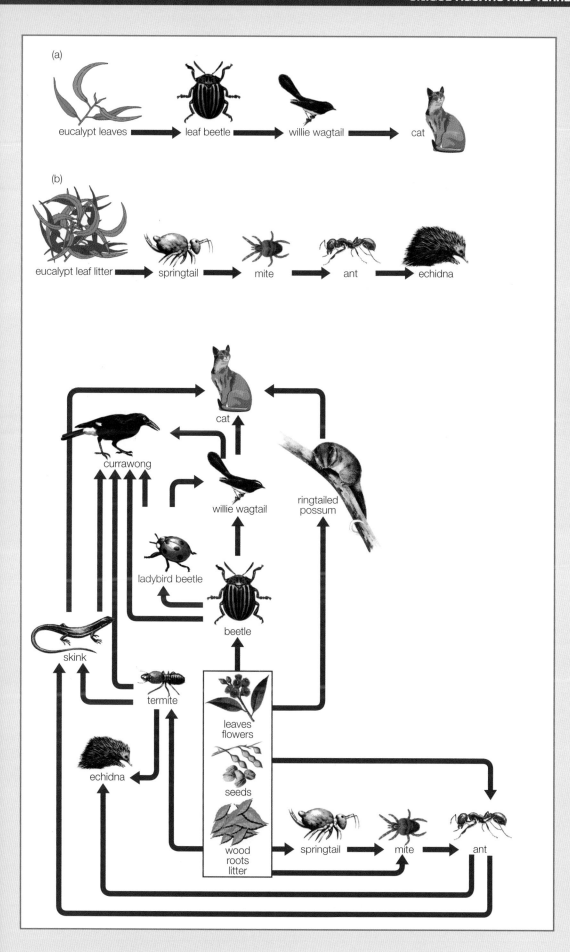

Figure 2.13
Examples of simplified food chains that might occur in eucalypt woodlands: (a) a grazing food chain; (b) a detritus food chain

Figure 2.14
A simplified food web, illustrating how food chains from Figure 2.13 may interconnect with other food chains in a eucalypt woodland

(a)

eucalypt leaves → leaf beetle → willie wagtail → cat

(b)

eucalypt leaf litter → springtail → mite → ant → echidna

numbers. In a pyramid of numbers, large numbers of herbivores (primary consumers) are consumed by smaller numbers of increasingly large carnivores (secondary and tertiary consumers) (see Fig. 2.15).

Pyramids of biomass

The amount of food at any trophic level depends, in part, on its biomass (total amount of living material present at any one time). A **biomass pyramid** shows the amount of biomass through each level of the food chain. At each level, energy (heat) and matter (food and wastes) are lost (90 per cent). Diagrams showing the biomass of trophic levels in an ecosystem are more frequently pyramidal in shape than diagrams of numbers of organisms. However, when an ecosystem is found to be unstable (where biomass from one level cannot support the next) then the biomass pyramid shape moves away from the

pyramidal shape (see Fig. 2.16). For example, if only 4 g of eucalypt leaves has to support a 20 g leaf beetle, then the biomass pyramid is unstable; the lower trophic level must have a larger biomass than the higher levels.

Pyramids of energy

Not all the energy and material taken in by one trophic group is passed on to the next, and not all organisms at one trophic level are consumed by the next; there are also losses of heat associated with cellular respiration and the elimination of wastes. Thus, there is a net loss of energy and materials as you move up each trophic level. Energy flow indicates the food value of trophic levels more accurately than either numbers or biomass. We can therefore represent this energy flow diagrammatically in **pyramids of energy** flow (see Fig. 2.17). A pyramid of energy is never inverted.

Figure 2.15
Ecological pyramid of numbers (individuals per 0.1 ha)—an oak forest in UK

Figure 2.16
Ecosystems as shown through biomass pyramid diagrams—stable ecosystems: (a) a grassy agricultural field in USA, (b) a tropical rainforest in Panama; unstable ecosystem: (c) plankton in the English Channel, North Sea

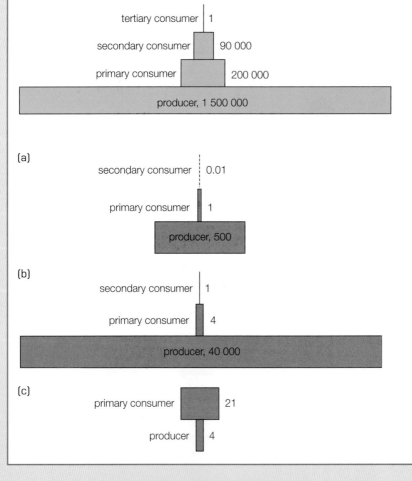

tertiary consumer | 1
secondary consumer | 90 000
primary consumer | 200 000
producer, 1 500 000

(a)
secondary consumer | 0.01
primary consumer | 1
producer, 500

(b)
secondary consumer | 1
primary consumer | 4
producer, 40 000

(c)
primary consumer | 21
producer | 4

Some ecologists illustrate ecological pyramids as a stepped shape (as represented in Figs 2.15, 2.16 and 2.17); however, others prefer to simplify it further to a triangular non-stepped shape. The following diagrams summarise the three different types of ecological pyramids illustrated as the simplified, triangular non-stepped shape (see Fig 2.18).

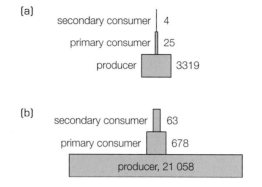

Figure 2.17
Ecological pyramids of energy flow (kJ m^{-2} year): (a) a desert grassland in New Mexico, USA; (b) a prairie in Oklahoma, USA

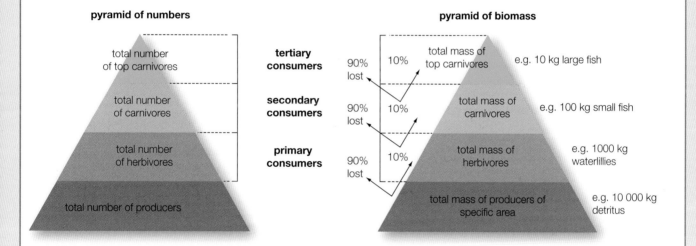

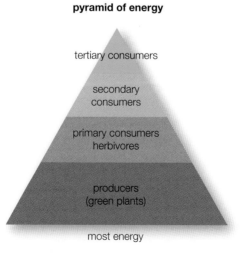

Figure 2.18 A summary of the three types of ecological pyramids

Constructing food chains and food webs

FIRST-HAND AND SECONDARY SOURCE INVESTIGATION

BIOLOGY SKILLS

P14

For recommended websites to obtain secondary source information

Table 2.1 Feeding information for organisms found in a simplified lake ecosystem

■ *gather information from first-hand and secondary sources to construct food chains and food webs to illustrate the relationships between member species in an ecosystem*

Background information

A food chain is a flowchart that shows the movement of energy from one living thing to another. It illustrates the feeding order or trophic levels. A food web is the combination of more than one food chain, representing the feeding relationships in a habitat or ecosystem.

Energy passing through from different trophic or feeding levels is illustrated by the direction of the arrows. Most food webs can be constructed from their composite food chains; however, most food webs you see illustrated are simplified. This is because most ecosystems have hundreds of organisms intertwined and interacting within the ecosystem. It would be very time consuming to attempt to draw every species in the food web so they are simplified to represent the major trophic levels and feeding relationships that occur.

Part 1: Constructing food chains and food webs from feeding information

Using the data provided in the following table (see Table 2.1), construct six food chains with at least three trophic levels. Once you have completed the food chains then construct one complete food web, start by combining the food chains you have constructed. Check for any organism relationships that you have missed and add these in; be careful with the direction that arrows are placed and that all necessary arrows are included. Note that **detritus** (accumulated debris of dead organisms) forms part of the food web but is not an organism. Use Figures 2.13 and 2.14 as a guide for how your food chains and web should be presented.

Lake organism	Trophic type	Food/prey
Human	■ Consumer —omnivore	■ Trout
Tadpole	■ Consumer —omnivore	■ Algae and zooplankton
Yabbie	■ Consumer —filter feeder —herbivore ■ Scavenger	■ Zooplankton ■ Large water plants ■ Detritus
Platypus	■ Consumer —carnivore	■ Insects, molluscs, worms and small invertebrates
Water bird	■ Consumer —herbivore	■ Large water plants
Silver perch	■ Consumer —omnivore	■ Aquatic insects, crustaceans and molluscs
Pelican	■ Consumer —carnivore	■ Small trout and silver perch
Large water plant	■ Producer	■ Photosynthetic, survives submerged or free-floating
Diving beetle (aquatic insect)	■ Consumer —carnivore	■ Aquatic insect larvae and adult insects blown into the lake
Water flea	■ Consumer —filter feeder	■ Planktonic algae
Trout	■ Consumer —carnivore	■ Zooplankton, yabbies, aquatic insect larvae, insects blown in to lake
Mosquito larva	■ Consumer —filter feeder	■ Planktonic algae
Planktonic alga	Producer	■ Photosynthetic
Kingfisher	■ Consumer —carnivore	■ Small fish, yabbies and tadpoles

Part 2: Constructing food chains and food webs for a chosen ecosystem

Method A: Gather information from first hand sources

From your first-hand experiences during the field trip (taken later in this chapter), you will be required to gather any information on different organisms and their feeding relationships within one particular ecosystem. Carry out Method B to construct food chains and food webs from secondary sources, in preparation for constructing your own from the field trip.

Method B: Gather information from secondary sources

Refer to the investigation on page 18 to recall the way to obtain secondary source information. Read the background information for this task and page 34 under 'Food chains and food webs' to find possible keywords to use in an electronic search. Also use the websites provided on the Student Resource CD. Don't forget to look at a variety of different information sources for this investigation and collate your findings.

Gather information on *one* ecosystem (perhaps use the ecosystem that you may be selecting to visit for your field study later

in this chapter) and attempt to create at least three food chains from the data you have. You should be able to then interconnect some of the completed food chains into food webs. If you are not sure of some feeding relationships or have gaps to fill in your food chains or webs, use information from secondary sources to complete the task. You should have at least three food chains and one food web which represent your chosen ecosystem. Arrows should be carefully placed representing the movement of energy. You do not necessarily need diagrams for each organism; using names only is satisfactory.

Analysis questions

Using your food web in Part 1:
1. Identify the longest food chain in your food web.
2. Explain why you think this is called a simplified food web.
3. Identify two producer organisms in your food web.
4. Identify two herbivores and two carnivores in your food web and list what they eat from the food web.
5. Describe the difference between food chains and food webs.

Adaptations

2.6

■ *define the term adaptation and discuss the problems associated with inferring factors of organisms as adaptations for living in a particular habitat*

Adaptations

An **adaptation** is any characteristic that increases an organism's likelihood of survival and reproduction relative to organisms that lack the characteristic. Simply, an adaptation is a feature of an organism that makes it suited to its environment. There are three types of adaptations:
■ *structural*—a physical characteristic
■ *physiological*—an organism's function or process
■ *behavioural*—the way in which an organism acts.

Problems associated with inferring factors of organisms as adaptations for living in a particular habitat

Adaptations are characteristics that an organism has inherited and that make it suited to its environment. An organism does not intentionally look at an environment and work on changing to suit it, nor does it try to produce offspring that have these changes. An adaptation is a result of a change occurring at random when organisms reproduce a new organism.

This random difference just so happens to benefit the organism by making it more suited to the environment it lives in. There are problems associated with inferring that a particular characteristic of an organism is a direct adaptation to its habitat.

Past environments

Characteristics of present-day organisms are a product of millions of years of change; ancestors have received adaptations to survive in different habitats. An organism's current characteristics may have been inherited a long time ago when the organism existed in a different habitat. The organism may still possess that characteristic (or adaptation) but it is now not of any use or related to its survival in its current habitat. Dolphins and whales are well adapted to life in water; however, they possess lungs which are characteristic of land-dwelling animals. If these animals occupied a very different environment to that of their ancestors, then we cannot infer that the lungs are any sort of adaptation to their current environment, but one inherited from a time when they could possibly have been land-dwellers.

Studying the environment

To be able to determine if a characteristic is an adaptation, biologists need to study the organism's environment. It is difficult to relate a characteristic to a specific feature of an organism's environment when we do not know the exact habitats it has lived in over generations.

Sometimes adaptations may be obvious (like the stick insect camouflaging itself within its environment) and sometimes not. Some characteristics may have no benefit to the organism in a particular habitat, or are just not adaptations at all. It may be difficult to be certain how one characteristic benefits the organism in a particular environment.

Using fossil evidence

Interpreting the characteristics of organisms from fossil evidence in particular may lead to incorrect assumptions. For example, the extinct organism stegosaurus possessed bony plates along its back (see Fig. 2.19a). Some suggest that this characteristic was an adaptation to its competitive environment and used for defence. Others suggest it was simply used to attract mates, or perhaps even used for thermoregulation. Another more recent example is the 375 million year old fossil, the *Tiktaalik*, discovered in Arctic Canada in April 2006 (see Fig. 2.19b). The research team suggests that it is technically a fish, complete with scales and gills, but it has a flattened head like a crocodile and unusual fins that have sturdy interior bones that may have allowed it to prop itself up. Without knowing the environments it lived in, it is very difficult to suggest that these characteristics are in fact adaptations. So, we must be careful not to assume that all characteristics of organisms are adaptations to their present day habitat or environment.

Figure 2.19: (a) Bony-plated stegosaurus; (b) *Tiktaalik* fossil found in 2006

(a)

(b)

Adaptations for survival in Australian ecosystems

2.7

■ *identify some adaptations of living things to factors in their environment*

Australian environments are varied and diverse with some tough conditions for organisms to survive in. The three main abiotic factors that affect survival in Australian ecosystems are water, temperature and sunlight. Australian organisms have adapted to survive harsh conditions such as lack of water, high temperatures and high exposure to sunlight. Some examples of such plant and animal adaptations are described below.

Plant adaptations

Adaptations for dry environments

Xerophytes are plants that have adapted structurally to dry environments by reducing the surface area of their leaves in order to minimise water loss.
For example:

■ cactus plants like the Mexican lime cactus have small spiky leaves to reduce the loss of water and shallow, widespread roots to catch surface moisture
■ the pigface, found on sand dunes, has fleshy stems which store water (see Fig 2.20a)
■ eucalypts have a waxy cuticle on their leaves which reflects heat and light to minimise water loss from evaporation
■ cyprus pines have tiny leaves reducing water loss through **transpiration**
■ porcupine grasses roll their leaves during the hottest part of the day to allow fewer **stomata** (leaf pores) to be exposed to the dry atmosphere, therefore less water is lost through evaporation
■ all spinifex species have tough, pointed and narrow leaves for reducing water loss.

Other plant adaptations

■ Sclerophyllous (hard) leaves minimise water loss with a waxy or hairy surface, sunken stomata or greatly-reduced leaves (e.g. desert oak)
■ Succulent leaves and stems or fleshy underground tubers store water (e.g. parakeelya)
■ Deep-root systems access deep-water supplies or shallow root systems to enable the rapid uptake of moisture when it suddenly becomes available after rainfall
■ Desert acacias (wattles) have leaflets that are vertically flattened and oriented towards the ground, reducing the amount of light and hence water loss.

Figure 2.20
(a) Pigface with its fleshy stems;
(b) Spinifex grass with pointed narrow leaves to reduce water loss

(a)

(b)

continued . . .

(c)

(d)

Figure 2.20
(c) Kangaroos lick their forearms to lose heat by evaporation;
(d) Spinifex hopping mouse produces concentrated urine

Animal adaptations

Lack of water

- Kangaroos do not sweat, so they avoid losing water through sweating.
- The bilby hides in burrows to reduce water loss by evaporation; most desert mammals are nocturnal to reduce exposure to daytime temperatures.
- The water-holding frog is a burrowing frog that spends the majority of its life underground. It seals itself in a waterproof cocoon made up of layers of shed skin. Water is stored in the bladder or in pockets under the skin.

- The fat-tailed dunnart (a carnivorous marsupial) does not need to drink water, it obtains it all from its food of invertebrates that it feeds on at night (it is nocturnal). Through the day it shelters in nests of grass or under logs or rocks.
- The desert mouse also does not need to drink as it gains moisture from food (it is currently on the presumed extinct list).
- The spinifex hopping mouse and some desert mammals reduce water loss from excretion by producing highly concentrated urine (see Fig 2.20b).

High temperatures

- Animals are generally small in size to reduce heat gain and loss.
- Kangaroos, such as the rat kangaroos, dilate or swell their blood vessels, bringing them close to the surface of the skin to lose heat more rapidly (called vasodilation).
- The bilby hides in burrows to reduce the temperature of the environment and lives nocturnally (it forages for food at night when it is cooler).
- The bilby has large vascular ears for extra surface area for heat loss.
- Kangaroos lick their forearms to lose heat as the evaporation of saliva draws heat from the surface (see Fig. 2.20c).

High exposure to sunlight

- Lizards constrict their bodies to reduce surface area exposed to the sun.
- Some lizards have a pale external colour to reflect sunlight, therefore reducing heat absorption.
- Kangaroos sit in the shade during the day to avoid the heat absorption from the sun (see Fig 2.20d).

Three types of adaptations

As you can see from the above adaptation examples, there is a large variety in adaptations for just one type of environment (arid), and of that variety of adaptations there are just three types: structural, physiological and behavioural. In order to understand the difference between the three adaptation types, examples are described below:

■ structural—plants that have long, narrow leaf structure in order to reduce water in a desert environment

■ physiological—animals that dilate or swell their blood vessels, bringing them close to the surface of the skin to lose heat more rapidly in a high-temperature environment (vasodilation)

■ behavioural—animals that burrow under the ground to avoid the sun in a desert environment.

Adaptations from the local ecosystem

2.8

■ *identify and describe in detail adaptations of a plant and an animal from the local ecosystem*

The local ecosystems that may be found near Australian schools can vary quite considerably. The abiotic factors of the surrounding environment play a large part in determining the ecosystems that may exist in the area.

Some detailed examples of plant and animal adaptations found in common Australian ecosystems are described below. Later in this chapter, students are required to observe specific plant and animal adaptations from an ecosystem they selected for conducting a field study.

Mangrove forest (wetland)

Plant adaptations—mangroves

Mangroves use their roots, leaves and reproductive methods in order to survive in a harsh, changing intertidal environment of low-oxygen (and soft) soils and saline conditions (see Fig. 2.21).

Roots

Mangroves live in shifting environments where tides and floods constantly move the mud in which they live, destabilising the trees. Some mangrove species have pneumatophores (aerial roots) which are filled with spongy tissue and small holes that provide structural support and transfer oxygen from the air to the roots trapped below the ground in low-oxygen soil. The roots are also adapted to prevent the intake of a high amount of salt from the water.

Leaves

Some types of mangroves have leaves with glands that excrete salt. Grey mangroves can tolerate the storage of large amounts of salt in their leaves which are later dropped when the amount of salt gets too high. Mangroves can restrict the opening of their stomata, pores in the leaves responsible for regulating the exchange of gases and water during photosynthesis. This conserves fresh water within the leaves which is vital for survival in a saline environment. Mangroves are also able to reduce the leaf surface exposure to the hot sun by turning leaves side-on. This reduces excess water loss through evaporation.

Mangrove adaptations—
field trip material

Figure 2.21
Mangrove adaptations:
(a) pneumatophores
(aerial roots);
(b) leaves release salt;
(c) seeds germinate
before dropping from
the parent tree, roots
establish once they
land in mud

Seeds

Some mangrove species are **viviparous**, meaning they retain their seeds until they have germinated (shoot). When dropped into the water from the parent tree, the seed is able to remain dormant (some surviving after a year at sea) until it finds soil when it is immediately ready to put out roots. Other mangrove species produce seeds that float, so the tide assists in the dispersal, avoiding the overcrowding of young plants.

(a)

(b)

(c)

Animal adaptations—mangrove crabs

Mangrove crabs burrow into the soft mud to gain protection from both dehydration and predators. They use the water in their burrows to keep their gills moist and keep away from the hot sun.

Figure 2.22 A male fiddler crab can be distinguished by its single large claw

Examples of mangrove crab species

The male fiddler crab has a distinctive single large claw (see Fig. 2.22). It burrows in the intertidal zone and as the tides recedes it comes out to feed on algae, microbes and organic matter. Sometimes it may drag leaf litter into its burrow to be eaten.

The red crab is nocturnal, leaving its burrow at night to feed on fallen mangrove leaves. The burrow leaves a large distinctive hole in the soft mud under trees at the back of the mangroves.

Competing for resources

■ *describe and explain the short-term and long-term consequences on the ecosystem of species competing for resources*

Competition

When in competition two organisms use one or more resources in common, such as food, shelter and mates. The competition is so the organism can acquire a limited factor in the environment. For example, plants compete for factors such as water, light, carbon dioxide and minerals. Organisms may compete with members of their own species or members of another species. Competition between members of the same species is called **intraspecific competition** (see Fig. 2.23). Competition between members of different species is called **interspecific competition** (see Fig. 2.24). Usually interspecific competition is less intense than intraspecific competition. This is most likely due to members of the same species having far more resource needs in common to compete for.

L. C. Birch, a zoologist from the University of Sydney ('The meanings of competition', *American Naturalist*, 1957), defined two types of competition: **resource competition**, where organisms utilise a resource that is in short supply; and **interference competition**, where organisms harm each other while obtaining a resource, even if that is not in limited supply.

Plants

Individuals compete for a range of resources. Plants compete with other nearby plants for soil nutrients, water and space or for access to sunlight. Some plants are better able to compete than others in certain parts of ecosystems. These species exclude their competitors from that part of the ecosystem. As discussed earlier (see page 28) some plants secrete allopathic chemicals into the soil to inhibit the growth of or kill other nearby plants.

Figure 2.23 Coral on the Great Barrier Reef—one species shades its neighbour as they compete for light

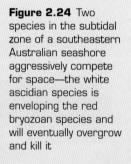

Figure 2.24 Two species in the subtidal zone of a southeastern Australian seashore aggressively compete for space—the white ascidian species is enveloping the red bryozoan species and will eventually overgrow and kill it

Animals

Animals compete for a number of different resources within an ecosystem. Animals may compete for mates from the same species. Animals also compete with the same and other species for:

- food
- shelter or hiding places to avoid predators
- shelter or hiding places in defence of territory or young
- shelter for nest sites.

Animals possess various defence mechanisms which may be used in intraspecific and/or interspecific competition. Some can attack intruders using teeth, claws, stingers and/or chemical means. Some use camouflage to hide such as the flower spider (see Fig. 2.25), while others use **mimicry** to resemble dangerous or unpalatable species (see Fig. 2.26). Noxious or unpalatable species, such as some frogs and butterflies, actually advertise that fact with warning colouration such as spots or stripes in bright colours (see Fig. 2.27).

Figure 2.27 (a) The Australian frog *Pseudophryne corroborree* is distinctly coloured gold with black stripes; (b) The monarch butterfly advertises its unpalatability by warning colouration

Figure 2.25 The flower spider (*Thomisus spectabilis*) hides through the use of camouflage

Figure 2.26 The blue poison arrow frog

(a)
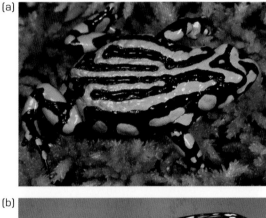

(b)

Effect of competition on populations

Organisms in competition will affect population numbers due to the impact on reproduction and survival rates. Population fluctuations can be directly linked to the competing species and their resource. Figure 2.28 illustrates the fluctuations in abundance that may occur in species competing for a resource. If the resource is a common food source, for example, as food sources become more readily available the abundance of both species increases (see Fig. 2.28: 1967–1968). As food sources decrease so may the abundance of both competing species (see Fig. 2.28: 1968–1969).

In some cases, some species may be better competitors than others. In the 1950s, Birch conducted an experiment observing the population sizes of two species of grain beetles. When the species were sharing the same environment, one species was always driven to very low numbers or became extinct (see Fig. 2.2a). Individuals of the less successful species were out-competed for food by individuals of the species that eventually replaced it. Interestingly, Birch was able to reverse this outcome simply by adjusting one aspect of the beetles' environment, temperature.

Consequences of competition on the ecosystem

Short-term consequences

When two species compete for a resource, the short-term effect is a decrease in population numbers. In most instances, one species is more successful than the other and so one species finds that their population numbers have dropped more significantly than the other (due to an increase in deaths and a decrease in reproduction rates). Depending on the continued success of this one species over the other, this trend may continue. However, depending on the supply of the resource they are competing for, the ability of the 'losing' species to adapt by occupying a different niche, or other environmental factors (i.e. temperature), this trend may change (or even reverse as Birch demonstrated).

Long-term consequences

If the trend of one species successfully out-competing another species continues, the long periods of decreased reproduction rates and increased deaths will eventually lead to the elimination of the 'losing' species in that area, and on the larger scale to possible extinction.

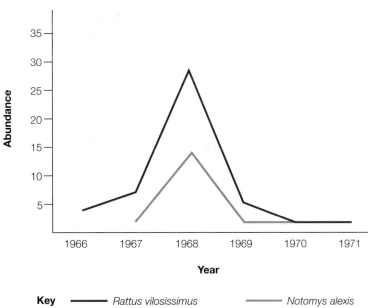

Figure 2.28
Fluctuations in populations of the long-haired rat (*Rattus vilosissimus*) and the spinifex hopping mouse (*Notomys alexis*)

Key —— *Rattus vilosissimus* —— *Notomys alexis*

Extension activity

47

2.10

The impact of humans

■ *identify the impact of humans in the ecosystem studied*

Student activity—
human impact

Figure 2.29 Examples of human impact:
(a) land clearing;
(b) oil pollution;
(c) salinity; (d) cane toad (*Bufo marinus*);
(e) prickly pear (*Opuntia*)

There are three broad types of ecosystems that vary in impact by humans: urban, rural and natural ecosystems. Of course in this chapter we have only been discussing natural ecosystems; however, the human impact on urban and rural ecosystems is significant. The ecosystem that you choose to study may be influenced by urban or rural development alone or perhaps even both. You may observe small pieces of evidence such as rubbish or erosion from walkers which have a small impact on the environment, but it may be distracting you from a larger impact that is not easily observed. Some or many of the examples of human impact below may exist in your ecosystem. You may not be able to see the impact directly; however, evidence of it may be easily observed when visiting the site. Further research into the area studied may provide examples that you have not been able to source directly from the site.

Examples of human impact

(See Fig. 2.29.)

■ Land clearance and habitat fragmentation (e.g. clearing of large areas of ecosystems)
■ Slash and burn agriculture (e.g. clearing with burning)
■ Integrated pest management (e.g. use of pesticides, biological controls)
■ Land and water degradation (e.g. poor waste management, dams, irrigation runoff, roads, mining)
■ Erosion (e.g. livestock, clearing/ploughing, roads, housing development)
■ Soil acidification (e.g. chemical runoff into soil water)
■ Soil and water salinity (e.g. irrigation runoff)
■ Polluting the atmosphere (e.g. industrial gases, vehicle emissions)
■ Introduced species (e.g. fox, rabbit, cane toad, lantana, Paterson's curse, prickly pear).

(a)

(b)

(c)

(d)

(e)

Application of a biological control and its affect on society and the environment

Table 2.2 describes an example of an application of biology, a biological control using cane toads, and its affect on society and the environment.

Using the same table format practise applying PFA P4 to another example such as the prickly pear (see Fig. 2.29e).

PFA
P4

Table 2.2 Application of biology affecting society and the environment

Example	Historical background	How increases in our understanding in biology have led to the development of useful technologies	Relevance, usefulness and the applicability of biological concepts and principles, and the affect on Australian society and the environment	Contribution made to Australian society
Cane toad (*Bufo marinus*) (see Fig. 2.29d)	Cane toads have been present in Australia for nearly 70 years. They were introduced to Queensland in 1935, when approximately 100 individuals were imported from Hawaii in an attempt to control cane beetles. The larvae of French's cane beetle and the greyback cane beetle (*Dermolepida albohirtum*) eat the roots of sugar cane and kill or stunt the plants; they are pests of the sugar cane industry.	Biologists discovered a way of avoiding the use of chemicals (pesticides) to eradicate pests through the use of biological controls. A biological control involves the use of natural predators to decrease a prey population. This particular application of a biological intervention, where the cane toad was thought to be a natural predator of the cane beetle, aimed to reduce the population of cane beetles in sugar cane crops.	Within six months of introducing the 100 cane toads, over 60 000 young toads had been produced. Having no natural enemies, the cane toad became a predator of Australia's native small mammals and thrived in its new environment, spreading rapidly to other areas of Australia (the Northern Territory and New South Wales). They are now a major threat to native animals on the far north coast of New South Wales. Estimated rates of spread vary from between 1 to 5 km per year in northern New South Wales to approximately 60 km per year in the Northern Territory. The cane toad produces in its glands a toxic poison which is detrimental to most animals when exposed to it. It can cause death in small animals (e.g. domestic cats). This has been commonly described as one of Australia's major biological disasters. Not only has it had a major adverse impact on the environment but also on the community—the cost of eradication, attempts at control, and the long-term damage to native species populations.	The failure of this biological application has increased our understanding of the potential dangers of implementing biological controls without extensive prior research into possible interactions with native populations. It has contributed to our understanding for future use of biological controls in Australia.

A blank copy of Table 2.2

Student worksheet— introduced species

Threatened species

The nationally endangered Baw Baw frog (*Philoria frosti*) (see Fig. 2.30) is only found in a small area comprising 135 square kilometres on the Baw Baw Plateau in the Central Highlands of Victoria. This native frog requires a special habitat, breeding in wet areas of subalpine heathland, montane wet forest and cool temperate rainforests. It lays a small clutch of unpigmented eggs within natural cavities under dense vegetation, soil, rocks or logs. In the non-breeding season the frogs move away from wet breeding habitats, sheltering in terrestrial habitats beneath dense vegetation, roots, logs, rocks and leaf litter. These non-breeding sites provide protection from extreme weather conditions. The species has suffered a significant decline in population numbers over the past 15 years, particularly from high-elevation habitats. Likely reasons for the frog's decline include the introduction of an exotic fungus and climatic change. Timber harvesting activities may also threaten remaining populations of the frog due to effects such as habitat destruction and fragmentation or pollution.

Biodiversity

Biodiversity refers to the variety of all forms of life, the diversity of the characteristics they contain and the ecosystems of which they are components. Characteristic diversity within a species is what allows populations to adapt to changes in the environment.

Globally, species are rapidly becoming extinct at the rate of 1000 to 10 000 times the natural rate and it has been estimated that 20 per cent of all species are likely to become extinct in the next 30 years. In Australia, 80 per cent of species are unique to Australia. Over 1150 plant species are endangered and about 145 species of birds, reptiles and mammals are endangered. In total, 27 Australian mammals have become extinct since European settlement. Many mammals were from arid areas but their populations were decimated by altered fire regimes and introduced species. It is predicted that if there is not a rapid change in patterns of land use then up to 50 per cent of bird species in Australia will become extinct within 100 years.

Most of the types of human impact listed above directly affect the biodiversity that exists in Australian ecosystems.

Conservation of biodiversity

Organisms and their roles are essential for ecosystem survival. Species interactions are complex and the loss of key species can have a substantial impact on ecosystems. Should one species disappear others which depend on it for food or shelter may struggle to survive, setting in motion a domino effect within that environment. For example, cassowaries are birds that have an important role in eating rainforest fruit so that rainforest tree species can be dispersed (see Fig. 1.6e on page 241). Cassowaries are threatened due to rainforest clearing and introduced species. This means that, if cassowaries disappear, some rainforest plant species will lack a medium for seed dispersal and struggle to survive.

Figure 2.30 Baw Baw frog (*Philoria frosti*)— a threatened species

Reasons for conserving biodiversity are:

- for bioresources such as food, fibre, medicines, timber etc. Australia has only identified 15 per cent of all its animals and plants. The potential of any undiscovered bioresources is significant. For example, ants possess specialised glands for producing antibiotics to reduce disease in their colonies. These discoveries hold great potential for therapeutic use

- humans enjoy the beauty of the natural environment and countries like to conserve their heritage to be passed down to future generations
- ethically all species have a right to exist just as humans do
- ecosystems underpin many of our natural resources and provide services such as clean water, healthy soil and pollination for crops.

Classroom activity— endangered and introduced species

Field study of a local ecosystem

Background

Choose a terrestrial or aquatic ecosystem that is convenient and easy to get to. This ecosystem may range from a woodland, rainforest, mangrove area or rock platform. Take extra care when studying tidal, or wave-exposed areas. The type of ecosystem chosen will affect the sort of equipment and resources needed to undertake the field study. To complete all of the data collection it is more efficient to work in small groups. Once the group is selected each group member should equally share the workload by dividing up the roles. Students are responsible for carrying out their own background research on the chosen ecosystem, and creating their own graphs after data collection. Follow-up work back in the classroom will be necessary. All the data from your group needs to be collated and distributed amongst the group members ready for individual presentation and analysis of the data. Materials usually needed for this type of field study are a map of the area, field guides of the area if available, lined and blank paper for taking notes and data collection, a clipboard or folder, pencils and pens, pre-drawn plots and tables for completion,

field study task information, and equipment for specific data collection.

Take care while in the field study area. Bring sunscreen, a hat, insect repellent and wear long sleeves to avoid insect bites.

Minimise environmental damage by collecting data in the field and returning to the lab to conduct chemical tests (e.g. pH and salinity tests). Also choose where you walk carefully and replace rocks or anything you disturb to minimise the physical damage and impact on the ecosystem.

FIRST-HAND INVESTIGATION

BIOLOGY SKILLS

P11.1; P11.3
P12
P13
P14.1

Figure 2.31 Students embarking on a rock platform field study near Sydney

Risk assessment sheet

Part 1: Preparation prior to field trip (refer to checklist on page 57, Table 2.5)

- *choose equipment or resources and undertake a field study of a local terrestrial or aquatic ecosystem to identify data sources and:*
  - *—identify data sources and gather data by:*
    - *—tabulation of data collected in the study*
    - *—calculation of mean values with ranges*
    - *—graphing changes with time in the measured abiotic data*

A. Measuring abiotic factors

—measure abiotic variables in the ecosystem being studied using appropriate instruments and relate this data to the distribution of organisms

A blank copy of
Table 2.3

Table 2.3 Measuring
abiotic factors

Once you have selected the aquatic or terrestrial ecosystem that you will study use Table 1.3 provided on page 6 to determine what possible abiotic tests you may need to conduct and list the equipment needed to carry this out in the field. Think about why you choose particular types of equipment over others. Write up the method you will need to follow to complete these tests when out in the field. If you have data loggers and sensors for each type of test then it will be much easier and quicker to collect data. However, data loggers and their probes/sensors are not essential for carrying out the field study data collection. Simple equipment from the lab is just as effective in obtaining data for this study. Streamwatch kits may also be useful.

Construct tables ready for entering data once out in the field. You must include a number of time measurements, the range (lowest to highest value) and mean values (average over a set time period). Choose time intervals depending upon time available at the site. Some suggestions are provided in Table 2.3.

Draft a graph to scale ready to plot points after data collection. Use time as the horizontal axis and each abiotic measurement scale as the vertical axis. Plot all zones for each abiotic factor on the one graph. Think of a title for each graph and label all axes with a name and units of measurement (see Fig. 2.32 on page 54).

Abiotic factor	Equipment used and units	Time measurements (0, 30 and 60 min)	Zone 1	Zone 2	Zone 3	Trend across zones
Air temperature	Thermometer (°C) or a data logger and sensor	1st (0 min—initial reading)				
		2nd (after 30 min)				
		3rd (after 60 min)				
		Range				
		Mean				
Soil temperature	Thermometer (°C) or a data logger and sensor	1st				
		2nd				
		3rd				
		Range				
		Mean				
Water temperature	Thermometer (°C) or a data logger and sensor	1st				
		2nd				
		3rd				
		Range				
		Mean				
Humidity	Hygrometer or a wet and dry bulb thermometer (°C) comparison	1st				
		2nd				
		3rd				
		Range				
		Mean				

Abiotic factor	Equipment used and units	Time measurements (0, 30 and 60 min)	Zone 1	Zone 2	Zone 3	Trend across zones
Air pressure	Barometer (hPa)	1st				
		2nd				
		3rd				
		Range				
		Mean				
Wind speed	Wind speed meter (ms^{-1} or an impression scale 1–5)	1st				
		2nd				
		3rd				
		Range				
		Mean				
Light intensity	Light meter (lux or an impression scale 1–5) or a data logger and sensor	1st				
		2nd				
		3rd				
		Range				
		Mean				
Soil pH	Universal indicator or a data logger and sensor	1st				
		2nd				
		3rd				
		Range				
		Mean				
Water pH	Universal indicator or a data logger and sensor	1st				
		2nd				
		3rd				
		Range				
		Mean				
Soil salinity	Silver nitrate (test for chlorides), hydrochloric acid (test for carbonates), barium chloride (test for sulfates) or a data logger and sensor	1st				
		2nd				
		3rd				
		Range				
		Mean				
Water salinity	Silver nitrate (test for chlorides), hydrochloric acid (test for carbonates), barium chloride (test for sulfates) or a data logger and sensor	1st				
		2nd				
		3rd				
		Range				
		Mean				

B. Estimating abundance and distribution

—estimate the size of a plant population and an animal population in the ecosystem using transects and/or random quadrats

Choose the plant and animal species that you will be attempting to estimate both the population size and distribution of (you may need to do some prior research or get assistance from your teacher).

Population size (abundance)

Using the information provided in Section 1.5 (page 13) select which technique you are going to use to estimate the size of the chosen plant and animal population in your ecosystem (e.g. random quadrats). Once the technique has been chosen list the equipment required to carry this out in the field and write up the method.

Construct transect and/or quadrat plots and tables ready for data collection out in the field. Use the information below as an example.

Plant population size

If you choose to use quadrats for estimating population size ensure that they are randomly placed in each zone and the size of each quadrat is recorded (e.g. 1 m^2). Use a key to indicate different species. Determine how many quadrats you will record for each zone. You may only have time for one or two. Draw up a table to include the number of quadrat recordings you will make with space allocated for the mean, range and total values (see Table 2.4). You will also need a rough estimate of the size of the area you have chosen.

Animal population size

Depending upon the ecosystem you have chosen you may be able to use quadrats again to estimate the abundance of sedentary or slow-moving animals such as barnacles on a rock platform. However, if you have chosen other methods it will involve prior preparation (discuss this with your teacher) and the use of the mark–release–recapture formula to assist in the calculation of abundance.

Figure 2.32 Temperature change over 60 minutes in the three zones of a mangrove ecosystem

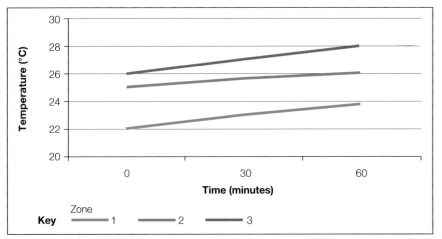

Table 2.4 Population size calculation table

	Plant Percentage cover/counts	Animal Percentage cover/counts
Zone 1 quadrat		
Zone 2 quadrat		
Zone 3 quadrat		
Zone 4 quadrat		
Trend across zones		
Mean value		
Range across zones		
Size of area (m^2)		
Total area abundance estimate (mean × size of area in m^2)		

Species distribution

—collect, analyse and present data to describe the distribution of the plant and animal species whose abundance has been estimated

Using the information provided in Section 1.4, select technique for describing the distribution of your chosen plant and animal species. Now list the equipment and resources needed to carry this out in the field and write up the method.

Construct the chosen plots (profile sketch or transect) that you will be using to record the distribution of a plant and animal species.

Suggestions

■ *Profile sketch* (see Fig. 2.33)—complete the profile sketch across all of the zones.

Carefully choose a key for the species and indicate the scale used (e.g. 1 cm = 1 m).

■ *Transect* (see Fig. 2.34)—mark the transect line with string, or preferably measuring tape, to pass through a representative area of the ecosystem (include all possible zones). Draw the organisms present in the 1 m width of the marked transect line across all of the zones. Carefully choose a key for the species and indicate the scale used (e.g. 1 cm = 1 m).

C. Trophic interactions

■ outline factors that affect numbers in predator and prey populations in the area studied
■ gather information from first-hand and secondary sources to construct food chains and food webs to illustrate the relationships between member species in an ecosystem
—describe two trophic interactions found between organisms in the area studied

Students will need to revise the material on trophic interactions (see Section 2.5 pages 33–9 and Section 2.2 pages 25–7) to make sure that they understand the different types of feeding relationships (in particular predator–prey relationships) and how to construct food chains and webs from feeding information. It is recommended that students research possible organisms and food webs that may exist in their chosen type of ecosystem. This will assist in providing feeding relationship information that may not be easily be sourced during data collection at the study site.

D. Adaptations

■ identify and describe in detail adaptations of a plant and an animal from the local ecosystem

Prior research will assist students in looking at the types of adaptations that may exist in their chosen ecosystem (see Sections 2.7 and 2.8 pages 41–4). With this in mind, students may find it easier to identify and describe in detail the adaptations of a named plant and animal once out at the field study site.

E. Human impact

■ identify the impact of humans in the ecosystem studied

A revision of examples of human impact and the need to look for evidence showing their occurrence (see page 48) will be of assistance. Preparing a list of possible evidence indicating human impact will assist you in knowing what to look for once out at the site. Prior research into the chosen study site may indicate more hidden but large-scale human impact factors, such as nearby industry. It may also provide the history of the area, examples of previous human impact and any changes occurring in recent years.

Figure 2.33 A blank profile sketch ready to use for your chosen ecosystem

A blank profile sketch

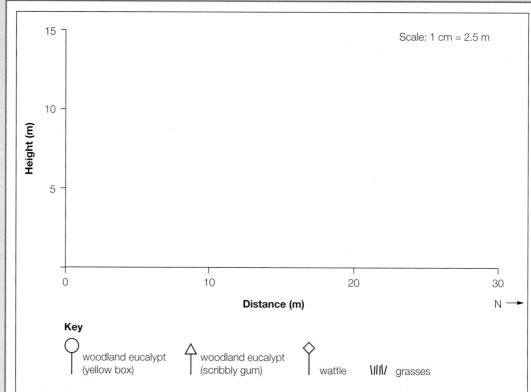

Figure 2.34 An example of transect plots

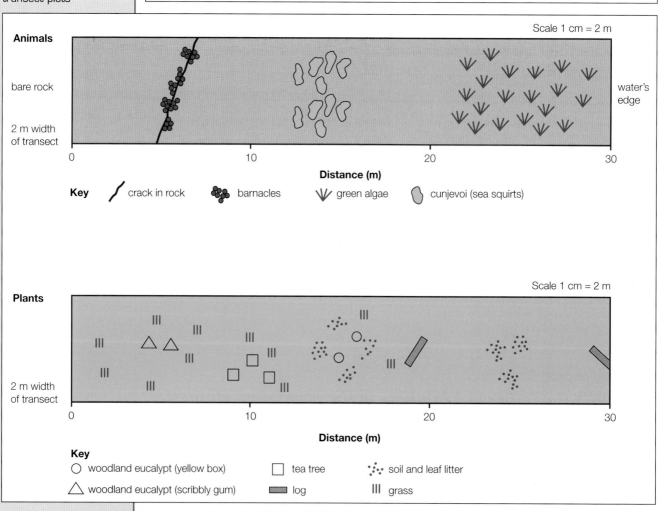

Checklist prior to data collection in the field

You should now have completed the following tasks prior to data collection in the field:

Part 1	Task	Check ✔
General information	■ Name of ecosystem chosen for field study ■ Map of the chosen area ■ Optional field guides to the area	
Method	■ Method written for conducting abiotic tests (including equipment used and units of measurement)	
Data collection preparation		
A. Measuring abiotic factors	■ Table constructed for measuring abiotic factors (including mean values and ranges) ■ Graphs drafted to scale using correct axes fully labelled and titles	
B. Estimating abundance	■ Method written including techniques chosen to estimate the size of the chosen plant and animal population and equipment required ■ Construct transect and/or quadrat plots and tables ready for data collection	
B. Estimating distribution	■ Method written including techniques chosen to measure distribution of the chosen plant and animal population and equipment required ■ Construct transect and/or profile sketches ready for data collection	
C. Trophic interactions	■ Revise Section 2.5 and the construction of food chains and webs ■ Research feeding relationships and food webs in your type of ecosystem	
D. Adaptations	■ Research the types of adaptations and examples that may occur in your type of ecosystem	
E. Human impact	■ Prepare a list of evidence that may indicate human impact to look for at the site ■ Research the actual chosen site and its exposure to recent human impact	

Table 2.5 Checklist for tasks to do prior to field trip

Table 2.5

Note: This is a rough guide in attempting to be as prepared as possible prior to visiting the field study site. If you are unable to complete all of the above tasks in preparation for the field trip much of this can be attempted by sourcing the information while on site. However, you must be aware of the time constraints during the actual site visit and be very organised in order to complete all data collection at the time. Remember, you may not get the chance to return to the site afterwards.

Part 2: Data collection out in the field

Once you have arrived at your field study site write a short description of the area (i.e. weather conditions, slope, aspect). Mark on your map of the area the position of the field study site. Study groups should make effective use of time by sharing the collection of data.

A. Measuring abiotic factors

Once you have marked out the site and determined the zones, start collecting abiotic data and completing your tables. Remember that these measurements are to be repeated a number of times at set time intervals while at the site. In between measurements other tasks can be completed such as estimating abundance and distribution. If you have time, plot a graph for each abiotic characteristic you measured showing the abiotic factor on the vertical axis (*y*-axis) and the time along the horizontal axis (*x*-axis). Time measurements for the different zones are to be plotted on the one graph. You will need to provide a key to identify each zone (see Fig. 2.32). Each axis must be to scale, fully labelled and the graph title must be provided. You may choose to hand draw each graph using graph paper or it can be easily plotted using Microsoft Excel, data loggers or other graphing programs.

B. Estimating abundance and distribution

Start collecting your abundance and distribution data for a plant and animal population.

Note: You may need to re-do some of your plots if messy or incomplete when back in the classroom. Present to scale (with a key for organisms) when drawing a quadrat, profile sketch or transect.

Throughout your data collection, note any observations or evidence of animals in the area (this obviously requires you to move and work quietly during data collection). After data collection, complete Table 2.6 for animal and plant species observed in the area. You may need to use relevant field guides for the area to identify some species. You may ask your teacher for assistance or write a brief description (or take a photograph) to use for identification when back in the classroom.

C. Trophic interactions

Select two examples of organisms interacting in a feeding relationship and complete Table 2.7 below. Select at least one predator–prey relationship.

Factors that affect numbers in predator and prey populations in the area

Use the lists of organisms and relationships that you have completed in Tables 2.6 and 2.7 to outline at least three factors that may affect the numbers of predators and preys in your area. Refer to page 27: 'Factors affecting numbers of predator and prey populations' for examples.

Food chains and web

Using your list of animal and plant species in the area, construct at least three possible food chains (include three organisms or trophic levels in each).

Using the food chains you have just completed, construct a simplified food web for the ecosystem you studied. Be careful to place all arrows correctly. You may need to conduct further research of secondary sources to complete this task.

D. Adaptations

Give two examples of plants and two examples of animals from any zone that show specific adaptation to your ecosystem. Outline the problem and how they appear to have adapted to it.

E. Human impact

Identify any evidence of human impact in the area (at least three) and describe the possible long-term effects of each of these on the ecosystem. Describe possible solutions for two of these problems.

Checklist at field study site

Before you leave your field study site, use the checklist (see Table 2.8) to ensure that all tasks have been completed.

Table 2.6: Animal and plant species in the area

Animal species	Plant species

TR

Mangrove adaptations— field study material

Table 2.7 Organisms in feeding relationships

Organism 1	Organism 2	Description of trophic (feeding) interaction
E.g. Kookaburra	*Worm*	*Predator-prey relationship: carnivore kookaburra consumes the worm*

Part 2	Task	Check ✔
General information	■ General description of the area ■ Mark area studied on site map ■ List members of group sharing data collection	
Data collection		
A. Measuring abiotic factors	■ Complete abiotic factors table (and graphs if time permits)	
B. Estimating abundance and distribution	■ Measure chosen plant and animal population size by conducting transect and/or quadrats plots ■ Measure distribution of the chosen plant and animal population by conducting transect and/or profile sketches	
C. Trophic interactions	■ Complete list of plant and animals species observed in your area ■ Describe two trophic interactions between organisms in your area ■ Outline three factors that may affect the numbers of predators and preys in your area ■ Construct three food chains (including at least three organisms) from your area ■ Construct a simple food web for organisms in your area	
D. Adaptations	■ Give two examples of plants and two examples of animals that show specific adaptation to your ecosystem	
E. Human impact	■ Identify evidence of human impact (at least three) in the area and describe the possible long-term effects of each of these on the area. Describe possible solutions for two of these problems	

Table 2.8 Field study site checklist

Table 2.8

Part 3: Back in the classroom—data analysis and report presentation

- *examine trends in population estimates for some plant and animal species within an ecosystem*
- *process and analyse information and present a report of the investigation of an ecosystem in which the purpose is introduced, the methods described and the results shown graphically and use available evidence to discuss their relevance*
  - *—collect, analyse and present data to describe the distribution of the plant and animal species whose abundance has been estimated*
  - *—identify data sources and gather, present and analyse data by:*
    - *—tabulation of data collected in the study*
    - *—calculation of mean values with ranges*
    - *—graphic changes with time in the measured abiotic data*
    - *—evaluating variability in measurements made during scientific investigations*

Sample field study
assessment task

Background

After each group has completed the data collection from the field study, and students have processed, presented and analysed data, each student must work individually in writing a report.

Process and analyse information

Process and analyse the information from the field study data collection so that it is ready to present in report format. You need to have completed all tasks from your field study (Parts 1 and 2) in order to complete this report.

Present a report of the investigation of an ecosystem

Part 3 is basically a presentation of your findings from Parts 1 and 2, looking at trends found and their relevance. Using scientific writing and format, write a scientific report for your investigation of your chosen ecosystem. Start with the subheadings of:

Aim

- Give a clear description of the aims of the field study.

Method

- Give a brief description of the methods used during the investigation.
- Use subheadings for method sections (e.g. measuring abiotic factors).

Results

- Write a concise paragraph describing the general characteristics of the area.
- Provide a map of the area with the position of your study area marked on it.
- Provide a clear and accurate profile sketch and/or transect drawn to scale with a key for identifying symbols.
- Provide clear and accurate tables of results from population size estimates using quadrats and abiotic measurements. Check that all mean value calculations have been fully completed, including the range.

- Plot a graph for each abiotic characteristic you measured showing any variation between zones.
- Provide a description of two trophic interactions found between organisms in the ecosystem.
- Outline factors that may affect the numbers of predators and prey in your area
- Draw three food chains from your ecosystem and a food web you have devised from all information collected on organisms in the area. Ensure that arrows are drawn in the right direction and clearly placed between all organism feeding interactions.
- Give two examples of plants and two examples of animals from any zone that show specific adaptation to your ecosystem. Outline the problem and how they appear to have adapted to it.
- Identify evidence for two examples of human impact in the area and the possible long-term effects.

Discussion/conclusion (discussion of trends found in results and their relevance)

- Describe trends occurring in each of the abiotic factors across the three zones.
- Describe what your data tells us about the distribution of the plant and animal species in your area.
- Discuss the possible relevance of abiotic factors in contributing to the population size of the plant and animal species you estimated and the distribution of these across the study area.
- Discuss how some organisms in your area have adapted to the conditions of the environment.
- Make two recommendations as solutions to identified problems of human impact on the area.
- Explain how these solutions may maintain the natural balance of the abundance and distribution of species in your ecosystem.

Analysis questions

■ *evaluating variability in measurements made during scientific investigations*

1. Do your abiotic measurements vary significantly over time? Look at your range of values and mean value.
2. (a) Are your group measurements the same as other groups?
   (b) Could this be due to other abiotic or biotic changes or do you think the trend is a valid one?
3. Was human error possibly involved? Discuss.
4. Were the instruments accurate and calibrated carefully?
5. Did you choose the most accurate equipment and method for the measurements you conducted?
6. Suggest how your methods could be adjusted to increase accuracy.

 http://members.optusnet.com.au/-janewest000/Mangrove/home.htm
A virtual field trip of mangroves

Websites with extension activities

REVISION QUESTIONS

1. Outline the factors that affect numbers in predator–prey populations in an area.

2. Identify one example of allelopathy, parasitism, mutualism and commensalism in an ecosystem and briefly describe the role of organisms in each type of relationship.

3. Describe the role of decomposers in ecosystems.

4. Identify the difference between food chains, food webs and pyramids of numbers, biomass and energy. Draw an example of each.

5. Define the term *adaptation* and discuss the problems associated with inferring characteristics of organisms as adaptations for living in a particular habitat.

6. Identify two named plants and two named animals with adaptations to factors in their environment.

7. Identify and describe in detail adaptations of a named plant and a named animal from the local ecosystem you studied.

8. Describe and explain the short-term and long-term consequences on the ecosystem of species competing for resources.

9. Identify the three impacts of humans in the ecosystem you studied.

10. Describe how you measured three different abiotic features in the ecosystem you studied using appropriate instruments and discuss how these factors may determine the distribution of organisms.

Answers for revision questions

PATTERNS IN NATURE

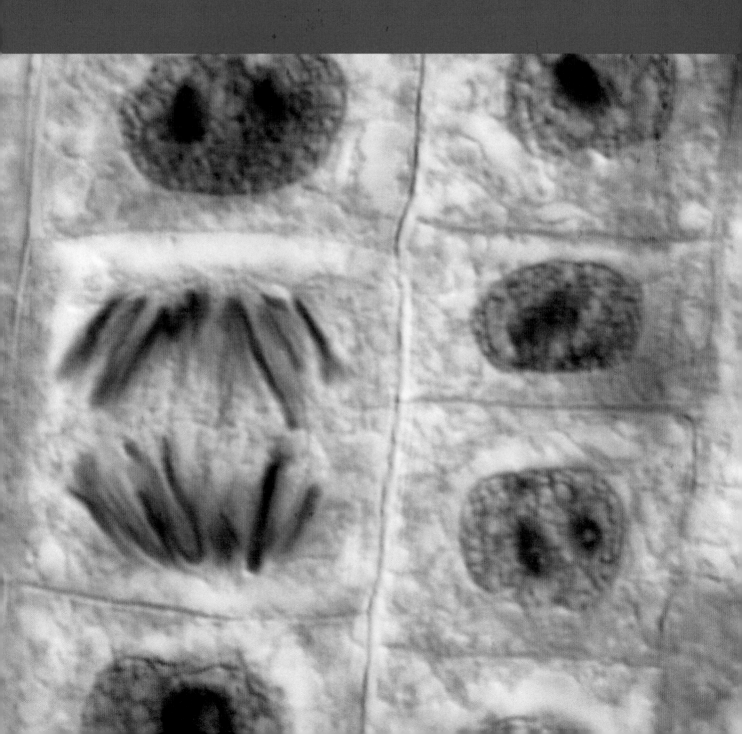

MICROSCOPE BEGINNINGS

THE SCIENTIFIC REVOLUTION

the cell theory

evolution

germ theory of disease

1590	*Hans and Zacharias Jansen* made the first **compound microscope** by placing two convex **lenses** in a tube.
1663	*Robert Hooke* introduced the term '**cell**' while observing cork under a light microscope. He also worked at improving a number of scientific devices, including the microscope, telescope and barometer.
1668	*Francesco Redi* conducted an experiment to challenge the theory of '**spontaneous generation**'.
1674–1683	*Anton Van Leeuwenhoek*, a Dutch lens maker: ■ produced *lenses* of higher quality, which allowed for greater magnification (up to 200 times). ■ described 'animacules' (unicells) ■ discovered bacteria.
1758	*John and Peter Dollard* (father and son), spectacle makers, produced the first achromatic (colour-free) *lenses*, making microscopes superior to hand lenses.
1796	*Edward Jenner* used cowpox in the first successful **vaccine** against the disease **smallpox**.
1801	*Robert Brown* a botanist and naturalist, first described the cell **nucleus** while observing plant cells in an orchid. He also noticed the random movement of pollen grains (Brownian motion).
1836	*Charles Darwin* arrived in Sydney Harbour aboard HMS *Beagle*.
1838	*Matthias Schleiden*, a **botanist**, stated that parts of plants are made of *cells* (not visible to the unaided eye).
1839	*Theodor Schwann*, a **zoologist**, stated that parts of animals are made of *cells*; agreed with Schleiden and they published **the cell theory** in a book, stating that the cell is the basis of the structure of all living things.
1843	*Robert Koch* studied the cause of the disease **anthrax**.
1855	*Rudolph Virchow* introduced the idea that cells reproduce by dividing, stating that all living cells can only arise from other living cells, further challenging the theory of 'spontaneous generation'.
1856–1858	*Gregor Mendel* began a series of controlled experiments with garden peas, to carry out a statistical study of **heredity**.
1858	*Charles Darwin* and *Alfred Wallace* presented a paper 'A Theory of **Evolution** by **Natural Selection**'.
1859	*Charles Darwin*'s book, *On the Origin of Species*, is published.
1860	The *Huxley–Wilberforce* debate takes place.
1861	*Louis Pasteur* published his experiments showing that **fermentation** was caused by something in the air, finally disproving 'spontaneous generation'.
1862	*Louis Pasteur*'s experiments with bacteria showed that infectious diseases are caused by micro-organisms, leading to the *germ theory of disease*.
1863	*Louis Pasteur* introduced **pasteurisation**, a practical application of what he had learnt through his fermentation experiments.
1866	*Gregor Mendel* published his work on *studying plant hybrids*.
1867	*Joseph Lister* made the connection between Pasteur's work on infection and introduced *antiseptic surgery* (published paper).
1880	*Charles Louis Alphonse Laveran* first identified cause of **malaria**: a microscopic organism.
1881	*Pasteur* developed a vaccine against anthrax.

disease	1882	*Walther Flemming* discovered nuclear material—termed '**chromatin material**'.
	1882–1893	*Koch* proposed *postulates*: 'rules of engagement' for bacteriologists.
	1885	*Pasteur* used a vaccine against **rabies** on humans for the first time, saving the life of a young boy who had been bitten by a dog.
	1891	*Robert Koch* concluded that malaria was transmitted by mosquitoes.
	1897	*Ronald Ross* demonstrated that female *Anopheles* mosquitoes were the **vectors** *(carriers) of malaria*, by showing that these mosquitoes carried malarial oocysts in their gut tissue.
genetics	1900	Significance of *Mendel*'s experiments in terms of heredity is noticed after three other scientists get similar results.
	1902	*Walter Sutton* and *Theodore Boveri* independently proposed and demonstrated a connection between **chromosomes** and inheritance. Sutton studied meiosis in grasshoppers. Boveri studied chromosome behaviour and inheritance in sea urchins.
	1911	*Thomas Hunt Morgan* studied *sex-linked inheritance* (Nobel Prize in 1933 for life's work).
	1909	*Wilhelm Johannsen* introduced the term 'gene'.
microscope advances, microbes and antibiotics	1928	*Alexander Fleming* noticed that the mould *Penicillium* killed bacteria in a petri dish.
	1933	*Ernst Ruska* built the first **electron microscope**.
	1935	*Howard Florey* began to search for a useful medicine to kill germs.
	1938	*Fritz Zernike* invented the **phase contrast microscope** which can be used to observe living, unstained cells.
	1939	*Howard Florey* extracted stable **penicillin** (the first antibiotic).
	1941	*George Beadle* and *Edward Tatum* published the results of their experiments with bread mould, in which they proposed the *one-gene-one-enzyme (protein) hypothesis*.
	1942	Viruses first seen under the electron microscope.
	1945	*Frank McFarlane Burnet* isolated influenza A virus (in Australia) and developed a vaccine.
	1945	*Howard Florey* and *Alexander Fleming* received the Nobel Prize for Physiology and Medicine for their work on penicillin.

molecular technology, biotechnology and health	1950	*Rosalind Franklin* and *Maurice Wilkins* made a crystal of **DNA** to study its structure.
	1953	*James Watson* and *Francis Crick* put together a model of DNA.
	1955	*Marvin Minsky* invented the *scanning* electron microscope.
	1960	*Frank McFarlane Burnet* and *Peter Medawar* received the Nobel Prize for Physiology and Medicine for their work in *immunology* and *organ transplants*.
	1962	*Vernon Ingram* did further work on genes and proteins leading to the change to the *one-gene-one-polypeptide hypothesis*.
	1962	*Watson, Crick* and *Wilkins* received the Nobel Prize for Chemistry for their discovery of DNA. (Rosalind Franklin died in 1958; her work was acknowledged, but Nobel prize nominations cannot be awarded posthumously.)
	1972	*Niles Eldridge* and *Stephen Jay Gould* put forward the theory of *evolution by punctuated equilibrium*.
	1980	**WHO** declared the disease smallpox eradicated worldwide.
	To present	Genetic and reproductive revolution: in-vitro fertilisation, genetic engineering, cloning and advanced biotechnology.

Note: Dates in many timelines show slight inconsistencies when compared. This is due to inconsistent record-keeping long ago. It is the *sequence of events* that is more important in reflecting the historical developments in science, than the absolute dates.

Cells and the cell theory

Organisms are made of cells that have similar structural characteristics

Introduction

Up until 400 years ago, objects that were too small to be seen with the naked eye could not be examined successfully. Magnifying glasses had been in use since the 13th century, but were still fairly ineffective instruments of observation because of the imperfect shape of the lenses and the low quality of the glass used to produce them.

The study of living things was popular, but at a **macroscopic** level, based on what could be viewed with the naked eye or with the lenses available—living organisms had certainly never been considered at a cellular level. Biologists at that time were called 'natural scientists', suggesting a broad study of nature. Today, biologists study living things not only at a macroscopic level, but also at a **microscopic** (cellular and sub-cellular) level and even at a molecular level. This progress began with the discovery of the microscope.

Characteristics of living organisms

From your studies in junior science, you will be familiar with today's accepted idea that all living things are made of one or more units called cells. This is the most basic characteristic of *living* things. How does one distinguish between something that is *living* and a *non-living* thing? All living things are made of cells, but based on everyday observations without using a microscope what tells us that, for example a beetle is alive but a stone is not?

Certain characteristics or life functions are common to all living things. Living things are made of one or more cells and they can:

- *reproduce*—produce offspring that resemble the parents
- *grow*–increase in size
- *move*—even plants can make some small movements such as opening and closing petals

Figure 1.1 Living or non-living? A beetle as opposed to stones

■ *respire*—produce chemical energy by taking in oxygen and combining it with sugar, giving out carbon dioxide as a by-product
■ *excrete*—get rid of wastes such as carbon dioxide
■ *respond to stimuli* in the environment—such as moving towards food or growing towards light
■ *obtain nutrients*
■ *die*—death is when all of the above functions cease.

To be *dead*, something must have once been living and when all of its life functions cease, death results. This differs from *non-living* things that are not alive and never were.

These functions of life are easy to picture in complex **multicellular** organisms, such as insects, sunflowers and humans. In **unicellular** organisms (microscopic living things made of only one cell), all of the life functions listed above still occur, but each single cell carries out every function.

The discovery of the cellular basis of living things

■ *outline the historical development of the cell theory, in particular, the contributions of Robert Hooke and Robert Brown*

Introduction

The statement that *all living things are made of cells* forms the basis of what is currently termed *the cell theory*. The historical development of the cell theory is interwoven with the story of the invention and development of the microscope. Improvements in the design and use of microscopes, as well as progress in techniques to prepare specimens for viewing, play a significant part in advancing our knowledge and understanding of cells.

To study the *historical development* of a theory (see PFA P1 on page ix), we need to know the currently accepted view ('now') and the views in the past ('then'). New ideas are often linked to *advances in technology*, which allow new discoveries to be made (see PFA P3 on page ix).

(The PFAs or Prescribed Focus Areas are different emphases in the Preliminary and HSC biology curriculum designed to increase students' understanding of biology as an ever-developing science. See page ix.)

The cell theory

The cell theory forms the basis of all biology. In its universally accepted form, it states that:

1. All living things are made of cells.
2. Cells are the basic structural and functional unit of organisms.
3. All cells come from pre-existing cells.

However, this has not always been the accepted biological view.

A **scientific theory** is a broad and general idea or *explanation* provided by scientists, and is related to observations and is supported by a large amount of evidence. It is not a fact and cannot be proved; it can only be supported or not supported by evidence. Since an explanation is a product of the mind, it is not a fact and therefore a theory may have to be modified if new evidence arises that no longer supports it.

Theories are tested by examining whether their consequences (predictions) are supported by observation and experiment.

The build up to the proposal of the cell theory is interesting and, in reading this historical account, we as scientists should look for evidence that has been gathered to validate the theory before we accept it (see PFA P2 on page ix).

Biological view prior to the proposal of the cell theory

Before the discovery of the microscopic world

Until the last decade of the 16th century, microscopes did not exist, cells had never been seen and so the living world had not been considered at a cellular level. One of the accepted views was the theory of **spontaneous generation**. This theory predicted that living creatures could arise from inanimate (non-living) material. This idea dated back to the time of Aristotle and the evidence was based on observation. For example it was noticed that maggots (fly larvae) appeared on rotting meat if meat was left exposed for a period of time. In the 1500s, this theory was being challenged, but it was not until the mid 1600s that scientists suggested that the flies that visited the meat contributed to the appearance of the maggots. Francesco Redi (1668) performed an experiment that tested this hypothesis successfully, showing that maggots only appeared in meat that had been exposed to flies in the environment— if the meat was covered, no maggots arose. This is one of the first recorded examples of experimentation being

used to oppose a theory—an example of the '**scientific method**' of today (see PFA P3 on page ix).

The idea that the meat 'spontaneously' gave rise to maggots may seem ridiculous now, but seems less so if one considers that people could not see fly eggs in those days. What happened between 1500 and 1668, to encourage people to think differently?

The invention of the compound microscope

In the late 1500s, scientists, who were using poor quality magnifying glasses to view small or 'minute' objects, tried many things to improve the images that they were viewing. The idea that led to the invention of the first compound microscope was that, to get a larger and clearer image, two convex lenses could be placed one above the other. The lower lens would produce a magnified image of the object and the upper lens would further magnify or enlarge the first image.

Two Dutch lens makers, a father and son named *Hans and Zacharias Janssen*, are credited with having made the first compound microscope in 1590. A **simple microscope** uses only one lens to magnify an object viewed, so the invention of the **compound microscope** relied on the principle of using *two* lenses, kept *a set distance apart*. It consisted quite simply of two convex lenses placed at either end of a wooden tube to keep them the ideal distance apart from each other. These tubes could magnify objects 3 to 9×, were held by hand and formed the basis of the first compound microscopes. (They had not as yet been named microscopes.)

Discovering a compound microscope

Figure 1.2 The first compound microscope (circa 1595)

Biological studies and technology that led to the proposal of the cell theory

Technology: improvements to the compound microscope

As people across Europe continued to use what are today known as compound microscopes, these instruments were being refined and improved upon all the time. By the early 1620s, most microscopes in use had a magnification of about 30×, but it is recorded that those used in Italy had magnifications of about 150×. This was probably due to the high quality glass that the Italians used, producing lenses of greater clarity. (Italy is still renowned for its high quality glass today.)

For a lens to be effective, it needs to do two things:
1. give an enlarged view of an object
2. make the detail appear clear, giving a precise (not fuzzy), outline to the parts of the object being viewed.

The ability to enlarge an image is termed **magnification**. The ability to show fine detail, distinguishing two very close objects as separate images, is termed **resolution**. A good quality lens is one that has *high magnification* and *high resolution*. The convex shape () of a lens enables it to magnify an object, but to get this shape, one has to grind the glass. Both the quality of the glass used, as well as the manner in which the glass is ground to minimise imperfections, play a role in determining a lens's resolution or resolving power.

During the 17th century, the hand-held tube designed by the Janssens was mounted onto a stand and the design of the microscope as we know it today began to take shape. *Robert Hooke* in England, *Anton van Leeuwenhoek* in Holland, and *Galileo Galilei* in Italy all made noted contributions to improving the design of microscopes.

Robert Hooke's compound microscope was progressive for its time because it used a fine adjustment knob to move the tube holding the lenses up and down. His microscope also had a light source to illuminate the specimen—another lens that concentrated the glow of a candle onto the specimen. Hooke probably had his microscopes built London, but he ground his own lenses. He gave the first demonstration of the use of his microscope to the Royal Society of London in 1663.

Biological view: understanding living things using a microscope

Robert Hooke

In 1665 Robert Hooke produced a book, the first recorded publication to describe observations of living tissue using a microscope. Entitled *Micrographia: physiological studies of minute bodies made by magnifying glasses*, his book included 57 diagrams. It was in this book that he used the term '**cell**' to describe the 'honeycomb' elements (units) of cork. He was looking at dead plant cells which had no contents and clearly resembled small *compartments,* similar to the cells of monks. Hooke's findings were respected, but not universally accepted by scientists at that time.

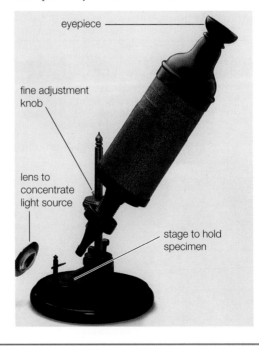

Figure 1.3
Hooke's compound microscope

The low quality lenses in use still distorted images and separated colours, giving a rainbow 'fringe' to the objects being viewed and so many scientists were sceptical about the 'artificial images' created.

Anton Van Leeuwenhoek

Anton Van Leeuwenhoek was a Dutch lens maker whose grinding technique was far superior to that of his contemporaries and so he was able to produce lenses of much higher quality. As a result of his work between 1674 and 1683 with crystal, quartz and even diamond lenses, he developed a **simple microscope** that used a *single*, powerful lens that could magnify up to 300×, perhaps more. The single lens used did not have the usual aberrations associated with lenses at that time. Unfortunately, Van Leeuwenhoek did not record his technique and so similar lenses could not be produced after his death. Starting off in the cloth trade, Van Leeuwenhoek used very many microscopes to study both fabrics and a great variety of living tissue. He discovered many single-celled living things, but because there was no cell theory at the time Van Leeuwenhoek had no framework in which to accurately name or describe his findings. When Van Leeuwenhoek

first presented his work to the Royal Society of London they asked Robert Hooke, a member of the society, to confirm these findings, which he did. Evidence of Van Leeuwenhoek's findings are documented in letters to the Royal Society, spanning 50 years. These letters have been translated from Dutch into English and Latin. Van Leeuwenhoek is credited with discovering bacteria, and from his descriptions, may have even seen nuclei.

Figure 1.5 Robert Brown

Robert Brown

Robert Brown, a Scottish botanist, is known for his discovery of the cell **nucleus** (plural *nuclei*). Although he first described nuclei seen in the outer layer of cells in orchid plant tissue, he discovered that nuclei were present in a wide variety of plant tissues that he studied. He had no idea at that time of the importance of the nucleus or its function in cells. (Robert Brown is famous in science for his diverse discoveries, including being the first person to observe and describe Brownian motion. He also travelled on a ship with Captain Mathew Flinders to Australia in 1801 and he identified many genera of Australian plants.)

Figure 1.4
Van Leeuwenhoek's simple microscope

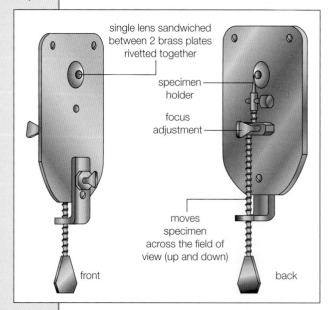

single lens sandwiched between 2 brass plates rivetted together

specimen holder

focus adjustment

moves specimen across the field of view (up and down)

front

back

Microscopes as valid scientific instruments

It was about 200 years after the discovery of microscopes, from the time when the first compound *optics* microscopes were in use (1824), that microscopes began to be acknowledged as useful scientific instruments. The lenses of these optics microscopes were achromatic (did not separate colours) and they no longer produced distorted images. There was the added benefit of powerful light sources and precise focusing screws, thereby increasing the precision of the instruments in general. With the improved technology, scientists became less suspicious of the 'artificial images' and observations made were accepted as valid scientific evidence. It is not surprising that, shortly after this time, the *cell theory* was proposed.

Schwann and Schleiden

In 1838, apparently over a cup of coffee after dinner, two German scientists—*Theodor Schwann* and *Matthias Schleiden*, were discussing the results of their microscopic studies of living things. As Schleiden (a botanist) described the regular placement of nuclei that he had observed in plant cells, Schwann (a zoologist) recognised a similarity to the animal cells that he had been studying and they both went right then to Schwann's laboratory to look at his slides. It was the first time that a *common basic structure for all living things* had become evident. A year later (1839) Schwann published a book on plant and animal cells, listing three main conclusions, two of which are still accepted today as the basis for **the cell theory**. Schwann's first two conclusions are summarised below.

1. *The cell is the unit of structure of all living things.*
2. *The cell exists as a distinct entity and as a building block in the construction of organisms.*

Further investigation led to evidence that his third conclusion, *cells form by free-cell formation, similar to the formation of crystals*, is not valid.

Rudolf Virchow

The accepted version of how cells arise is attributed to a German medical scientist, *Rudolf Virchow*. In 1855 his studies led to his statement that: 'Where a cell arises, there a cell must have previously existed'. From this is derived the accepted third statement of the cell theory:
3. *All cells come from pre-existing cells.*
Virchow had not only discovered cell division but, by implying that living things could not arise from non-living elements, had convincingly refuted spontaneous generation. In 1879, *Walther Fleming* confirmed Virchow's observations and named the process of division **mitosis**.

Effect of microscope on disease theory

From the time of Hippocrates until the discovery of cells, it was believed that disease resulted from 'imbalance in body humors'. This was replaced with a cell-based theory of disease—look at the timeline (see pages 61–5) to discover the close relationship between the discovery of the cell theory and advances in the understanding of disease.

Collaboration in science: the importance of the contributions of Hooke and Brown

Although the work of other scientists was not formally acknowledged by Schwann in his book, the basic cell theory is today attributed to both Schleiden and Schwann and significance is given to the work of previous scientists such as Hooke and Brown. It was the regular placement of nuclei in plant and animal tissue which suggested to Schleiden and Schwann that all living tissue has a similar, compartmentalised basis.

This compartmental nature of tissue led them directly to the idea that cells are the basic unit of living things. Without the work of Hooke (who, more than 150 years before, had recorded the compartmentalised nature of cork and named these compartments 'cells') and Brown (who had discovered the nucleus six years before Schwann's book was published), Schleiden and Schwann could not have built their cell theory. It is noticeable that it is often the collaborative work between scientists, as well as their building on the work of previous scientists, that leads to a new theory in science.

The cellular basis of life was a major breakthrough in biological thinking and led not only to further studies of cells, but also to a cell-based theory of disease. You will notice in the timeline summary (see pages 64–5) that both the discovery of cells and progress in the study of disease coincided with advances in microscopy (the history of the discovery of disease forms part of the HSC course).

Figure 1.6
Photomicrographs:
(a) plant and (b) animal cells seen under a light microscope showing the compartmental nature of cells

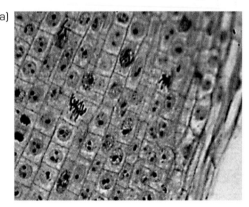
(a)

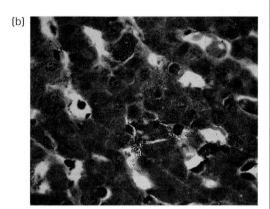
(b)

Evidence to support the cell theory

■ *describe evidence to support the cell theory*

(PFA P1) The evidence to support the cell theory has been described in detail, along with the historical development of the cell theory on pages 67–72. Table 1.1 provides a summary of these findings.

Table 1.1 Summary of evidence for the cell theory

Time frame and/or person	Contribution (discovery or proposal)	Evidence to support finding	Response of scientific community (acceptance or rejection and grounds)
Past: time of Aristotle (380 BC) until the Renaissance (14th to 16th century)	Spontaneous generation: belief that creatures could originate from inanimate (non-living) material.	People relied on observation with the naked eye and drew inferences from what they saw (e.g. rotting meat left exposed developed maggots—fly larvae).	Theory accepted, but was being challenged. (Cells not yet discovered.)
1663 Robert Hooke	Introduced the term 'cell'.	Observed units seen in thin slices of cork using a compound microscope (published in Hooke's book *Micrographia*).	Not well received at first— believed distorted images and colour separation may have given 'artificial images'. Later accepted.

Time frame and/or person	Contribution (discovery or proposal)	Evidence to support finding	Response of scientific community (acceptance or rejection and grounds)
1674–1683 Anton Van Leeuwenhoek	Discovered bacteria. May have seen cells or nuclei.	Viewed microscopic 'animalcules' ('tiny beasties living all around us'), viewed 'globules' in tadpoles and eggs. Findings recorded in letters to the Royal Society of London.	Royal Society asked Robert Hooke to verify these findings, which he did.
1801 Robert Brown	Discovered the nucleus in cells.	Microscopic studies of plants (orchids) and later many other plant tissues revealed that each cell had a nucleus.	Discovery of nucleus noted, but were not aware of its importance.
1838 Schleiden and Schwann	Proposed the cell theory: 1. *All living things are made of cells.* 2. *Cells are the basic unit of organisms.*	Microscopic examination of plant tissue (Schleiden), and animal tissue (Schwann), revealed a common cellular basis for all living tissue. Findings published in Schwann's book, *Microscopic investigations on the accordance in the structure and growth of plants and animals.*	Two of three statements accepted by scientific community and still hold true today.
1855 Rudolf Virchow	Proposed cell theory: 3. *All cells come from pre-existing cells.* This disproved the theory of spontaneous generation.	Studies of living tissue using a microscope showed that cells only arise if other cells are present to give rise to them.	(1879) Walther Fleming confirmed Virchow's observations and named process 'mitosis'.

Technological advances and the development of the cell theory

1.2

■ *discuss the significance of technological advances to developments in the cell theory*

Continued advances in light microscope technology

Improvements to the light microscope continued and, in the 1870s, **oil-immersion lenses** were introduced by Zeiss and Abbe, enabling a good image of up to 1500× magnification to be seen. By the 1890s the top-level microscopes of the time were fairly similar in their viewing capacity to the current senior school microscopes. Over the next 100 years improvement to the quality of images produced by microscopes has resulted from ongoing research into the technology (see pages 74–8).

Because the microscopes at this stage of the advance in technology were similar to those that you currently use at school, this is an appropriate time for you to become acquainted with the workings of a compound microscope (see classroom activity on next page).

Figure 1.7 The compound microscope

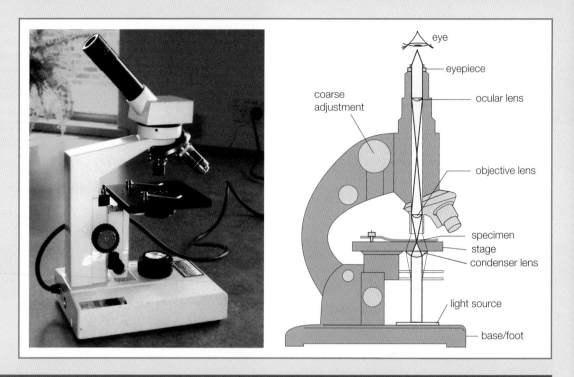

CLASSROOM ACTIVITY

This classroom activity practical is continued on the Student Resource CD and the Teacher Resource CD

Practical introduction to using a microscope

This investigation, although not specified by the Preliminary Course syllabus, is recommended to guide students in the correct use of a compound light microscope. It should also assist their understanding of the size of microscopic fields, magnification and resolution, and the importance of introducing contrast to improve the image that is being viewed.

The microscope is the main technology used to investigate cell structure and functioning. Three main attributes of a microscope that allow you to clearly view a specimen are the *magnification*, *contrast* and *resolution* of a microscope. It is also important for you to understand *measurement* under the microscope. In this introductory practical we will:

■ identify parts of a microscope and investigate the functions of each part, including the diaphragm (for contrast)
■ investigate magnification and become familiar with microscopic units of measurement
■ estimate/calculate the diameter of the fields of view of a microscope
■ investigate resolution.

The invention of the electron microscope

■ By the end of the 19th century compound light microscopes had been developed to a point where they were no longer limited by the quality of the lenses—their main limiting factor had become the wavelength of light. The wavelength of visible light (0.5 μm) limits the resolving power of microscopes so that objects closer together than 0.45 μm are no longer seen as separate objects, even if the shortest wavelength of light is used. The best optical microscope cannot effectively magnify larger than 2000×. This led scientists to begin experimenting with forms of energy other than light.

■ The next big breakthrough in our knowledge of cells was as a result of the invention (1933) and advancement of the **electron microscope**. With this technology images are produced using a *beam* of electrons—electrons that are made to behave like light (waves). In 1928, Ernst Ruska and his supervisor Max Kroll built the first electron microscope, but it only had a magnification of 17×. Ruska continued working on the device and by 1933 he had built the first transmission electron microscope that could magnify up to 12 000×. Ruska's team continued working on the electron microscope during the second world war, achieving a magnification of one million times.

■ The basic principle of the *transmission electron microscope* is similar to that of the compound light microscope, except that the energy source transmitted through the specimen is a beam of electrons instead of a beam of light. Modifications to the design have had to be made because electrons do not normally travel in a manner similar to light, but bounce off anything that they hit, such as air. The electrons must therefore pass through the specimen in a *vacuum*, making it possible to view only non-living, preserved tissue. The electrons are focused by electron *magnets*, rather than by glass lenses, and the image is produced on a screen where it shows up as *fluorescence*, or it may be projected onto a photographic plate.

■ The invention of the *scanning electron microscope* followed in 1955. The electron beam causes the specimen to emit its own electrons, producing a three-dimensional image (but it has a low resolution). The picture on the front cover of this textbook was produced by a modern day, scanning electron microscope.

Advantages

The main advantage of the transmission electron microscope is the high magnification and resolution which show an enormous amount of detail. The electron microscope reveals structures at not only the cellular level, but also the sub-cellular level. Many parts of cells (organelles) were seen for the first time after the invention of the electron microscope. Other parts previously seen with the light microscope can be seen in far greater detail, providing increased knowledge of their internal structure. This has led to an understanding of their functions in cells.

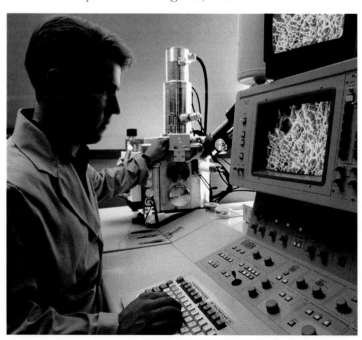

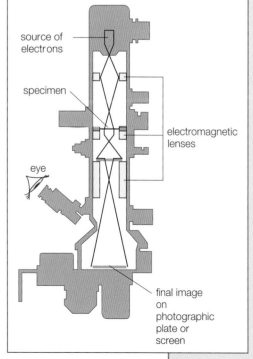

Figure 1.8 The electron microscope

source of electrons

specimen

eye

electromagnetic lenses

final image on photographic plate or screen

Disadvantages

The main disadvantage of the transmission electron microscope is that the specimen must be placed in a vacuum for viewing, as air would interfere with the flow of electrons. As a result, living tissue cannot be viewed. This leads scientists to question how different the preserved specimens are from living tissue, as a direct comparison cannot be made.

Another difficulty is the size, expense and maintenance: electron microscopes are very large (one microscope fills a small room), must be kept at constant temperature and pressure, and are extremely expensive. As a result, they are not accessible to the general public or to schools. The biology department of a university usually has an electron microscope, but it is in high demand and researchers would need to book time to use it.

Techniques for preparing specimens for viewing

The preparation of tissue for viewing under microscopes has become an integral part of microscopy—as microscopes improved, technology for specimen preparation has had to keep up.

Two main criteria must be met when preparing tissue for viewing under the microscope:

1. The sections must be *thin* enough to allow light or electrons to pass through them.
2. Very thin sections of living tissue are mostly transparent, so the structure is difficult to observe unless some *contrast* is created between the tissue and its background.

While preparing the tissue for viewing, the technique should minimise the alteration of tissue from its living form, otherwise what we view under the microscope may be an *artefact* (aberration or 'artificial image').

To meet these criteria, a four-step process is used to prepare slides involving *fixation, embedding, slicing* and *staining*:

1. fixation: the tissue is placed into a preservative substance that kills it and preserves it, as closely to the living from as possible. In some cells chemical fixation disrupts the cell and its contents, so it is important to study cells prepared in a variety of ways
2. embedding: tissue is embedded in a hard medium such as wax (or an even harder substance such as resin for electron microscopy), to overcome the difficulty of cutting soft tissue into very thin sections
3. slicing or sectioning: a machine called a *microtome* was invented, which could cut much thinner sections of tissue more smoothly than could be done by hand. An *ultramicrotome* has been invented more recently to allow ultra-thin specimens to be cut, suitable for viewing under the electron microscope. The thinner the section, the greater is the clarity of the image being viewed
4. staining: *colour* is produced by a variety of stains to create a contrast between the transparent material and its background or heavy metals may be used to stain tissue for viewing under the electron microscope.

Historical evidence of specimen preparation

Robert Hooke noticed that he could get a clearer view of his cork cells if he cut a section very thinly to allow the light to pass through it.

The use of dyes to stain tissue and improve visibility in specimens began in the late 1770s, but it was in the 1880s that Walther Flemming, using synthetic dyes, named the material that became most strongly stained **chromatin material.** And in 1888 Wilhelm Waldeyer-Hertz named the shortened threads of chromatin, **chromosomes** (chromo = coloured; soma = body).

Use a table to compare (show the differences and similarities between) the light microscope and the transmission electron microscope. Headings that may be useful as points of comparison are suggested below:
- Energy source
- Focus
- Specimen preparation
- Magnification
- Resolution
- Can live specimens be viewed?
- Image—colour or black and white?
- Advantages and disadvantages

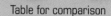

Table for comparison

By using advanced preparation techniques to view tissue under the microscope, our knowledge and understanding of cell structure is further increased.

Current biological research, technology and the cell theory

The electron microscope and further developments in the cell theory

The development of the electron microscope has allowed scientists to study the ultrastructure of cells (parts smaller than can be seen with a light microscope). Electron microscopes are now also linked to computers; this allows the study of sub-cellular structures in enormous detail, providing evidence of their functioning. This technology is also used in the areas of genetics and ecology, providing evidence which has resulted in modern biologists adding a further three statements to the original cell theory. The modern day additions are that:

4. *Cells contain hereditary information which is passed on during cell division.*
5. *All cells have the same basic chemical composition.*
6. *All energy flow (resulting from chemical reactions) of life occurs within cells.*

Further advances in microscopy

Phase contrast microscopes

These microscopes use an alternate way of creating contrast that does not involve altering the specimen. They take advantage of the fact that when light passes through structures of different densities, it changes *phase* because of the wave-like nature of light. A *phase contrast* in the incoming light is created by the different optical system of the microscope.

Cutting edge technology—contemporary light and electron microscopes

- Current developments in compound light microscopes include link-ups with computers, where the image can now be digitally enhanced. *Confocal microscopes* use laser light to allow a three-dimensional view of a specimen to be built up, similar to medical scans. This has the advantage that the specimen no longer needs to be sliced into sections to be viewed.
- *Synchrotrons* are very recent microscopes that accelerate electrons to a speed close to that of the speed of light. They can be used to study structure at the atomic level and, like most electron microscopes, they control the direction of movement of the electrons with magnets.

At the time of writing, an Australian synchrotron is being built at Monash University and will be about the size of a football field.

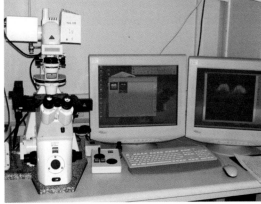

Figure 1.9 A confocal microscope

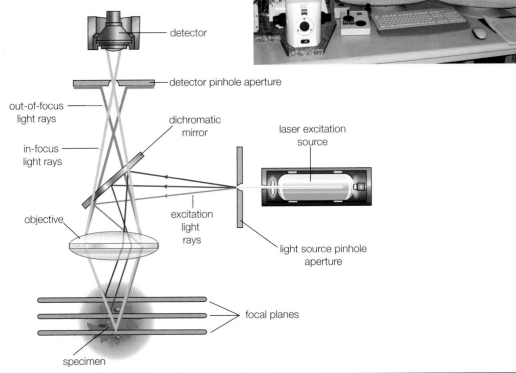

The impact of technology on the development of the cell theory

SECONDARY SOURCE INVESTIGATION

PFAs

P3

BIOLOGY SKILLS

P11.1
P12.3; 12.4a—f
P13.1a—e
P14.1; 14.3b, d
P15

■ *use available evidence to assess the impact of technology, including the development of the microscope, on the development of the cell theory*

Scientists in the past were limited in their research by the technology available to them. As equipment and techniques became more sophisticated, they could collect new evidence, leading to new biological views/theories.

Task

This is a complex task that requires high-order thinking skills from students, so a suggested method of tackling this task is given below.

1. Collect relevant information about two things:
■ the advances in technology (such as microscopes and techniques for preparing specimens for viewing)

■ the improvement in understanding of a biological concept (the cell theory) over time. Plot the relevant information on a timeline.

2. Analyse information to enable you to answer the dot point. Using both your timeline and information in the textbook, answer a series of questions which should improve your understanding of the links between the history of the cell theory and the history of the invention and improvement of the microscope.

3. Answering the dot point: a 'scaffold' has been provided on page 80 to assist you with this step.

Introduction

The history of the development of the cell theory is closely linked to the invention and improvement of the microscope. The PFA P1: *Outline the historical development of major biological principles, concepts and ideas* is a precursor to PFA P3: *assessing the impact of technological advances on understanding in biology*, and so we begin our research on the history of the development of the cell theory by using a timeline. The suggestions below will help you to streamline your search for information, so that it is concise and relevant.

Timeline activity

Read pages 65–78 and any additional secondary source information, and then produce a timeline as outlined below. Remember to use a variety of sources and to crosscheck any uncertainty in the accuracy of your information.

- Research the main contributions to the cell theory that each of the following scientists made and the dates of these contributions:
  —Robert Hooke
  —Hans and Zacharias Janssen
  —Anton Van Leeuwenhoek
  —Matthias Schleiden
  —Theodor Schwann
  —Rudolf Virchow
  —Ernst Ruska.
- Draw a timeline showing the chronological order of the historical development of the cell theory: draw lines on the upper side of the line to list the invention of the technologies; below the line, list the discoveries made that led to the proposal of the cell theory. The earliest date should be on the left of the timeline and the most

recent date on the right. Be sure your spacing shows a reasonable approximation of the amount of time elapsed between dates.
- Label the timeline in a logical and legible manner:
  —record the name, date and contributions of each scientist to the cell theory below the timeline. Some dates may vary slightly in different sources: evaluate the source and use the one you think is most accurate
  —record the advances in technology made (e.g. improvements to the microscope and specimen preparation techniques) above the timeline.

Answering the dot point

(PFA P3) Use the information summarised in your timeline as a guide to answering the questions below. You will also need to refer to more detailed information (see pages 65–78) to answer some of the questions. You may answer these questions as a rough draft on A4 paper first and then transfer your final answers to the scaffold provided, or you may write them straight onto (or type them on the computer into) the scaffold provided.

1. Technology

1.1 Identify the technology available PRIOR to the proposal of the cell theory and outline its uses and limitations.

1.2 Identify the technology (microscopes and specimen preparation) available AT THE TIME of the proposal of the cell theory and outline its uses and limitations.

1.3 Identify the most advanced CURRENT technology available and describe

PFA P3

Figure 1.10 Example of how to draw a timeline

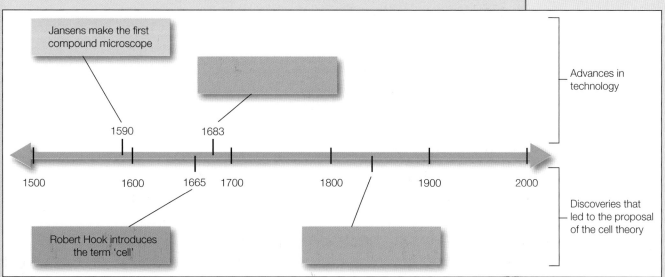

Jansens make the first compound microscope

1590

1683

Advances in technology

1500   1600   1665   1700   1800   1900   2000

Robert Hook introduces the term 'cell'

Discoveries that led to the proposal of the cell theory

four ways in which this technology (microscopes and specimen preparation) is an improvement on the past technology.

2.   Knowledge and understanding

2.1   Outline any areas of knowledge and understanding of the cell theory that came about as a result of PAST technology.

2.2   Outline FURTHER areas of knowledge and understanding of the cell theory that resulted from the use of CURRENT technology.

3.   Putting the two together

3.1   Explain how the advance in technology allowed the progressive accumulation of knowledge and understanding of the cell theory. (Remember, *explain* means 'relate cause and effect'; that is, show the relationship between the improvement of the microscope and the increased knowledge and understanding about the cell theory.)

3.2   Assess the impact of (1) on (2); that is, 'make a judgment of the value' of the advance in technology on the development of the cell theory.

Table 1.2 The impact of technology on knowledge and understanding

A blank copy of Table 1.2

For a sample answer of Table 1.2

TECHNOLOGY
(Identify technology and outline its uses and limitations)

THEN	NOW
PRIOR to: the proposal of the cell theory: *PAST*: at the time of the proposal of the cell theory	*CURRENT*: technology (outline *four ways* in which microscopes and specimen preparation have improved)
IMPROVEMENT (advance) in technology:	

KNOWLEDGE AND UNDERSTANDING

THEN	NOW
PRIOR to: the cell theory *PAST*: at the time of the proposal of the cell theory	*CURRENTLY*:
IMPROVEMENT in (progressive accumulation of) knowledge and understanding:	

PUTTING IT ALL TOGETHER

Explain how the *advance in technology* allowed the *progressive accumulation of knowledge and understanding* of the cell theory.
[Show the relationship between the improvement of the microscope and the increased knowledge and understanding about the cell theory.]

Assess the *impact of technology* on the *development* of the cell theory.

Cell structure and functioning

1.3

Introduction: levels of organisation

Most living organisms that are seen every day consist of many cells and are termed **multicellular**. However, some living things consist of only *one* cell that carries out all of their life functions. These are said to be **unicellular** and can be seen with a microscope (e.g. Protists such as *Euglena*, *Amoeba* and *Paramecium*, living in pond water,

and some disease-causing organisms such as bacteria) (see Fig. 1.11).

The term 'cell' is therefore used to describe the basic unit of any organism, whether it is the only unit or one of many units making up an organism. Table 1.3 shows how the concept of 'cells' fits into the overall organisation of living things.

Table 1.3
An introduction to structural organisation in living things

Level of organisation	Examples	Diagrams:
*Multicellular organism	Plant, animal	
Systems	Transport system, respiratory system	
Organs	Heart, lungs, muscles, roots, stems, leaves	
Tissues	Blood tissue, lung tissue, photosynthetic tissue	
Cells	Nerve cell, blood cell, muscle cell, *unicellular organism	
Organelles	Nucleus, chloroplast, mitochondria, ribosomes	
Molecules	Water (H_2O), protein, sugar, lipid, chlorophyll	
Atoms	Carbon (C), hydrogen (H), oxygen (O), nitrogen (N)	

* terms are explained in the following text

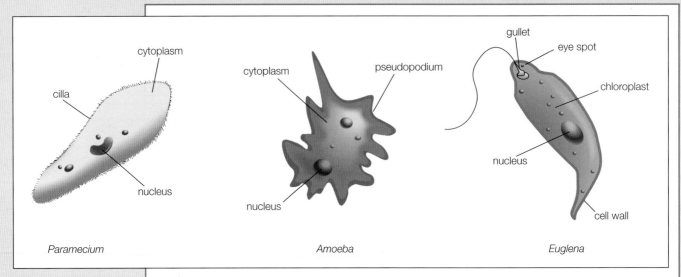

Paramecium   Amoeba   Euglena

Figure 1.11 Common unicellular organisms that may be seen in a drop of pond water using a light microscope

Teaching analogy

Figure 1.12 Cells of multicellular organisms that can be seen using a compound light microscope: (a) non-photosynthetic plant cells (onion)

The structure of cells (as seen using a light microscope)

■ *identify cell organelles seen with current light and electron microscopes*

The *general contents* of cells can be studied using the *light microscope*, but if more detailed information is required an *electron microscope* must be used.

Cells vary greatly in shape, size, structure and function. There is, in reality, no such thing as a 'typical' cell. The majority of cells that form the tissues and organs of an organism become highly specialised for particular functions, for example lung tissue and blood tissue. To allow an understanding of the general structure and functioning of cells, a hypothetical or 'typical' cell of plants and one of animals is often studied, as shown in Figures 1.12 and 1.13 (light microscope) and 1.16 (electron microscope).

Cells that are found in plants and animals have the same basic features, with some variations. The cell parts discussed below are those that are visible with a light microscope:

■ It is in the **protoplasm** of cells that the functions essential to life, such as growth and respiration, are carried out. The **cytoplasm** (that part of the protoplasm outside of the nucleus) consists of a liquid-based background, the cytosol, in which there are *dissolved* chemical substances (e.g. ions such as chloride ions), *suspended* organelles and insoluble granules. Approximately 90 per cent of the cytoplasm is water—the medium in which all cell chemicals are dissolved or suspended.

(a)

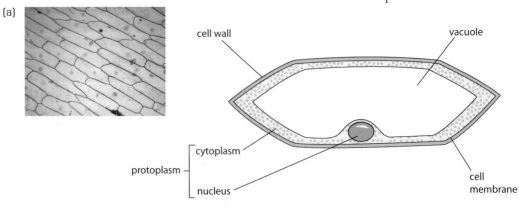

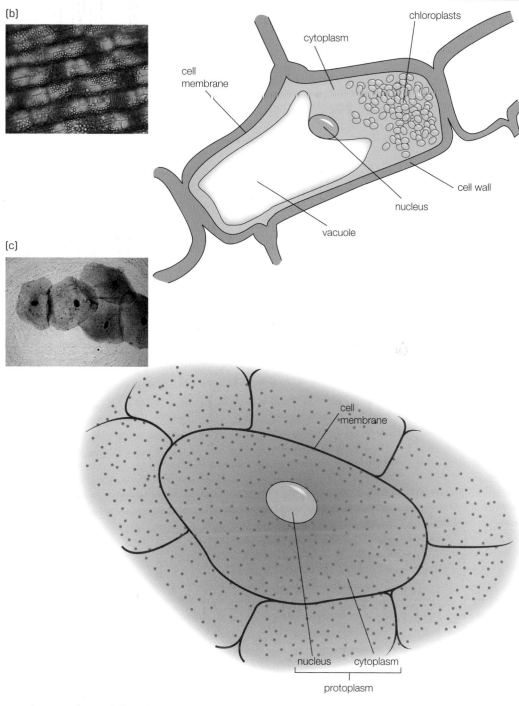

(b)

(c)

Figure 1.12 Cells of multicellular organisms that can be seen using a compound light microscope: (b) photosynthetic plant cells (pond weed); (c) animal cells (cheek cells)

- The **nucleus** (plural *nuclei*) appears as a large, spherical, oval or sometimes elongate structure in the cytoplasm. It is colourless, transparent and slightly more jelly-like than the rest of the cell. Most organisms have one nucleus per cell.

- The **cell membrane** (alternate names are the *plasma membrane, cytoplasmic membrane* or *plasmalemma)* surrounds the cell contents in all cells and separates the cell contents from its surroundings. It controls the passage of water and other chemical substances into and out of cells.

Plant cells

Plant cells have some additional structures which can be viewed under a light microscope. These structures are exclusive to plant cells and therefore not usually found in animal cells.

■ **Chloroplasts** are organelles that are green in colour, due to the presence of a pigment called *chlorophyll.* Chloroplasts are responsible for **photosynthesis**—the manufacturing of sugar in plants, using the energy of sunlight. Chloroplasts are not present in all plant cells, they are only found in the green tissue of plants that can photosynthesise. Under the light microscope, they appear as green, disc-shaped structures, smaller than the nucleus. An electron microscope is needed to see the detailed interior.

■ **Vacuoles** in plant cells are large, permanent, fluid-filled sacs in the cytoplasm of mature cells. Each vacuole consists of a watery solution called **cell sap**, surrounded by single membrane, the *tonoplast.* Cell sap contains substances such as mineral salts, sugars and amino

acids dissolved in water. It may also contain dissolved pigments that give cells their colour, for example the reds, pinks and purples seen in some flower petals. Besides having a storage function, vacuoles play a very important role in providing *support* to plant cells. By filling up with water, the vacuole pushes outwards with the cytoplasm, exerting a pressure on the cell wall, keeping it firm. As a result of the outward pressure of the cell contents and the resisting pressure of the cell wall, the cell becomes firm or **turgid**. (Small, temporary vesicles may sometimes be found in animal cells, but these do not play a role in cell support, so permanent vacuoles that give turgidity are considered to be a feature excusive to plant cells.)

Figure 1.13 is a comparative diagram of a plant and an animal cell. To compare two things, both the similarities and differences must be

Figure 1.13
Comparative diagram of typical plant and animal cells as seen under a light microscope

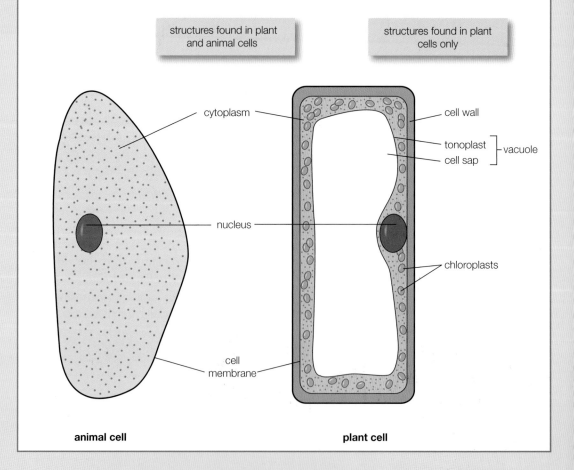

structures found in plant and animal cells

structures found in plant cells only

cytoplasm

cell wall

tonoplast
cell sap
vacuole

nucleus

chloroplasts

cell membrane

animal cell

plant cell

examined. When this is done using diagrams, the features common to both are labelled down the centre and the features that are different are labelled on the outside.

The nucleus is one of the largest organelles visible with the light microscope. Chloroplasts and vacuoles in plant cells are also large enough to be seen with a light microscope, but all other organelles are much smaller and appear as granules of various sizes in the cytoplasm, if viewed under a light microscope.

Table 1.4 Comparison of organelles visable with light and electron microscopes

School light microscope (10–200x)	Top technology light microscope (800–2 000x)	Electron microscope (60–2 000 000x)
■ Cell wall ■ Cell membrane ■ Nucleus and nuclear membrane ■ Chloroplast ■ Vacuole: tonoplast and cell sap ■ Cytoplasm	All structures in previous column as well as: ■ Golgi body ■ mitochondria ■ nucleolus ■ (special staining required for all)	All structures in previous two columns as well as: ■ endoplasmic reticulum ■ ribosomes ■ lysosomes ■ centrosome ■ cytoskeleton (special staining needed for this)

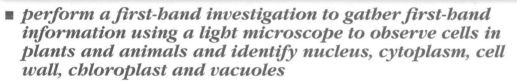

Observing plant and animal cells using a light microscope

■ *perform a first-hand investigation to gather first-hand information using a light microscope to observe cells in plants and animals and identify nucleus, cytoplasm, cell wall, chloroplast and vacuoles*

FIRST-HAND INVESTIGATION

BIOLOGY SKILLS

P 12.1; 12.2; 12.4
P 13.1

Background

The microscope

The practical introduction to using a microscope (see page 74) helped students to review parts of a microscope, safe work practice in microscopy, inversion of image, as well as giving them a clearer understanding of magnification, calculating the size of a microscope's field of view and the use of the diaphragm. This knowledge should be applied when completing this investigation.

The specimens to be viewed

You are required to examine both plant and animal cells under the microscope and gather first-hand information. This investigation involves:

■ preparing your own slides for the plant tissue to be viewed, using the wet mount technique described and illustrated below

■ using permanent prepared slides of animal cells such as cheek cells and blood cells, that have been made in commercial laboratories for you. (As a matter of OH&S, we no longer prepare our own slides of this tissue, to avoid the risk of transmitting infections between students.)

(*Note to teacher*: A demonstration of how to prepare a wet mount can be done on a plastic sheet on the overhead projector, as all materials are transparent and can be viewed easily.)

Preparing a wet mount

The technique will be described on the next page, in the *method* of the practical and your teacher may also demonstrate this technique to you. Figure 1.14 illustrates the technique.

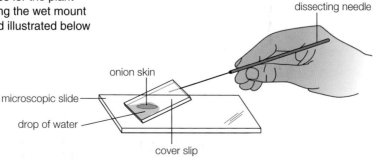

Figure 1.14 Diagram showing technique for lowering coverslip on to a wet mount

dissecting needle

onion skin

microscopic slide

drop of water

cover slip

General resources

Recording your results

■ Accurately present your information by selecting and drawing two to three cells of each type viewed under the microscope. Remember to calculate and state the magnification for each diagram. It is not essential to draw the circular field of view around the cells, but this sometimes helps to remind you that this is a representation of a microscopic section.

■ The photomicrographs and diagrams provided (see Fig. 1.12) should help you to find and recognise the tissues that you are looking for.

Drawing cells seen under the microscope

Scientific drawing skills apply—use a sharp pencil and draw single, solid lines. Each diagram should be large enough (approximately 6 to 7 cm in size) to clearly show all structures visible inside the cells, have detailed and accurate labels (see Table 1.4). Label lines should be parallel if possible, should never cross each other and should have no arrowheads, but touch the actual structure being labelled (see Figs 1.12 and 1.13).

Practical task

Aim

To investigate the structure of plant and animal cells under a light microscope.

Equipment

■ One compound light microscope per student if possible

■ Glass microscope slides, coverslips, dissecting needle, razor blade or scalpel

■ 50 mL beaker, water, dropper, stains such as iodine or toluidine blue, paper towel, lens tissue, oil for oil immersion

■ Onion, elodea (water plant), leaves of agapanthus

■ Prepared slides of cheek cells, blood cells

Method

1. Plant cells

Working in pairs, one student prepares a wet mount of a section of onion tissue, while the other student prepares a wet mount of a piece of pond weed (elodea). Follow the procedure for preparing a wet mount as demonstrated by your teacher (see Fig. 1.14).

A. Onion cells (see Fig 1.12a)

■ Remove the onion skin and carefully lift a thin section of onion tissue from the surface of one of the layers.

■ Cut a piece about 1 cm² in area and place this in a drop of water plus iodine (stained) or water (unstained) on the glass microscope slide.

■ Carefully lower the coverslip using a dissecting needle, to avoid the formation of air bubbles.

■ Place a piece of paper towel over the coverslip and slide to dry any excess water or stain. (*Note*: There are no chloroplasts in white onion cells, but you should be able to view all other plant cell structures visible with a school microscope listed in Table 1.4 on page 85.)

B. Pond weed—elodea (see Fig 1.12b)

■ Follow the instructions above for an unstained wet mount. Pond weed is thin enough for light to pass through and you should be able to view chloroplasts clearly.

■ View under low power, then under higher power. You will need to place drop of oil on top of the coverslip to view the cells under the highest power. Remember that your microscope is **parfocal**, so it is not necessary to adjust the focus before changing to a higher magnification, but make sure you use the fine adjustment knob only, when bringing the specimen into focus.

Draw the cells seen in both parts A and B.

2. Animal cells

C. Cheek cells

■ Using a prepared slide of stained cheek cells, observe and draw these as seen under high power or under oil immersion, as instructed by your teacher.

■ Label the cheek cells (see Table 1.4 for suggested labels).

■ Record your magnification.

(a)
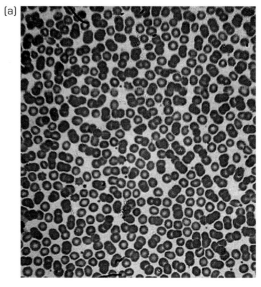

(b)
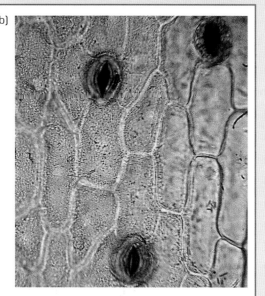

Figure 1.15
Photomicrographs taken under a compound light microscope: (a) blood cells; (b) agapanthus—surface view of epidermal cells and stomates

Further investigations for students who finish early:
(a) blood cells
(b) surface view of leaf of epidermal cells showing guard cells and stomates an agapanthus leaf: scour the surface of the leaf into 1 cm² sections with a scalpel blade and then place transparent sticky tape over the leaf surface to lift the uppermost layer (epidermis). Place these cells attached to the sticky tape in a drop of water on a microscope slide, as described

for the unstained onion and complete the wet mount as described. Draw and label this surface view of epidermal cells of agapanthus. Breathing pores called stomates should be visible amongst the epidermal cells
(c) pond water: place a drop of pond water on a microscope slide. Cover with a coverslip and view under low power to find cells and then under high power to draw (see Fig. 1.11). Record your magnification.

The ultrastructure of cells (electron microscope)

1.4

■ *describe the relationship between the structure of cell organelles and their function*

In order to look at cells and their organelles in detail, photographs that have been produced using an electron microscope are studied. These photographs are called **electron micrographs** and they reveal the structure of these sub-cellular components which have been greatly magnified—termed the *ultrastructure* of cells.

The structure of each organelle is closely related to its function within the cell.

In any cell, membranes are extremely important structures which not only separate one cell from another, but may also separate organelles within a cell from the surrounding cytoplasm.

(a)

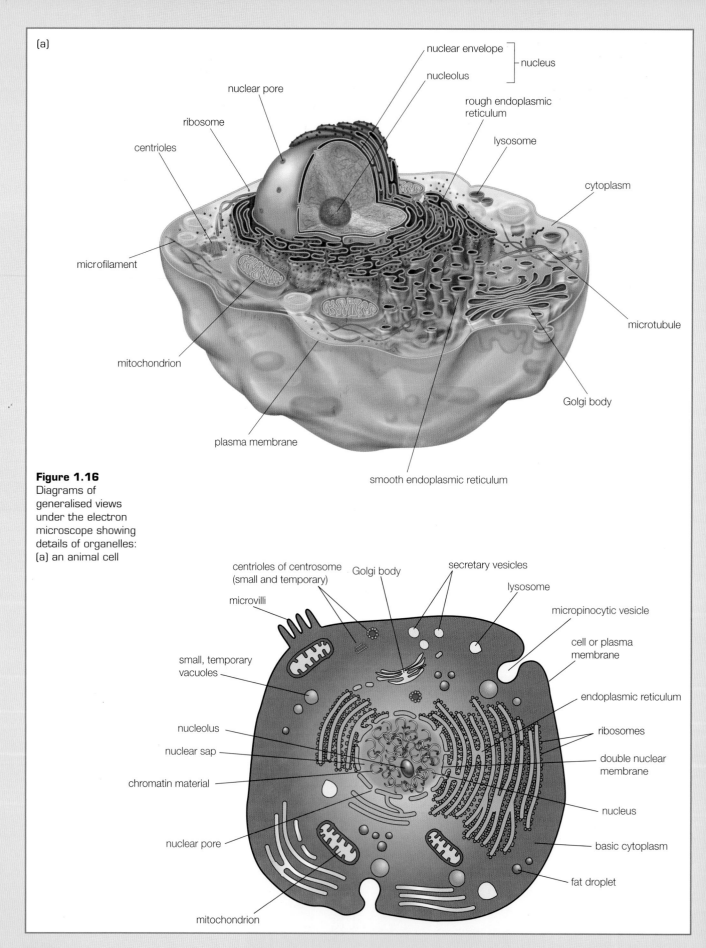

Figure 1.16
Diagrams of
generalised views
under the electron
microscope showing
details of organelles:
(a) an animal cell

Labels (top diagram):
nuclear envelope
nucleus
nucleolus
nuclear pore
rough endoplasmic reticulum
ribosome
lysosome
centrioles
cytoplasm
microfilament
mitochondrion
microtubule
plasma membrane
Golgi body
smooth endoplasmic reticulum

Labels (bottom diagram):
centrioles of centrosome (small and temporary)
Golgi body
secretary vesicles
microvilli
lysosome
micropinocytic vesicle
small, temporary vacuoles
cell or plasma membrane
nucleolus
endoplasmic reticulum
nuclear sap
ribosomes
chromatin material
double nuclear membrane
nucleus
nuclear pore
basic cytoplasm
mitochondrion
fat droplet

(b)

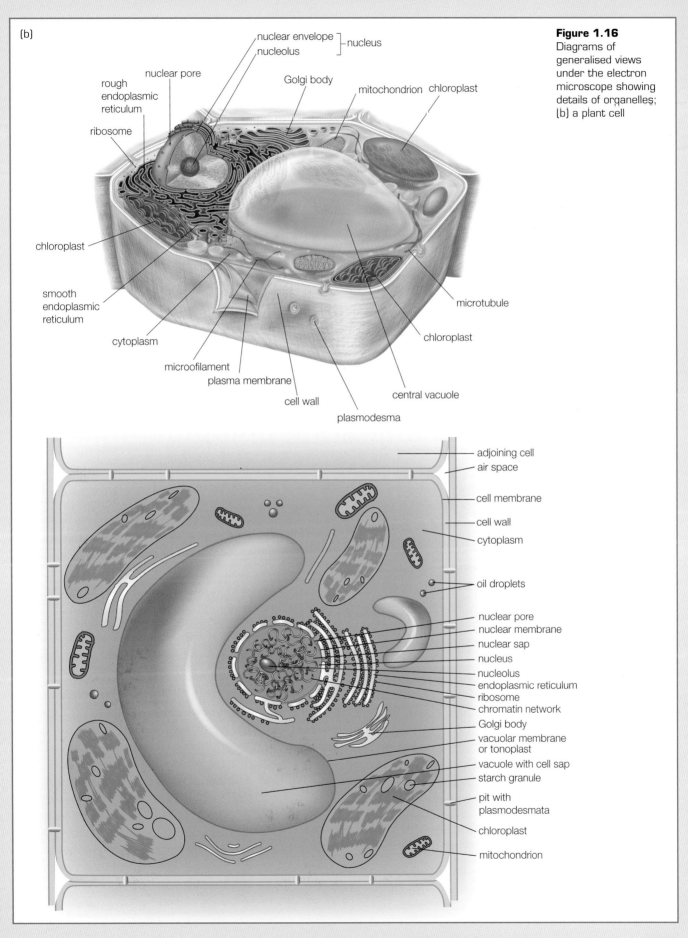

Figure 1.16
Diagrams of generalised views under the electron microscope showing details of organelles; (b) a plant cell

nuclear envelope
nuclear pore
nucleolus
nucleus
rough endoplasmic reticulum
Golgi body
mitochondrion
chloroplast
ribosome
chloroplast
smooth endoplasmic reticulum
cytoplasm
microfilament
plasma membrane
cell wall
plasmodesma
central vacuole
chloroplast
microtubule

adjoining cell
air space
cell membrane
cell wall
cytoplasm
oil droplets
nuclear pore
nuclear membrane
nuclear sap
nucleus
nucleolus
endoplasmic reticulum
ribosome
chromatin network
Golgi body
vacuolar membrane or tonoplast
vacuole with cell sap
starch granule
pit with plasmodesmata
chloroplast
mitochondrion

In the cells of multicellular organisms and many unicellular organisms, each organelle within a cell is surrounded by its own membrane, which may be either a single or a double membrane. *Cells* with membrane-bound organelles are termed **eucaryotic cells** (alternate spelling *eukaryotic*) and the *organism* is classified as a *eucaryote*. This name refers to the fact that the genetic material is contained within a nucleus, separated from the cytoplasm by a double nuclear membrane. (Derived from the Greek '*eu*' meaning '*true or proper*' and '*karyon*' meaning '*nucleus*').

With the use of the electron microscope, primitive cells called **procaryotic cells** (alternate spelling *prokaryotic*) were discovered. (Greek '*pro*' = before, '*karyon*' = nucleus). These organisms, for example bacteria, do not have their genetic material separated from the cytoplasm by a membrane, they simply have a strand of genetic material floating within the cytoplasm. Further studies with the electron microscope showed that procaryotes do not have any membrane-bound organelles, their sub-cellular components float within the cytoplasm.

Membranes—selective boundaries

The cell membrane is a *selective* barrier, permitting the passage of only certain molecules into or out of cells. This property gives the cell membrane the feature of being *selectively permeable*. (This will be dealt with in greater detail in the following chapter). Both plant and animal cells have a cell membrane. The membranes surrounding organelles are also selective in allowing only certain substances to pass between the cytoplasm and the organelle.

The nucleus—the control and information centre

- The nucleus stores the information needed to control all cell activities. It is therefore essential for the nucleus to be able to communicate with the surrounding cytoplasm.

- Electron micrographs reveal that the nucleus is surrounded by a *double* **nuclear membrane** or **nuclear envelope**, pierced by tiny **pores**. These pores regulate the passage of substances between the nucleus and cytoplasm, allowing communication between them.

- The **nucleoplasm** or **nuclear sap** is the liquid background of the nucleus in which the *chromatin* material is found. Chromatin is made up of *protein* and *nucleic acid*.

- The nucleic acid **DNA** is a very large chemical that holds, in a coded form, all the genetic information (the 'blueprint') necessary to control the cell's functioning. It is this DNA which contains the hereditary information of an organism that gets passed from one generation to the next. Within one organism, the information stored in the DNA of each cell is the same. Before a cell divides, the information on the DNA must be copied so that it can be transmitted (passed on) to newly formed cells.

- The chromatin material separates into short, thick separate rod-shaped structures called **chromosomes,** which become visible in dividing cells. ('chromo' = coloured, 'soma' = 'body'—so named because chromosomes take up stains easily when being prepared for microscopy.) Each species of organism has its own particular number of chromosomes; for example, humans have 46 chromosomes, a platypus has 52, a lettuce has 18 and a camel has 70!

- The **nucleolus** is a dense, granular region commonly seen within the nucleoplasm and it contains a large amount of nucleic acid—some DNA, but mostly RNA. The nucleolus is responsible for the manufacture of organelles called ribosomes, essential 'machinery' of the cell.

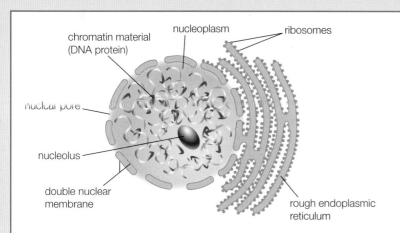

chromatin material
(DNA protein)

nucleoplasm

ribosomes

nuclear pore

nucleolus

double nuclear
membrane

rough endoplasmic
reticulum

Figure 1.17 Diagram of nucleus and associated ER and ribosomes

It is important to remember that the view of cell organelles revealed by a transmission electron microscope is a section cut through the organelle. From these micrographs it is possible to build up a *three-dimensional* view of each organelle (see Table 1.5 on page 95). Scanning electron microscopes give an *external* view of the surface of these organelles.

Endoplasmic reticulum—transport and processing of proteins and lipids

The outer nuclear membrane is usually continuous with a network of flattened, interconnected membranes—the **endoplasmic reticulum (ER)**. The ER provides a connection of pathways between the nucleus and the cell's environment, allowing intracellular transport (transport within a cell). The immense folding of the sheets of membrane increases its surface area. ER may have ribosomes attached (rough ER) or may have no ribosomes (smooth ER). The main function of ER is *transport*, but it also plays a role in *processing* cell products: rough ER folds and processes proteins products made by the cell and it can also synthesise lipids; smooth ER is the main site of lipid production, essential for membrane repair and manufacture. ER may also transport substances from one cell to another in plant cells, passing through channels in the cell wall called cell pits.

Ribosomes—protein synthesis

These small organelles appear as dense granules in electron micrographs of cells. Their small size and rounded shape increase their surface area for easy interaction with chemicals during their functioning. Each is made of the chemicals RNA and protein, and their function is protein synthesis. Ribosomes are the 'machinery' that carries out the genetically coded instructions of DNA to *produce* any *proteins* necessary for cell functioning and structure. Ribosomes may be found free in the cytoplasm or scattered over the surface of ER. Newly synthesised proteins pass from the ribosomes into the ER where the folding of the protein occurs.

Golgi bodies—packaging and sorting the products

Although the Golgi body is also made of flat membranes, it differs from ER in that it does not have ribosomes attached and the membranes are often in stacks of four to ten. The Golgi body is easily recognisable by its curved shape on one surface, where *vesicles* can be seen budding off. This surface is called the *forming face* and the vesicles are evidence of the secretory function of Golgi bodies. The opposite surface may be convex or flat in shape.

Golgi bodies process, package and 'sort' cell products. They are involved in adding proteins and carbohydrates to cell products and they also provide a membrane around the cell products to 'package' them. The membranes provided by Golgi bodies vary and serve as a 'packaging label'. Features of the membrane are used to 'sort' these products, determining where they will end up—they may be transported within the cell to wherever they are required or they may be secreted out of the cell.

Figure 1.18
Golgi body and vesicles

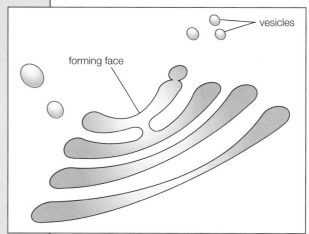

Lysosomes—digestion and destruction

One example of the products of Golgi bodies are lysosomes. These are little fluid-filled sacs, most commonly seen in animal cells. They are surrounded by a single membrane and the vesicle is filled with digestive enzymes, for intracellular digestion. Lysosomes commonly break down worn out cell organelles, so that the materials can be recycled and used to make new organelles. The membrane is essential to prevent the enzyme from digesting the normal cell contents.

Mitochondria—cellular respiration: production and storage of energy (ATP)

- Mitochondria are the 'powerhouses' of a cell, producing energy by the process of **chemical respiration**.
- Mitochondria (singular = *mitochondrion*) are usually rod-shaped, but may be round,

they vary in both shape and size. Mitochondria are smaller than the nucleus and chloroplasts, but larger than ribosomes.

- The number of mitochondria in a cell depends on how much energy the cell needs to carry out its functions. Less active cells contain few mitochondria, whereas very active cells have hundreds or even thousands of mitochondria (e.g. active liver cells contain 1000 to 2000 mitochondria).
- Just as machines in a factory need electrical energy in order to work, so cells need energy, in the form of a chemical called ATP (adenosine triphosphate), to work. Mitochondria combine oxygen with sugars during the process of chemical respiration to release energy in a form (ATP) that the cell can use.
- Each mitochondrion is surrounded by a **double membrane**: the *outer membrane* gives the mitochondria its shape and allows the passage of small substances into and out of mitochondria. The *inner membrane* is folded into fine, finger-like ridges or **cristae**—this increases the surface area for the attachment of groups of enzymes that are responsible for making energy for the cell. The groups of enzymes appear as knob-like particles on the inside of the cristae.
- The central space in a mitochondrion is filled with fluid and is termed the **matrix**. It contains mitochondrial DNA and enzymes that give mitochondria the unusual feature of being able to replicate (make copies of) themselves. The mitochondria divide by pinching off and then growing, something that usually occurs in very active cells or cells that are about to divide. This ability of mitochondria to reproduce themselves is extremely useful in evolutionary studies.

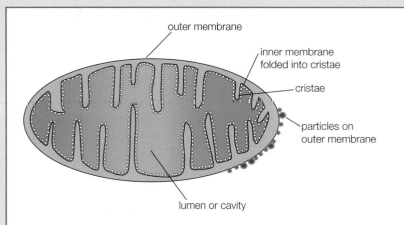

Figure 1.19
Simplified scheme of mitochondrion in longitudinal section

outer membrane

inner membrane folded into cristae

cristae

particles on outer membrane

lumen or cavity

Chloroplasts—photosynthesis

■ Chloroplasts belong to a group of organelles called *plastids* which are biconvex in shape and vary in colour. Plastids that are red, yellow and orange are called *chromoplasts* and they contribute to the colour of some flowers and fruit. Some plastids are white (e.g. leucoplasts in potatoes). Chloroplasts are green plastids which carry out the process of **photosynthesis**.

■ Chloroplasts are larger than mitochondria, but they are similar in that they also contain their own DNA and the number of chloroplasts per cell varies.

■ Chloroplasts are not found in all plant cells, only in green cells that photosynthesise; for example, they are not present in cells of roots, but are common in leaves.

■ Chloroplasts are surrounded by a *double membrane* which allows substances to pass between the cytoplasm and the chloroplast but, unlike mitochondria, the inner membrane of the chloroplast is not folded (see Fig. 1.20).

■ The liquid background of the chloroplast is called the stroma and it is here that stacks of membranes called **thylakoids** are found. Each stack or *group* of thylakoids is termed a **granum** (plural: *grana*) and the green pigment, **chlorophyll**, is found on these membranes.

■ The layering of the membranes increases the surface area over which chlorophyll occurs, allowing a large amount of sunlight to be absorbed for the process of photosynthesis. This captured energy of sunlight is then used by the plant to make food. All the enzymes needed for photosynthesis are present in the stroma and food made during photosynthesis is stored in the stroma as starch grains.

Cytoskeleton—keeps organelles in place

Organelles are not randomly scattered within a cell, their distribution is organised and they are held in place by a network of tiny microtubules and microfilaments called the cytoskeleton, which extends throughout the cytoplasm.

Figure 1.20
Simplified scheme of chloroplast seen in longitudinal section

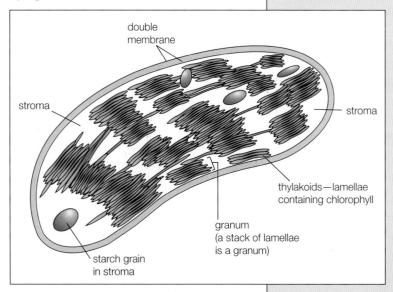

double membrane

stroma

stroma

thylakoids—lamellae containing chlorophyll

granum (a stack of lamellae is a granum)

starch grain in stroma

Centrioles—spindle production in cell division

A dense, granular structure, the centrosome, is often found near the nucleus in animal cells. It consists of two centrioles, which play an important role in the formation of spindle fibres when a cell divides (this will be referred to in Chapter 5).

Plant cell wall—shape and support

The cellulose cell wall that surrounds plant cells differs from the cell membrane beneath it, because the cell wall is not really selective in terms of what substances it does or does not allow into the cell. Its structure allows it to provide strength and support. The cell wall is built up of strands of cellulose fibres which have a little elasticity and are somewhat flexible, giving cell walls their characteristic feature of being able to resist the outward pressure of the vacuole and cell contents in a well-watered plant cell, conferring turgidity to the cell. Some cell walls are thickened with additional chemicals such as *lignin*, which could make the walls hard and woody (e.g. in tree trunks), or the chemical *suberin* for waterproofing (e.g. in cork or the waxy cuticle of leaves).

Figure 1.21
Summary of structures found in plant and animal cells

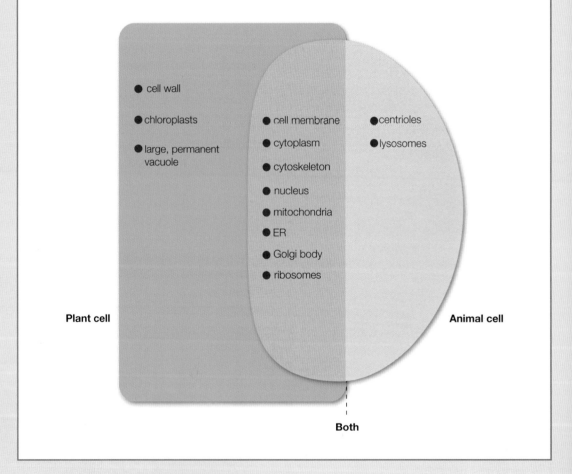

Electron micrographs of cell organelles

SECONDARY SOURCE INVESTIGATION

BIOLOGY SKILLS

P12.3; P12.4

■ *process information from secondary sources to analyse electron micrographs of cells and identify mitochondria, chloroplasts, Golgi bodies, lysosomes, endoplasmic reticulum, ribosomes, nucleus, nucleolus and cell membranes*

Table 1.5 shows a series of micrographs of the organelles of plant and animal cells, as seen using a transmission electron microscope. Draw a scientific diagram of the organelles shown in each micrograph, label the parts listed in the column next to the micrographs and then select any two features labelled on your diagram and describe how their structure relates to the function of the organelle of which it forms a part.

Table 1.5 Identifying cell organelles and relating structure to function

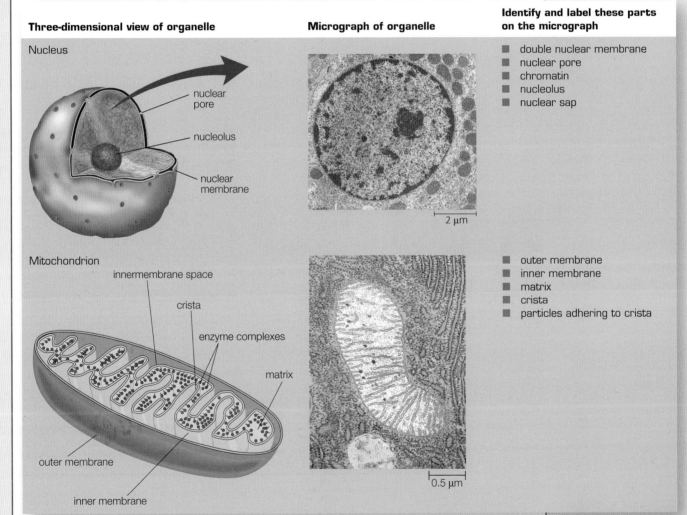

Three-dimensional view of organelle	Micrograph of organelle	Identify and label these parts on the micrograph
Nucleus nuclear pore nucleolus nuclear membrane	2 µm	■ double nuclear membrane ■ nuclear pore ■ chromatin ■ nucleolus ■ nuclear sap
Mitochondrion innermembrane space crista enzyme complexes matrix outer membrane inner membrane	0.5 µm	■ outer membrane ■ inner membrane ■ matrix ■ crista ■ particles adhering to crista

continued . . .

 SR
 TR

Student worksheet

Three-dimensional view of organelle	Micrograph of organelle	Identify and label these parts on the micrograph

Chloroplast

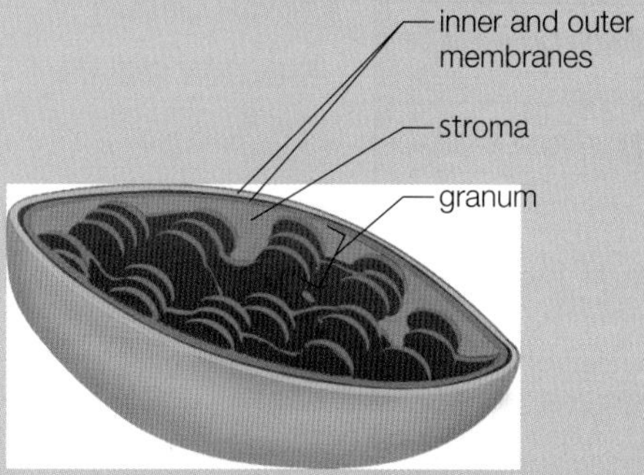

inner and outer membranes
stroma
granum

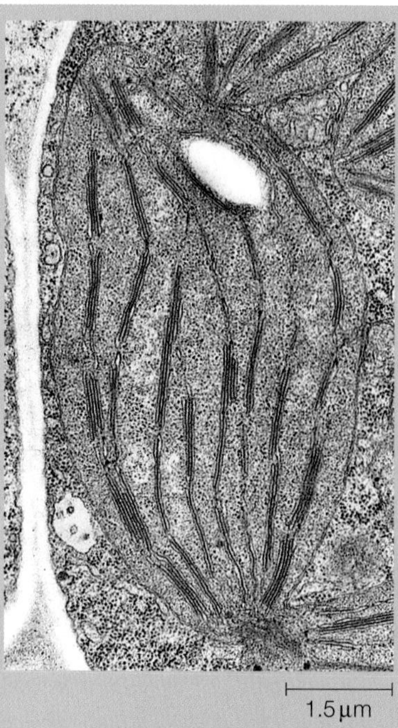

1.5 μm

- double membrane
- stroma
- lamella
- granum
- cholorophyll
- starch grain

Endoplasmic reticulum and ribosomes

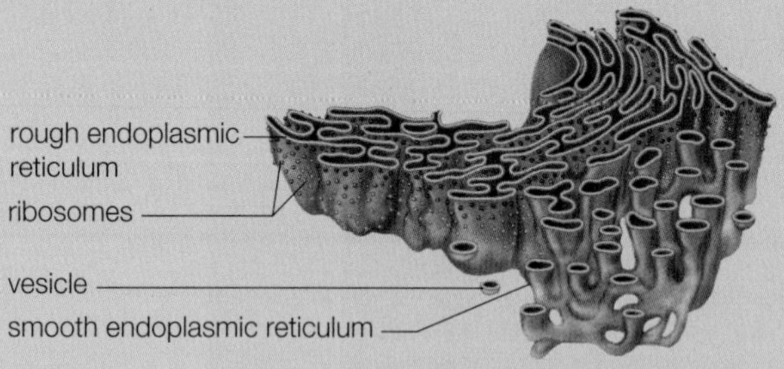

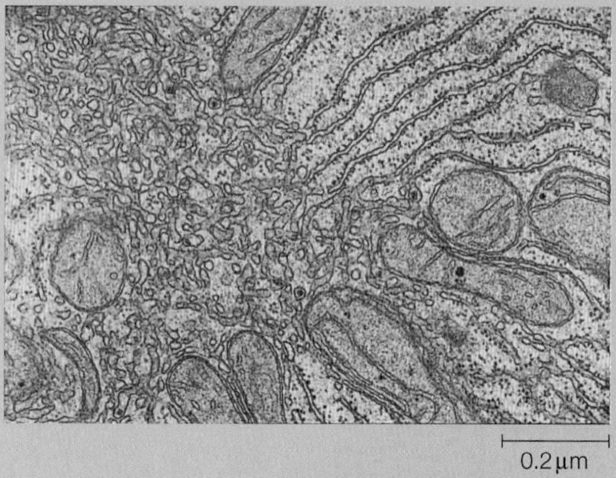

rough endoplasmic reticulum
ribosomes
vesicle
smooth endoplasmic reticulum

0.2 μm

- flattened membranes
- ribosomes
- rough endoplasmic reticulum
- smooth endoplasmic reticulum

Three-dimensional view of organelle	Micrograph of organelle	Identify and label these parts on the micrograph

Golgi body and lysosomes

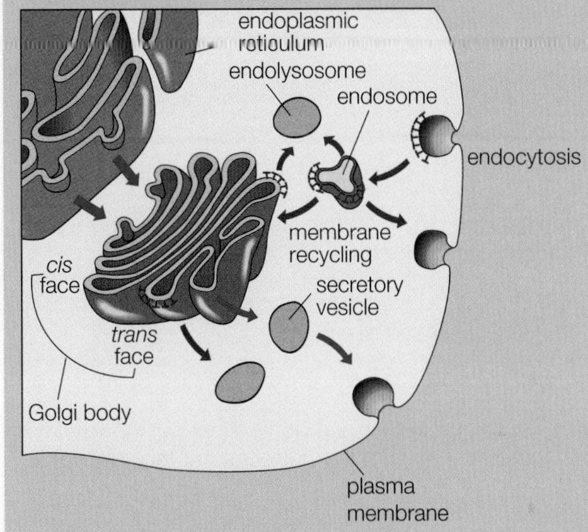
endoplasmic reticulum
endolysosome
endosome
endocytosis
membrane recycling
cis face
secretory vesicle
trans face
Golgi body
plasma membrane

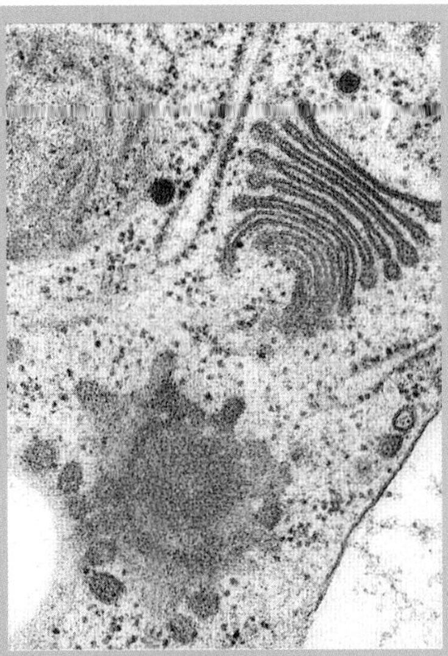

- membranes of Golgi
- convex forming face
- vesicles budding off
- lysosome

Cell membrane and cell wall

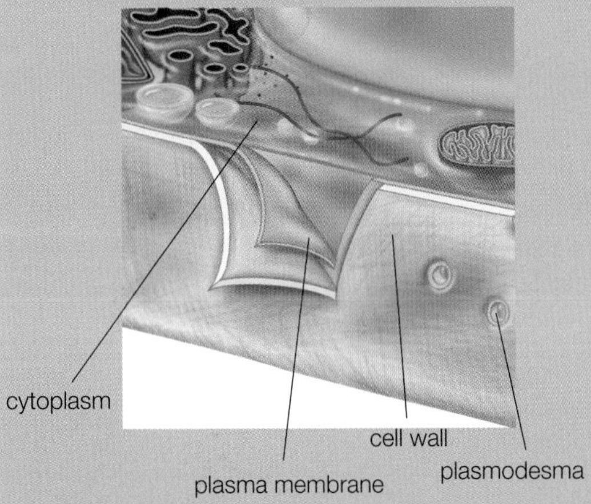
cytoplasm
cell wall
plasma membrane
plasmodesma

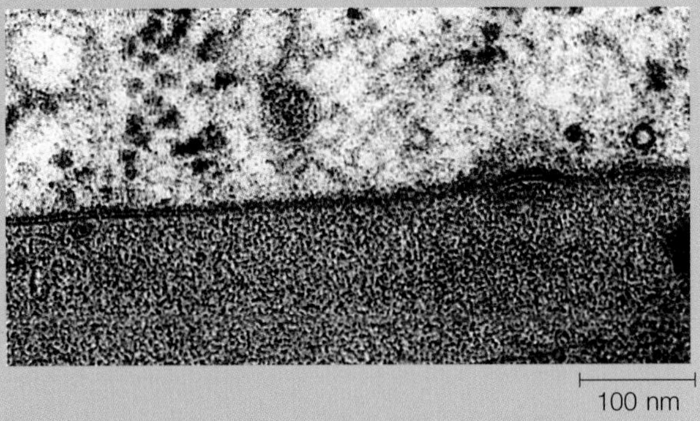
100 nm

- cell membrane
- cell wall
- cytoplasm

REVISION QUESTIONS

1. (a) Clarify what is meant by a scientific *theory*.
   (b) Describe how you would go about validating a scientific theory.
   (c) State the cell theory.

2. Before the development of the cell theory, it was commonly believed that living organisms could arise by spontaneous generation.
   (a) Outline the theory of spontaneous generation.
   (b) Describe experimental evidence that was used to discount this theory.
   (c) Explain the role that the invention of the microscope played in the dismissal of the theory of spontaneous generation.

3. Describe the contributions of Robert Hooke and Robert Brown in the development of the cell theory. (Find a reliable way of remembering which 'Robert' did what!)

4. Discuss why biologists have continued to use light microscopes since the invention of the electron microscope.

Table 1.6

5. Put the following words into order of size, from smallest to largest: organelles, molecules, cells, atoms and organisms.

6. Compare the detail seen with a light microscope in plant and animal cells. As a guide use the table below. (Copy out a larger version or print it from the Student Resource CD.)

Table 1.6 Parts of cells visible under a light microscope

Part of cell	Plant cell	Animal cell
Boundary		
Organelles		
Cell shape		
Vacuoles		

7. State whether the each of the following photographs shows plant or animal cells. Justify your answers.

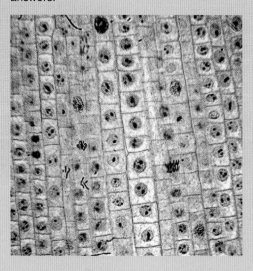 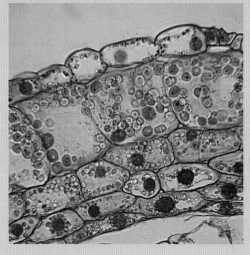

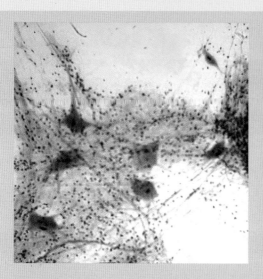

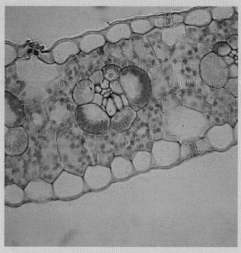

8. Using the method described for comparative diagrams on page 84, draw a comparative diagram of:
   (a) light and electron microscopes
   (b) chloroplast and mitochondria.

9. Identify the two types of nucleic acid found in cells.

10. Analyse the micrograph below and assess whether it was viewed under a light or an electron microscope and whether it shows plant or animal cells. Justify your choice.

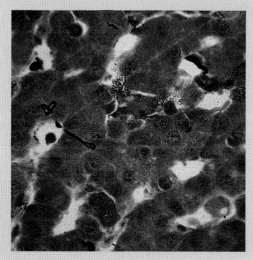

Answers to
revision questions

Membranes, chemicals and movement in and out of cells

Membranes around cells provide separation from and links with the external environment

2.1 Chemicals in cells

Studies of the ultrastructure of cells reveal a variety of organelles in the cytoplasm that function together so that a cell can carry out life processes such as respiration, growth and many others. Our studies now proceed to a chemical level—looking at what chemical substances make up these cell organelles.

■ *identify the major groups of substances found in living cells and their uses in cell activities*

The chemical substances found in cells fit into two main groups: *organic* substances and *inorganic* substances.

Organic compounds are chemical substances that are synthesised by *living* things and contain atoms of carbon and hydrogen. Carbon atoms bond very strongly with other carbon atoms and form long chains—the basis of the large organic molecules made by cells. (*Note*: The term 'organic', used to refer to chemical compounds (as above) should not be mixed up with the modern day use of the term 'organic' meaning a product of a farming method that avoids the use of pesticides, artificial fertilisers and herbicides.)

Examples of organic compounds are:
■ *carbohydrates* such as sugars and starch
■ *lipids* such as fats and oils
■ *proteins*
■ *nucleic acids* (DNA and RNA)
■ *vitamins*.

Inorganic compounds are part of the inanimate, *non-living* world. These substances do not contain the element carbon combined with hydrogen and do not have long chains. Examples of inorganic compounds are:
■ *mineral salts* such as calcium salts, sodium chloride (ordinary table salt) and phosphates
■ *water* (H_2O)
■ some *gases* such as carbon dioxide (CO_2) and oxygen (O_2)—although carbon dioxide contains the atom carbon, it has no hydrogen and so it is inorganic.

Organic and inorganic compounds are important both in forming structural parts of cells and in cell functioning.

Inorganic compounds

Table 2.1 Inorganic compounds in cells

Inorganic compound	Position in cells	Uses in cell activities
Water: ■ chemical elements: oxygen and hydrogen ■ chemical formula: H_2O.	90% of the protoplasm (cytoplasm and nucleus) is water	■ Water is the **transport medium** in cells and in organisms ■ Water is an important **solvent** for many molecules (gases, sugars, some amino acids and organic acids) inside cells. (*Note*: Molecules in cells that are *insoluble* in water are starch granules, lipids and some parts of proteins) ■ Water is the medium in which all **chemical reactions** in cells take place; water may be used in the reactions (e.g. it is a reactant in photosynthesis, the food-making process in plant cells) ■ Water is used to *regulate temperature* in organisms, as it needs to be heated to change to water vapour. This heat energy comes from the body, cooling it down.
Mineral salts: chlorides, nitrates, phosphates and carbonates of sodium, magnesium, calcium, potassium and ammonium (e.g. sodium chloride, NaCl; and calcium carbonate, $CaCO_2$).	Dissolved as ions in the cytoplasm and in vacuoles in plant cells	Generally: ■ mineral salts assist all chemical reactions in cells by helping enzymes to function (they are then called *co-enzymes*) ■ mineral salts are used in the synthesis of many macromolecules and body tissues (e.g. calcium for bones and teeth, and iron in blood cells) ■ sodium ions (Na^+) and chloride ions (Cl^-) assist in water balance in cells and are essential for cell membrane functioning and the function of nerve and muscle cells.
Gases: ■ carbon dioxide (CO_2) ■ oxygen (O_2).	Dissolved in the protoplasm; used and/or produced in chloroplasts and mitochondria	Carbon dioxide: ■ used by plant cells within chloroplasts during the process of photosynthesis (food production) ■ released as a product of aerobic respiration (plants and animals) ■ can react with water to form bicarbonates, a buffer limiting changes in acidity or alkalinity in cells. Oxygen: ■ used by all living organisms during aerobic cellular respiration to release energy for cells to function ■ released as a product of photosynthesis.

Organic compounds

Most organic compounds are very large molecules, termed **macromolecules** or **polymers**. These include carbohydrates, lipids, proteins and nucleic acids as well as vitamins. Macromolecules are built up from smaller organic molecules called **monomers**, which are joined together in a repetitive manner (see Table 2.2). Macromolecules are assembled into a variety of structures and make up the components of cells and their organelles.

Table 2.2 Organic compounds in cells

Polymer	Diagrams
Carbohydrates: ■ **monosaccharides**—simple sugars consisting of single units ■ **disaccharides**—complex sugars consisting of two units ■ **polysaccharides**—complex polymers consisting of more than five (up to hundreds) of units. *Note*: monomers = simple sugars	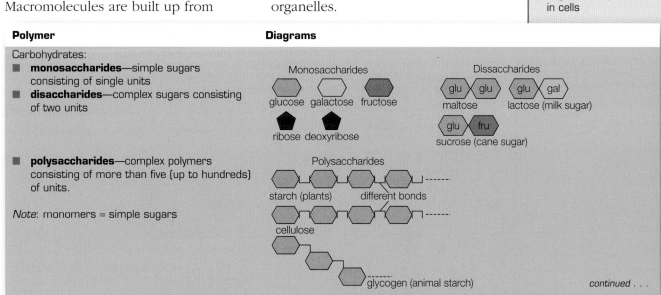

continued . . .

Polymer	Diagrams

Lipids:
- glycerol
- fatty acids.

Note: Three fatty acids attach to one glycerol molecule to from a lipid molecule.

Proteins:
- amino acids
- dipeptides
- polypeptides.

Note: monomers = amino acids

Nucleic acids:
- DNA
- RNA.

Note: monomers = nucleotides: sugar – phosphate – base

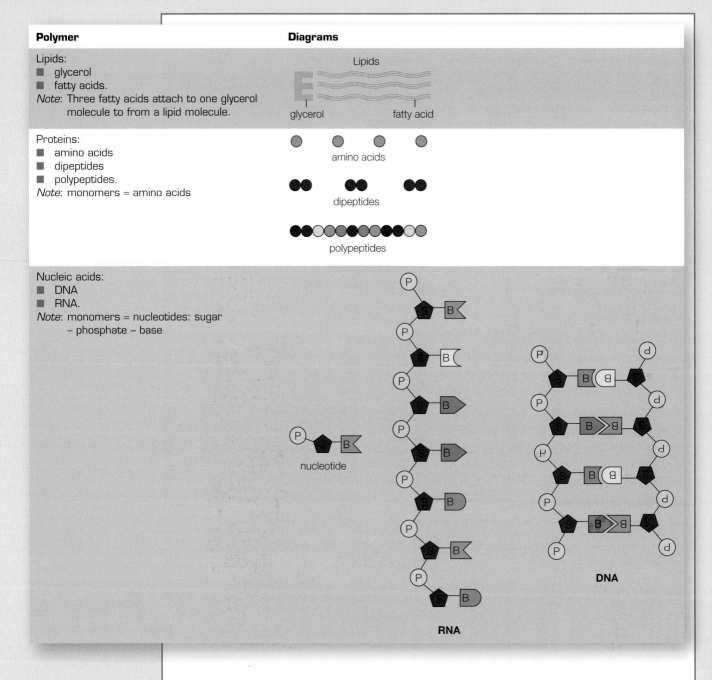

Carbohydrates

Structure

Carbohydrates are a group of organic molecules made up of carbon, hydrogen and oxygen atoms. They have a high proportion of oxygen atoms, having a ratio of one oxygen atom for every two hydrogen atoms. The general chemical formula for a carbohydrate is $Cx(H_2O)y$ where x is equal or approximately equal to y. For example the sugar glucose has the formula $C_6H_{12}O_6$; the sugar ribose is $C_5H_{10}O_5$; and sucrose is $C_{12}H_{22}O_{11}$ or $C_{12}(H_2O)_{11}$. In all of these, the ratio of H:O is 2:1.

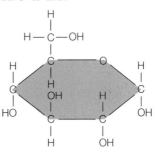

Figure 2.1 Structural formula of carbohydrate (e.g. glucose)

Uses of carbohydrates in cells

- **Sugars** are a source of *quick energy* in both plant and animal cells. Mitochondria use sugars, together with oxygen, to make a form of chemical energy called ATP (adenosine triphosphate) during chemical respiration. ATP is the form of energy that a cell needs in order to carry out its life functions. In some plant cells sugars may be stored dissolved in the vacuole, but most carbohydrates are stored in cells as polysaccharides.

  There are three main types of polysaccharides found in cells:
  1. **cellulose**, which forms a *structural part of cell walls* in plant cells, giving them strength and support
  2. **starch**, a form of *stored energy* in plant cells. It is most commonly stored as starch grains in the chloroplasts or in the cytoplasm
  3. **glycogen**, a form of *energy stored* as granules in the cytoplasm of animal cells.

- Starch and glycogen are too large to dissolve in water and so these polysaccharides are a useful form of *stored energy* in cells, as they do not affect the water balance of a cell.

- Plants *manufacture carbohydrates* in a process called photosynthesis. The carbohydrates made are an energy source that can be used by the plant itself and it can also be passed on in food chains to animals.

- Animals *consume plants* or other animals to obtain carbohydrates in their diet, gaining both energy (from starch and glycogen) and fibre or roughage (from cellulose).

- Cellulose in cell walls is often associated with a non-carbohydrate substance called **lignin**, an organic compound that helps walls of plant cells to bind. It also makes cell walls strong, rigid and impermeable to water. It is commonly found in the wood of trees.

Lipids

Structure

Lipids are macromolecules that contain many carbon and hydrogen atoms, but few oxygen atoms. Lipids are organic compounds that have an oily, greasy, or waxy consistency. They are relatively insoluble in water and tend to be water-repellent (e.g. have a waxy cuticle on leaf surfaces). Most lipids are made of a glycerol molecule to which fatty acids attach. Triglycerides have three fatty acids attached to each glycerol.

Lipids may be *fats* (solid at room temperature, e.g. butter) or *oils* (liquid at room temperature, e.g. olive oil). Most *fats* are *animal products* and *oils* are *plant products* (some exceptions do exist, e.g. fish lipids are oils, not fats).

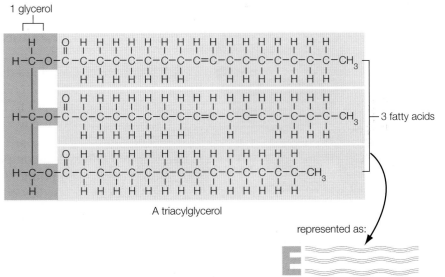

Figure 2.2 Structural formula of a lipid (triglyceride)

Uses of lipids in cells

- Lipids form an extremely important *structural part of all membranes* in cells.
- Lipids are important *biological fuels*, storing large quantities of energy for both plant and animal cells. Lipids are stored in the cytoplasm of cells—as oil droplets in plants and as fat droplets in animal cells. Carbohydrates can be converted into fats by enzymes and these fats are stored within cells such as adipose (fat) tissue beneath the skin of many animals (e.g. mammals and birds). When food is abundant, this store of fats is increased to be used during times of food shortages.
- Some lipids are essential *structural parts of hormones* which are chemical messengers produced by cells (e.g. steroids).

Proteins

Structure

Proteins are large, complex macromolecules and they are the second most abundant chemical in cells. Proteins are made up of one or more long chains of nitrogen-containing amino acids. Each chain is called a **polypeptide**. Proteins contain the chemical elements carbon, hydrogen, oxygen and nitrogen and some may contain sulfur and phosphorus. The amino acids in each linear sequence or polypeptide chain are held together by chemical bonds (forces of attraction) known as **peptide bonds**. *One or more polypeptides* can be twisted together into a particular shape, resulting in the overall structure of a *protein*. There are about 20 different amino acids and these can be put together in chains of up to 300 amino acids. The *sequence and arrangement of the amino acids* determines the type of protein, just like sequences of the letters of the alphabet can be used to make words and then sentences.

Uses of proteins in cells

- Proteins form *structural* components in cells and tissues; together with water, they form the basic structure of protoplasm (the cytoskeleton) in cells and form part of tissues such as bone, hair and nails.
- Proteins are an important structural part of **cell membranes** and, together with lipids, they regulate the passage of substances across cell membranes.
- Proteins may be important in conjunction with other molecules, such as DNA, to keep them compactly packaged inside cells.
- Some proteins have a *functional* role in cells; for example, *enzymes* are proteins that control all the metabolic (chemical) reactions of the cell.
- Proteins occur suspended in the protoplasm of cells or combine with other macromolecules to form an important structural part of all cell membranes.
- Nitrogen-containing compounds (nitrogenous compounds) cannot be stored in the bodies of animals and so they are often broken down and excreted as nitrogenous wastes.

Figure 2.3
Protein structures:
(a) structural formula of amino acids proline and glycine;
(b) two polypeptide chains making up a protein (e.g. insulin)

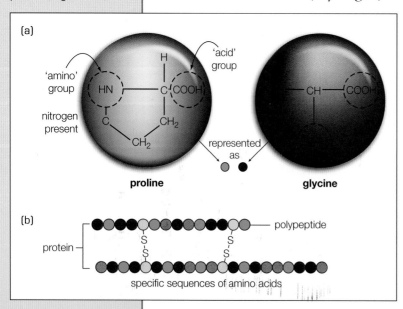

(a)

'amino' group

'acid' group

HN

H

C—COOH

nitrogen present

C

CH₂

CH₂

represented as

proline

CH—COOH

glycine

(b)

protein

polypeptide

specific sequences of amino acids

Nucleic acids

Structure

Nucleic acids contain the elements carbon, hydrogen, oxygen, nitrogen and phosphorous. All nucleic acids are made up of simple repeating units (monomers) called **nucleotides**, linked together to form single chains (RNA) or double strands (DNA), often of great length (see page 102). Each nucleotide is made up of a simple sugar (ribose or deoxyribose), a phosphate and a nitrogenous base. It is the sequence of bases which differs, providing the 'genetic code' for a cell.

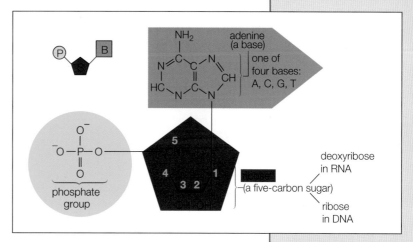

Figure 2.4 Structural formula of a nucleotide

Uses of nucleic acids in cells

- DNA (deoxyribonucleic acid) stores the information that *controls the cell* and thereby the whole organism. DNA is the main chemical making up chromatin material in the nucleus, although small amounts are also found in mitochondria and chloroplasts.

- DNA is responsible for *transmitting inherited information* from one cell to another during cell division and from one generation to another during reproduction.

- RNA (ribonucleic acid) is a nucleic acid found in small amounts in the nucleus, but in larger amounts in the cytoplasm; associated with ribosomes. There are three types of RNA. One type, the messenger RNA, is involved in passing on information that is stored in DNA, transporting a transcribed copy from the nucleus to the cytoplasm. The other types of RNA assist the message to be *'translated' into proteins*. (You will learn more details of this in your HSC course.)

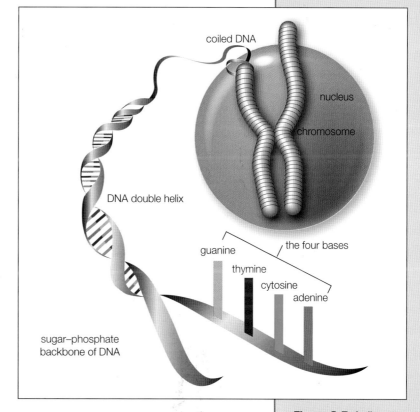

Figure 2.5 A diagram of the unravelling of a chromosome that shows the way that DNA is packaged in a cell

FIRST-HAND
INVESTIGATION

BIOLOGY SKILLS

P11.1; P11.2; P11.3
P12.1; P12.2; P12.3
P13.1

Investigating chemicals in cells

■ *plan, choose equipment or resources and perform a first-hand investigation to gather information and use available evidence to identify the following substances in tissues:*
—*glucose*
—*starch*
—*lipids*
—*proteins*
—*chloride ions*
—*lignin*

In this investigation, you will learn the standard tests to identify particular chemical substances and will then apply them to test cells or products of living cells for the presence of these chemicals.

Background information

All tissue that is living, or once was living, is made up of cells and these cells contain chemical substances. The chemicals may form part of the structure of these cells or may be held in a storage form within the cells.

Handy hints

■ Read the section on standard tests for organic chemicals that follows (see Table 2.3).
■ You are going to design an experiment to test for the presence of the *cellular* chemicals: glucose, starch, lipids, proteins, chloride ions and lignin. *Plant tissues* that could be tested include any type of fruit or vegetable matter and *animal tissues* such as meat or fish, but it is sometimes simpler to test the *products* that have been produced by animal cells (e.g. milk, butter or eggs) or the products of plant cells (e.g. grape juice or apple juice).
■ It is best if the cells are macerated (broken up) to allow the reagents to interact with the cell chemicals, so you may want to place the tissue of your choice in a blender or grind it using a mortar and pestle.
■ Remember that many of the tests rely on observing a colour change and so you should select your tissue or cell products carefully to give the best results.

Planning a valid scientific investigation

When planning the method, remember to ensure that your investigation is a 'fair' test.
■ Use a *control* to ensure the validity of the experiment and to ensure that it tests what it sets out to test. To design a control remove the factor that you are testing for and compare the results with the experiment when the factor was present. This comparison should show that if that factor is missing (control), you will not get a positive result, but if the factor is present (experimental), it is that factor which gives a positive result. In each chemical test, you will need to set up two sets of apparatus:
—*experimental*—the factor being tested is the only variable and all other factors are kept constant
—*control*—all factors are controlled and kept constant as for the experimental test and the (variable) factor being tested in the experimental is left out.
For example if you are using Benedict's reagent to test for the presence of glucose in a solution, your control test tube should contain Benedict's solution and water only, but no glucose. Your experimental test tube should contain Benedict's solution and *glucose* in water. This allows you to make a fair comparison between the results of the experimental and the control. Design similar controls for all your chemical tests.
■ Identify the **independent variable** (the one that you, as the experimenter, change) and the **dependent variable** (the one that you measure/observe and record in your results). It is termed *dependent* because its outcome depends on changes that you make in the experiment.
■ List all other variables that you should keep constant.
■ Identify any risks or dangers that could be encountered in this practical and state what measures you will implement to ensure safety.

Task

Aim

To test for the presence of cellular chemicals. Design an experiment to test whether any of the following chemicals are present in cells of your choice: carbohydrates (starch or glucose), proteins, lipids, chloride ions and lignin.

Planning the experiment
Preparation
1. List all the plant and/or animal tissues or products that you have selected to test and name the organic substance(s) for which each will be tested. A large table would be a suitable format to record this.
2. Read each standard test as described in Table 2.3 and record the equipment you will need for your test.

Recording your investigation
3. Write an aim, hypothesis, materials and methods for the investigation, and include safety precautions.
4. Draw up a table in which you can record your results. It should have a suitable number of rows and columns, each with an appropriate heading so that you simply have to fill in the results when you conduct the investigation.
5. Conduct the first-hand investigation, recording your results in the table as you proceed. Read the aim again and then write a general conclusion for the overall investigation. If your experiment was not successful, state this in the conclusion.
6. In your discussion, describe how you would modify your method, materials or equipment to improve your investigation if you were to repeat it. Compare your results with expected results and discuss any discrepancies here. Answer the discussion questions that have been set. (Refer to details on these *investigative skills* on page x.)

Ensuring validity, reliability and accuracy
Read the skills section on pages x–xii and take note of 12.4e and f; write a brief note in your discussion to assess the **validity**, **reliability** and *accuracy* of your results.

The **scientific method** involves doing a 'fair' test:
- validity improves with multiplicity (many sets) of results (i.e. use *large sample sizes* or *many trials*). Validity also relies on using a *controlled* experiment which measures only *one variable at a time*; all others should be kept constant
- reliability—the same method should yield the *same results when repeated by other people*
- accuracy—the results should comply with *similar scientific information* (e.g. data in scientific journals); also involves selecting the most *precise measuring equipment* you have available and *correct use* of the equipment to avoid human error.

Standard tests for organic chemicals

Test each substance (excluding the lipids) by placing a known amount of the chemical into water in a test tube and then add a set amount of indicator. A second test tube will be required for each control. Some samples may have to be ground up using a mortar and pestle before being placed in the test tube. See details in Table 2.3.

Safety considerations
Some indicators and materials that you will use are irritants to skin, eyes and mucous membranes of the nose. Identify which chemicals are irritants. Describe ways to avoid contact with, or the inhalation or ingestion of any chemicals or materials used. Running water should be used as a rinsing agent.

Detailed information on *validity*, *reliability* and *accuracy*

Results table— chemicals in cells

Table 2.3 Standard indicators used to test for the presence of organic chemicals

Chemical being identified	Indicator used	Heat/no heat	Positive result	Negative result
Starch	Iodine	No heat	Blue–black/purple	Yellow–brown, no colour change
Glucose	Benedict's solution	Heat in a water bath using a Bunsen burner	Orange–red	Pale blue, no colour change
Protein	Biuret's test: ■ 5 drops dilute sodium hydroxide ■ 3 drops copper sulfate	No heat	Purple	Pale blue, no colour change
Lignin	3 drops of toluidine blue	No heat	Blue–green	Pale blue, no colour change
Chloride ions	3–5 drops of silver nitrate	No heat	A white, precipitate (white solid material that settles to the bottom of your test tube)	No precipitate

2.2 Cell membranes, diffusion and osmosis

■ *identify that there is movement of molecules into and out of cells*

For any cell to function it must interact with its surrounding environment and with the cells which surround it. Substances required by cells for their functioning need to move into cells and waste substances need to pass out of cells. These substances enter or exit cells through the cell membrane.

Substances that enter and leave cells

Substances needed by cells are *gases* (oxygen and carbon dioxide), *nutrients* (sugars, amino acids, glycerol and fatty acids) and water, the main solvent in cells, as well as mineral salts dissolved in water

Substances that must leave cells are *wastes* (urea, uric acid and excess carbon dioxide) and *products* secreted by cells that may be needed to coat the outside of cells (e.g. mucus) or may pass to other cells (e.g. hormones).

Boundaries

The movement of these chemicals occurs across the cell boundary which may be a cell membrane only (animal cells) or a cell membrane and a cell wall (plant cells). In both plant and animal cells, the **cell membrane** is in direct contact with the cytoplasm and it controls the passage of *water* and *other molecules* (many in a dissolved form) into or out of living cells.

Cell membranes are **selectively permeable** (sometimes referred to by the simplified term **semi-permeable**) because they allow only certain molecules to pass through them. Microscopic pores that exist in the cell membrane determine what molecules may or may not enter a cell. (See Table 2.4.) The pores in the cell membrane work in a manner similar to the gates in the boundary or fence surrounding a school—people who are associated with the school are allowed in and strangers are refused entry. People walking into the school use pedestrian gates, but if people want to drive into the school in motor vehicles they need to do so through special gates, since the pedestrian gates are too small. Similarly, the pores of a cell membrane may restrict molecules because of their size and larger molecules will need special assistance to enter cells.

Plant cells may have a cell wall in addition to this membrane. In contrast to the selectively permeable nature of a cell membrane, a cellulose cell wall is **permeable**—it is a non-selective boundary that allows water and most molecules to pass freely inwards or outwards. The movement of these molecules is restricted only when they come into contact with the cell membrane. Some cells on the outside of plant parts (such as the surface cells of leaves) need to restrict the loss of water so that they do not dehydrate. Because cellulose is permeable these cells need an extra layer called a waxy cuticle, on the outside of their cellulose walls, to reduce water loss.

Movement across membranes of cells

Movement across the *cell membrane* occurs when substances need to pass from one cell to another or when substances are exchanged between cells and their surrounding environment.

Movement across *organelle membranes*: some intracellular transport involves substances passing between the cytoplasm and membrane-bound organelles such as chloroplasts or mitochondria. The passage of molecules through these membranes is similar to that of molecules moving through the cell membrane, as they share the same basic structure.

To understand how membranes control and regulate the movement of substances, biologists examine the structure of membranes and try to relate this to their functioning.

Current model of the cell membrane

■ *describe the current model of membrane structure and explain how it accounts for the movement of some substances into and out of cells*

The detailed chemical structure of a cell membrane cannot be seen, even with an electron microscope. Many years of research have gone into attempting to accurately describe the structure of cell membranes and our current accepted understanding is based on a model called the **fluid mosaic model** of cell membranes.

Models in science

PFA
P2

(PFA P2) *Why do we use models in science?* Models are used in science to help our understanding of things:

■ to represent something too large or too small to be seen

■ to explain something complex in a simple manner

■ to make predictions of expected results.

More importantly, how do we know that the model that we use is correct? Before a model is accepted, it needs to be *validated*—that is, certain predictions should be made and, when tested using the model, should hold true. Often models are accepted for a period of time until a prediction is made that is no longer supported by that model. At that point the model must be adjusted or changed to include the new prediction. Therefore, as scientists, we need to always be aware of the limitations of any model and be conscious of what is based on scientific fact and what is based on scientists' explanations of how they understand phenomena. Theories and models tend to change as science progresses, technology improves and more information becomes available. The current model for cell membranes, proposed in 1972, replaces the previous model which was proposed in 1935.

The current model of membrane structure: the fluid mosaic model

The structure of the selectively permeable cell membrane is described as a *fluid-mosaic*, based on a model proposed by J. Singer and G. Nicholson in 1972. This model proposes a 'lipid sea' with 'many and various proteins floating on and in it'. The model has been accepted because the behaviour of membranes, estimated surface area, chemical analysis and electron microscope studies are all compatible with the model and it accounts for most functions associated with cell membranes.

Lipid component

The 'fluid' part of the cell membrane is composed of two layers of **phospholipids**, in which:

■ their hydrophobic ('water-hating') tails are positioned inwards (towards each other)

■ their hydrophilic ('water-loving') heads are positioned facing outwards (towards the cytoplasm on one side and to the outside of the cell on the other side, see Fig. 2.6a).

(a) — hydrophilic head (attracted to water)

— hydrophobic tail (repels water)

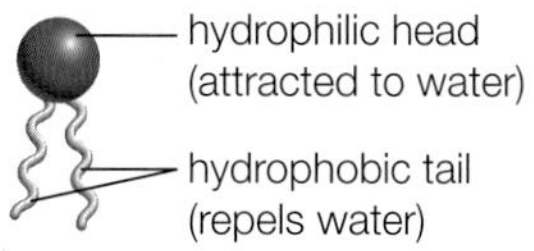

Figure 2.6
Phospholipids: (a) a phospholipid molecule

Figure 2.6
Phospholipids:
(b) a lipid bilayer
showing phospholipid
alignment

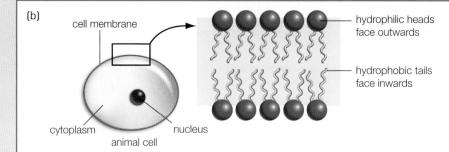

Figure 2.7
(a) Diagram
representing the fluid
mosaic model of a cell
membrane;
(b) A 3-dimensional
view of the fluid
mosaic model of
a cell membrane

This layering is termed a **bilayer** and it is not rigid in structure, hence the term 'fluid' mosaic. This structure forms the basis of the cell membrane and all other membranes within cells. Proteins are then interspersed throughout this structure. The diagram above would be termed a *unit membrane* and would be presented in a diagram of a cell by a single line (see Fig 2.6b).

Protein component
Protein molecules are scattered throughout the lipid bilayer, suspended in it. Some proteins penetrate all the way through the bilayer forming channels that allow some materials to cross the membrane. Other proteins may be partly embedded in the membrane and it seems that some proteins are fixed in place, but others travel about freely. The proteins are described as 'floating' in the lipid bilayer 'like icebergs in a lipid sea', giving a *mosaic* effect.

Some proteins function as pores (temporary or permanent), others form active carrier systems or channels for transport, while others (glycoproteins) have carbohydrates attached for cell recognition.

Other components
■ *Carbohydrates* may be attached to the outer surface of the cell membrane and play a role in cell recognition. They are termed glycolipids and glycoproteins, depending on the part of the membrane to which they are attached.
■ *Microtubules* may be attached to the inner surface of the cell membrane, forming the anchorage of the cytoskeleton (internal skeleton of the cell).

The type of carbohydrates and proteins attached to a cell surface are important for the immune system in mammals to recognise whether cells belong to that organism ('self') or if foreign or invading cells have entered the body ('non-self'). These surface

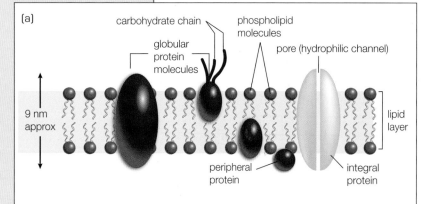

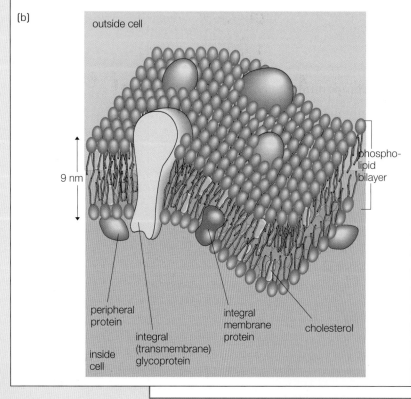

molecules therefore act as a 'uniform' by which cells can be recognised. They are the reason why bacteria are attacked and killed and why tissue matching is so important in organ transplants. (You will learn about these markers called *antigens* in more detail in your HSC course.)

Evidence for the fluid mosaic model

- The membrane allows lipid-soluble substances to pass through easily, suggesting a lipid basis.
- The behaviour of membranes, for example how they reseal themselves when punctured with a fine needle, led to the idea that they are not rigid, but are more like a fluid.
- Monolayer or bilayer—when the expected total area of a monolayer of cell membrane was estimated for a particular cell, it turned out to be twice the surface area of the cell, suggesting that it is arranged as a bilayer.

- An earlier model of the cell membrane, which suggested more simply that it consisted of lipids coated by proteins, was supported by electron microscope work; staining for particular chemicals showed three layers—a lipid centre with a protein layer on either side. The currently accepted fluid mosaic model is a progression from the earlier model because it accounts for the fact that not all membranes are identical (e.g. when a mouse membrane was combined with a human membrane, mouse proteins moved to become dispersed within the human membrane).

The permeability of membranes

The **permeability** of a membrane refers to its ability to allow substances to pass through it. Table 2.4 is far more simplistic than the workings of a cell membrane, but it gives some understanding of the terms permeable, impermeable and selectively permeable barriers.

Table 2.4 A simple explanation of the permeability of barriers

Terminology	Description	Diagram
Impermeable	If a barrier is impermeable it does not allow any molecules to pass through it (e.g. a plastic bag). It does not have any pores through which the molecules can pass.	no pores particles can not pass through
Permeable	If a barrier is permeable it will allow all molecules to pass through it and it probably has large spaces or pores.	large pores particles pass through
Selectively permeable (semi-permeable or partially permeable)	Some barriers are selectively permeable—they only allow certain molecules to pass through and others are held back. These barriers may be thought of as having small pores.	small particles pass through some particles too large to pass through

The movement of molecules across cell membranes

The permeability of a membrane to a molecule depends on the molecule's:

- size
- electrical charge
- lipid solubility.

Small molecules move across membranes fast. Water-soluble (hydrophilic) molecules have difficulty penetrating a membrane, whereas lipid-soluble molecules do not; for example, urea and ethanol have *high permeability*.

Electrically-charged molecules are not very soluble in lipids and therefore have *low membrane permeability*, whereas neutral molecules (e.g. gases such as carbon dioxide and oxygen) have a high permeability.

Although water is a charged molecule, membranes are *highly permeable* to it. Water moves through special tiny hydrophilic pores in the membrane by a process called **osmosis**.

Molecules that have low permeability rely on carrier proteins to transport them across membranes in cells (this will be dealt with in more detail later in this chapter).

The selectively permeable nature of cell membranes

FIRST-HAND INVESTIGATION

PFAs

P2

BIOLOGY SKILLS

P12.1; P12.2; P12.4
P14.2; P14.3

- *perform a first-hand investigation to model the selectively permeable nature of a cell membrane*

Introduction

Use two tea strainers to model the selectively permeable nature of cell membranes.

Prepare a mixture of icing sugar and lollies such as Smarties. Place two teaspoons of each into one tea strainer and then, using an elastic band, tie the handles together. Shake the tea strainer 'cell' over a sheet of coloured paper and record in a table which substances pass through. Validate your model by using a variety of other substances (these may include salt, sugar granules and any other substances such as tea leaves and water).

Discussion questions

1. Explain what you think is the purpose of the syllabus requiring you to model the selectively permeable nature of a cell membrane.

2. Draw and label your model and, using a different coloured pen, write next to the appropriate label what structures of a selectively permeable membrane are represented.

3. Describe in what ways your model successfully shows the selectively permeable nature of a membrane.

4. Discuss any limitations in the working of your model, when compared with the functioning of a selectively permeable cell membrane.

5. Justify the validity of your model.

6. The description of a cell membrane being like a school fence (see page 108) could also be considered a model that shows the selectively permeable nature of membranes. Discuss an aspect of membrane functioning described in that model which is not shown in the tea-strainer model.

Osmosis, diffusion and active transport

■ *compare the processes of diffusion and osmosis*

The movement of materials into and out of cells takes place either *passively* or *actively*. **Passive movement** includes the processes of **diffusion** and osmosis, types of movement requiring no energy input from the cell. **Active transport** requires an input of cellular energy to actively

Figure 2.8 Diffusion, osmosis and active transport across cell membranes

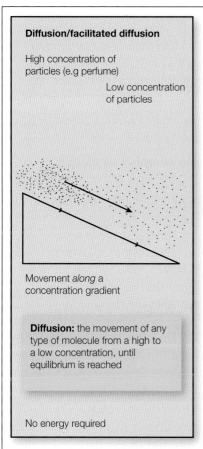

Diffusion/facilitated diffusion

High concentration of particles (e.g perfume)

Low concentration of particles

Movement *along* a concentration gradient

Diffusion: the movement of any type of molecule from a high to a low concentration, until equilibrium is reached

No energy required

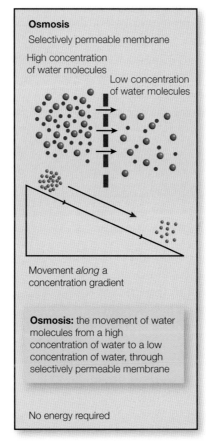

Osmosis
Selectively permeable membrane

High concentration of water molecules

Low concentration of water molecules

Movement *along* a concentration gradient

Osmosis: the movement of water molecules from a high concentration of water to a low concentration of water, through selectively permeable membrane

No energy required

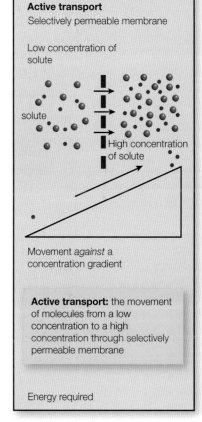

Active transport
Selectively permeable membrane

Low concentration of solute

solute

High concentration of solute

Movement *against* a concentration gradient

Active transport: the movement of molecules from a low concentration to a high concentration through selectively permeable membrane

Energy required

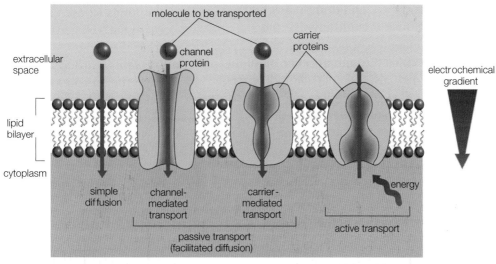

move molecules. Diffusion and osmosis result from the random movement of particles, called Brownian motion, whereby particles continually collide and move randomly. When particles are in a higher concentration in one region than other regions, their constant movement slowly results in the particles spreading out from this highly-concentrated region and eventually becoming evenly distributed within the space.

Diffusion

Diffusion is the movement of *any* molecules from a region of high concentration to a region of low concentration of that substance, until **equilibrium** is reached. This does not require an energy input.

Simple diffusion

For example, if deodorant is sprayed at the front of the classroom, its smell becomes evident to the nearest students in the front row at first and then slowly, as the particles spread out, to students further away. The particles of deodorant move from the front of the room, where they were in the highest concentration, to the rest of the room, where they were in low concentration, until they are equally spread throughout the room (equilibrium is reached).

Diffusion could involve the movement of solid, liquid or gas molecules through another medium that may also be solid, liquid or gas. Examples are the movement of perfume through the air (gas through gas), milk in tea (liquid through liquid), sugar in tea (solid through liquid) and ink up filter paper (liquid through solid).

Movement from a high to a low concentration is described as movement along a **concentration gradient** (a gradient is a slope). Picture the movement of a rock rolling down a hill (molecules moving down a

concentration gradient), needing no energy (see Fig. 2.8). The rate of diffusion changes, depending on the concentration gradient. If there is a greater difference in the concentration of substances, the concentration gradient will be steeper and the diffusion will occur faster. Diffusion can also speed up or slow down, depending on the temperature: heat increases the rate of diffusion because the kinetic energy of the particles speeds up.

When equilibrium is reached, the molecules continue to move randomly, but do not move in any particular direction.

Biological examples of simple diffusion

Lipid soluble molecules and small, simple molecules diffuse across cell membranes. Living organisms rely on diffusion to function in a large number of instances. Try to think of examples of substances that move by simple diffusion in animals (including humans) and plants. Some examples are listed below to help you.

- Oxygen diffuses from the air (where it is in higher concentration) into cells in the lungs of humans (where it is in lower concentrations).
- Oxygen diffuses from the cells of the lungs into the blood capillaries in the lungs (these capillaries can then carry the 'oxygenated' blood away from the lungs).
- In plants, carbon dioxide diffuses from the surrounding air into cells in the leaves of plants for photosynthesis.
- Some 'unnatural' solutes that diffuse across membranes in our bodies are anaesthetics (into the lungs and then the bloodstream), alcohol (from the digestive tract into the bloodstream) and pesticides such as DDT (across the skin).

Some larger molecules and electrically-charged molecules move

Demonstration of simple diffusion—diffusion of a liquid in a liquid

Aim
To demonstrate the diffusion of a liquid through a liquid and to investigate if it diffuses at the same rate at different temperatures.

Materials
Food colouring liquid, drinking straw, two beakers, hot water and cold water.

Method
Label one beaker 'cold' and the other 'hot', then three quarters fill each with cold and hot water as labelled. Place a drinking straw into the beaker, touching the bottom, and pour one drop of food colouring into the straw. Remove the straw, leaving the food colouring in the centre of the beaker (try not to disturb the water as the straw is withdrawn). Observe the movement of the colour through the water over the next half hour and then again after 24 hours.

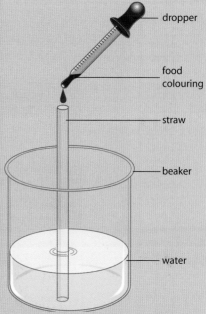

Figure 2.9 Experiment to demonstrate diffusion

too slowly by simple diffusion to satisfy the needs of cells. To overcome this problem, proteins in the cell membrane may act as *membrane transporters*, accelerating the movement of molecules across the membrane. If this movement is along a concentration gradient, no energy input is required and the process is termed **facilitated diffusion**.

Facilitated diffusion (passive protein-mediated transport)
Facilitated diffusion is similar to simple diffusion where molecules move along a concentration gradient, but in facilitated diffusion this movement is assisted (facilitated) by carrier proteins in the membrane. Each protein that acts as a membrane-transporter is specific to one solute or several similar solutes. It allows the movement of a large or charged molecule to occur more rapidly than would be expected as a result of simple diffusion for that molecule. The protein may act in one of two ways— as a **channel protein** or a **carrier protein**. (See Fig. 2.8.)

A *channel* is the fastest form of transport. A channel is similar to a drinking straw, it is formed by a protein that spans the cell membrane and allows direct passage from one side of the membrane to the other. Channels work like gates and they have 'open' and 'closed' states, determined by electrical or physical signals (comparable to electronic gates or those that rely on physical opening and closing).

Solution[a]	Solution A	Solution B	Solution C
	100% water	5% salt 95% water	20% salt 80% water
Water concentration	Highest	Lower than A	Lower than A and B
Solute concentration	Lowest	Higher than A	Higher than A & B
Relative solute concentration = tonicity[b]	Hypotonic to B and C	Hypertonic to A and hypotonic to C	Hypertonic to both A and B
Concentrated/dilute solution[c]	Dilute	More concentrated than A	More concentrated than A and B
Osmotic potential/osmotic pressure[d]	Lowest	Higher than A	Higher than A and B
Direction of osmosis	A → → → B and C	B → → → C	- - - - - -

Table 2.5 Terminology related to osmosis

[a]A solution is a mixture made up of a *solvent* (e.g. water) in which a *solute* (e.g. salt or sugar) is dissolved.

[b]The *solute concentration* determines the *tonicity* of a solution. It is used to *compare* how concentrated two solutions are in relation to one another. The term *hyper*tonic is used to describe the solution with the *larger* number of *dissolved substances* (hyper = large or over abundant, e.g. hyperactive = overactive) and the term *hypo*tonic describes the solution with fewer dissolved substances (hypo = small/under). If both solutions have the same concentration of solutes, they are termed **isotonic**.

[c]To remember the meaning of the terms *concentrated* and *dilute*, think in terms of mixing cordial with water—too much cordial and it is too concentrated, too little and it is too dilute. These terms refer to the concentration of *solute* (cordial) in water. If cordial has a very high *water concentration*, is it *dilute* or *concentrated*? Check the use of these terms in relation to water concentration in the table above.

[d]Osmotic potential or pressure describes the ability of a solution to draw water towards itself by osmosis. That is, the greater the concentration of dissolved substances, the lower the water concentration and the more likely the solution will attract water by osmosis (movement from a high to a low *water* concentration).

A *carrier protein* binds with a solute and then the protein changes shape or conformation to move the solute to the other side of the membrane. Once it releases the solute on the other side, the carrier returns to its original shape. Because the carriers can become saturated, the process is slower than that of channel-mediated transport, but quicker than the expected rate for simple diffusion. Biological examples of facilitated diffusion in cells include the uptake by plant roots of mineral salts from the soil and the absorption of mineral salts in the large intestine of animals.

The movement of water through cell membranes involves protein channels and so it is considered to be a specialised form of facilitated diffusion known as osmosis.

Osmosis

Osmosis is the movement of *water* molecules from a region of high water concentration to a region of low water concentration through a selectively permeable membrane. It does not require an energy input.

The movement of water through cell membranes is passive but, because water is a charged molecule, the movement is not directly through the bilayer of lipids. Water moves through special tiny protein channels in cell membranes called **aquaporins** ('water pores'). As a result, the movement of water across membranes is much more rapid than would otherwise be expected. Tissues with a high-water permeability have a greater number of aquaporins in their cell membranes (e.g. kidney cells).

The rate of water movement through the membrane is affected by two things:
1. solute concentration—the concentration of dissolved substances in the solution
2. opposing physical pressure or tension exerted on the water (how much it is being squeezed or pulled) by the space that it is moving into.

The effect of solute concentration on osmosis (see Fig. 2.5)

To understand how *water* concentration is affected by the number of solutes dissolved in it, think how distilled water (100 per cent water) has a greater water concentration than a solution of 5 per cent salt and 95 per cent water. The water concentration in a solution of 20 per cent salt and 80 per cent water would have an even lower water concentration. So, the higher the solute concentration in a solution, the lower the water concentration will be. Table 2.5 illustrates this and introduces some of the terminology that is often used in relation to water concentrations and osmosis.

Osmosis and cells

When a cell has a higher solute concentration than the water surrounding it, water drawn towards the cell can only continue to move into that cell if the cell is able to physically expand to allow space for the incoming water (see Fig. 2.10). (This would also apply to the osmosis experiment conducted on page 119, where the movement of water from the beaker into the sucrose solution was limited by the available space in the dialysis tubing.) The cellulose wall of a cell is only slightly elastic. When the cell

wall is distended by incoming water and stretches to capacity, it exerts an opposing force or *pressure* on the incoming water. If this opposing force, known as **wall pressure**, is equal to or greater than the force of the water moving in (**osmotic pressure**) water will no longer be able to enter the cell. The expanded cell is then said to be in a state of **turgor**. Plants rely on turgid cells for support.

Plant cells that are placed in a solution that is more concentrated than the cell contents, or plant cells that are exposed to air, will lose water to their surroundings by osmosis. Water may continue to leave the cells until the cell contents (cytoplasm and cell membrane) begin to shrink, leaving a gap between the cell membrane and wall. This state is termed **plasmolysis**. Plasmolysed or **flaccid** cells no longer provide sufficient support to the plant and the plant wilts. Wilting may be temporary (e.g. watering a plant may allow it to recover and become upright once again) or permanent. (See Fig. 2.10.)

Diffusion and osmosis both rely on a concentration gradient to direct the flow of substances and are examples of passive movement with no input of energy. Sometimes in living things, a chemical may need to be moved *against* the concentration gradient and to do this requires energy.

Figure 2.10 Turgor pressure in plant cells: (a) turgid cell placed in a hypotonic solution; (b) and (c) plasmolysed or flaccid cells in hypertonic solutions

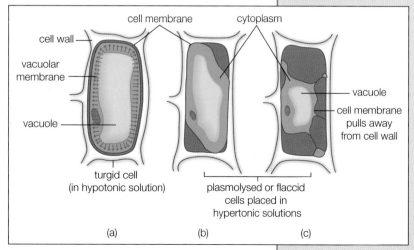

turgid cell (in hypotonic solution)

plasmolysed or flaccid cells placed in hypertonic solutions

(a) (b) (c)

Active transport

Active transport is the movement of molecules from an area of *low concentration* to an area of *high concentration*, requiring the input of energy.

Active transport requires a carrier protein that spans the membrane to actively move chemicals from a *low* to a *high* concentration, utilising cellular energy. An example of active transport is when glucose and amino acids are reabsorbed by kidney cells so that they are not lost in urine or when the end products of digestion are loaded from the intestine into the bloodstream to absorb food after eating.

Endocytosis and exocytosis—transport of large molecules across membranes

When particles larger than individual molecules need to enter or leave cells, the cell membrane of an animal cell is able to surround and engulf the particle in a process known as endocytosis. ('endo' = into/inside; 'cyto' = cell). A typical example would be when a white blood cell in the human body engulfs and digests bacteria to fight infection. The membrane enfolds and encloses the particle in a vesicle (sac) which pinches off and the particle enters the cell (see Fig. 2.11). If a solid particle is engulfed, the process is termed **phagocytosis** ('cell eating'). Sometimes fluid is engulfed and the process is then called **pinocytosis** ('cell drinking').

In a reverse situation, membrane-like vesicles can be transported to the surface of the cell and emptied out, a process called exocytosis. This could happen when a cell secretes hormones that it has made (see diagram on page 97).

All of these processes are made possible by the fluid mosaic nature of the cell membrane, because of its ability to flow around a particle and seal up when it joins with other membranes.

Table 2.6 Movement of substances into and out of cells—comparing active and passive transport

	Active transport	Passive transport
Entry point	Through carrier proteins	Through carrier proteins (facilitated diffusion) or between phospholipids (osmosis and diffusion)
Rate of movement	Fast	Slow
Concentration	Low to high	High to low
Energy	Energy required	Energy not required
Examples of molecules transported	Sugar and amino acids	Water, oxygen, carbon dioxide, sugars and amino acids

Figure 2.11 Phagocytosis—a white blood cell engulfing bacteria

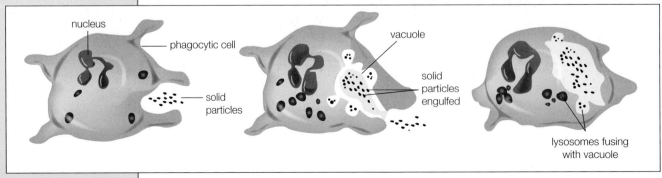

nucleus

phagocytic cell

solid particles

vacuole

solid particles engulfed

lysosomes fusing with vacuole

Demonstrating osmosis and diffusion

FIRST-HAND INVESTIGATION

BIOLOGY SKILLS

P11.3
P12.1; 12.2; 12.4
P13.1
P14.1; 14.2; 14.3

■ *perform a first-hand investigation to demonstrate the difference between osmosis and diffusion*

Suggestion for co-operative learning: allow students to work in pairs, each partner labelling themselves A or B. Split the class into two—all students labelled A do the osmosis experiment and students labelled B carry out the diffusion experiment. Students need to answer Question 1 individually before explaining their experiment to their partners and then they can jointly discuss and answer Questions 2 and 3, comparing osmosis and diffusion.

Aim 1—osmosis

To demonstrate osmosis using dialysis tubing as a selectively permeable membrane.

Materials—osmosis

Two 200 mL beakers, distilled water, sucrose solution, cotton thread, two glass rods, two 15 cm strips of dialysis tubing, 25 mL measuring cylinder, filter funnel, marking pen and plastic cling wrap to prevent evaporation.

Aim 2—diffusion

To demonstrate diffusion using dialysis tubing.

Materials—diffusion

Two 200 mL beakers, distilled water, dilute iodine solution, a mixture of starch (cornflour starch or rice water) in water, cotton thread, two glass rods, two 20 cm strips of dialysis tubing, 25 mL measuring cylinder, filter funnel, marking pen and plastic cling wrap to prevent evaporation.

Read the method below, identifying all the variables that you will need to keep constant in this experiment. (List these in your discussion—see Question 1.1.)

Method

■ Cut two strips of dialysis tubing of the same length.

■ Tie one end of each strip securely with a piece of cotton, leaving the other end untied.

■ Open the untied end of the tubing (this is made easier if you wet it under a running tap). Using a funnel and a measuring cylinder, pour a measured volume (about 15–20 mL) of sugar solution (osmosis) or starch mixture (diffusion) into the tubing so that the tubing is two-thirds full. (Record the exact volume of sugar solution used in each.)

■ Tie the top of the tubing with cotton. Attach it as shown in Figure 2.12 to a glass rod. This is your *experimental* apparatus.

■ Repeat the previous two steps, this time filling the tubing with distilled water instead of sugar (osmosis control). Keep the control for the starch solution the same as the experiment, except place no iodine in the beaker.

■ Suspend the *experimental* and the *control* apparatus each in a beaker of distilled water (osmosis experiment) or in a beaker of distilled water with iodine solution (diffusion experiment) as shown in Figure 2.12. Ensure that the tied ends of the dialysis tubing are just above the distilled water in the beaker to prevent leakage. Do not have the tubing too far out of the water as evaporation may occur and this will interfere with the accuracy of your results.

■ Mark the level of water in the beaker with a marking pen.

■ Cover the top of the experiment with plastic wrap to prevent the water in the beaker from evaporating. Leave the apparatus to stand for several hours or overnight.

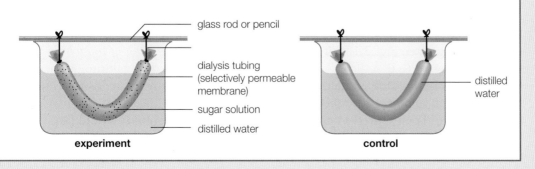

glass rod or pencil

dialysis tubing (selectively permeable membrane)

sugar solution

distilled water

experiment

distilled water

control

Figure 2.12
Experiment to demonstrate osmosis

- Record your results in the form of a table. Draw up a table and remember to include headings.
  Osmosis experiment—record the volumes of solutions in each dialysis bag and in each beaker (both the experimental and the control) at the start of the experiment, and then again after 24 hours. You should also record observations such as changes in the fullness of the dialysis tubing. (Remember to note the exact number of hours that the experiment was left standing.)
  Diffusion experiment—record the volumes of solutions in each dialysis bag and in each beaker (both the experimental and the control) at the start of the experiment and then again after 24 hours. You should also record observations such as the colour of solutions in the dialysis bag and in the beaker at the start of the experiment, at the end of the lesson and after 24 hours.
- Your results will be both **quantitative** (measured, e.g. using a measuring cylinder) and **qualitative** (e.g. a description of how full the bags are, drawing diagrams at different stages of the experiment, or a description of any colour change). Think about the advantages of qualitative versus quantitative results.

Results

Record results in a table and/or use diagrams.

Conclusion

It can be concluded that .... has taken place and .... has moved from a high concentration to a low concentration through a selectively permeable membrane.

Discussion

Individual

1. Answer the following questions before discussing the experiment with your partner:
   1.1. List all the variables in this experiment that should be kept constant.
   1.2. Identify the dependent and independent variable.
   1.3. Justify the conclusion given above, using evidence from your experiment.
   1.4. Describe ways that you could modify this experiment to improve its accuracy.
   1.5. Discuss the advantages and disadvantages of qualitative versus quantitative results.

Partners

2. Record the following as a table of comparison using the results form each experiment:
   2.1. Compare the variables that you kept constant.
   2.2. Compare your dependent and independent variables.
   2.3. Validate your experiments (i.e. explain how you tested what you set out to test).
3. In a written paragraph, describe how the experiments *distinguish* between osmosis and diffusion.

SR **TR**

Results tables for diffusion and osmosis experiments

2.5 Surface area: volume ratio and rate of movement

- *explain how the surface area to volume ratio affects the rate of movement of substances into and out of cells*

What is surface area?

The surface area of an object is that part of the object exposed to its surroundings (see Fig. 2.13). If the object is a cube (e.g. a die):
- the *area of one side* is calculated as the length multiplied by the width, $l \times w$, and is measured in cm^2.
  For example in a cube that is 2 cm in width and 2 cm in height, each side will have an area of 2 cm × 2 cm = 4 cm^2
- the *surface area of the whole cube* is calculated by taking into account that there are *six* sides to a cube (think of a die): $6 \times (l \times w)$.
  In a 2 cm cube it would be:
  $6 \times (2 \times 2)$ cm^2 = 24 cm^2

What is volume?

The volume of an object is the amount of space that it occupies. To calculate the volume of a cube, the length, width and depth must all be multiplied together: $l \times w \times h$ (measured in cm^3). In a 2 cm cube, this would be: 2 cm × 2 cm × 2 cm = 8 cm^3

What is surface area to volume ratio?

A ratio is a fraction that allows us to compare things. It is obtained by dividing the two things to be compared. For example the ratio *surface area:volume* is the same as obtained from the fraction: $\dfrac{\text{surface area}}{\text{volume}}$

In the example above, the cube would be: $\dfrac{SA}{V} \dfrac{24 \text{ cm}^3}{8 \text{ cm}^3} = \dfrac{3}{1}$

Therefore the ratio would be 3:1.

How the surface area to volume ratio affects the rate of movement of substances into and out of cells

Substances that enter cells must travel from the outside environment across the surface of the cell and then diffuse inwards until they reach the centre of the cell. If a cell has a large volume, the organelles in the centre of the cell are further from the outside (see Fig. 2.14). If a cell is flatter or smaller, the organelles in the centre are close to the outer surface. This increases the efficiency of substances diffusing into or out of a cell due to the cell having a larger surface area to volume ratio. The folding of a cell also increases the surface area to volume ratio, increasing the efficiency of substances entering and leaving the cell.

Extension activity— to demonstrate surface area

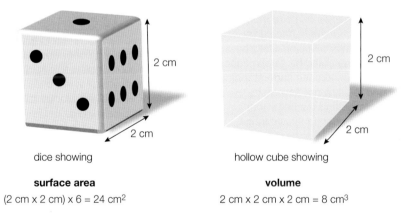

dice showing

surface area
(2 cm x 2 cm) x 6 = 24 cm^2

hollow cube showing

volume
2 cm x 2 cm x 2 cm = 8 cm^3

Figure 2.13 Surface area (e.g. a die) and volume (e.g. a hollow cube)

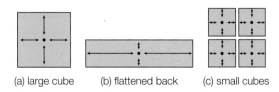

(a) large cube   (b) flattened back   (c) small cubes

Figure 2.14 Distance over which diffusion occurs to reach the centre in cells of different shapes and sizes: (a) a large cube, (b) a flattened block, (c) small cubes

121

Investigating the surface area to volume ratio and rate of movement

FIRST-HAND INVESTIGATION

PFAs

P2

BIOLOGY SKILLS

P12.1; P12.2; P12.4
P1.31
P14.1; P14.3

■ *perform a first-hand investigation to demonstrate the effect of surface area to volume ratio on rate of diffusion*

Introduction

Understanding the relationship between the size of cells (volume) and their surface area helps us to realise how the physical limitation of the size of cells is important if they are to function effectively. In order to carry out necessary chemical activities (metabolism) cells need to absorb substances that they require from the surroundings. This uptake is often dependent on the process of diffusion. Diffusion is a slow process and so, to be more effective, the number of molecules that can be taken up in a particular space of time should be increased. The **hypothesis** to be tested is that, by increasing the surface area over which uptake occurs (for an object of a particular volume), diffusion can be increased and molecules can reach the centre of an object *sooner* than if the object has a lower surface area to volume ratio. Therefore, although the rate of diffusion is still the same, by increasing the *amount* of diffusion occurring, the *overall rate of movement* increases.

This investigation makes use of a model to investigate the effect of the surface area to volume ratio on cellular diffusion. Cells of different sizes are represented by gelatin cubes, a simplification to try to increase understanding of the concept and to allow predictions to be made. When interpreting the behaviour of substances in the model (in this instance, the movement of the chemicals by diffusion) the limitations of the model must be taken into account. For example one limitation is that the jelly cubes do not have a selectively permeable boundary such as a cell membrane. The validity of the model should also be assessed (see PFA P2 on page ix). For example the lack of a cell membrane and an energy source suggests that no active transport can take place, indicating that only diffusion, as stated in the hypothesis, is being investigated.

Aim

This model will be used to investigate:
■ the change in the surface area to volume ratio with 'cell' size
■ the effect of a change in the surface area to volume ratio on the rate of diffusion.

Equipment

Agar-phenolphthalein-sodium hydroxide cubes (gelatin blocks) (three sizes: sides = 1 cm, 2 cm and 3 cm), 0.1 M hydrochloric acid (approx 150 mL), ruler (with centimetre and millimetre markings), knife or razor blade, paper towel, a 250 mL beaker, a plastic spoon and a stopwatch.

Formulae to be applied

Surface area (SA) of 1 side = length^2
Total surface area = $6 \times$ length^2
The units for SA are cm^2
Volume (V) = length^3
The units of volume are cm^3
Work out the SA to V ratio
$$\frac{SA}{V} = \frac{\_\_\_\ cm^2}{\_\_\_\ cm^3}$$

Safety

Wear a laboratory coat as phenolphthalein can stain clothing. Take care not to get HCl on skin or in eyes—use a dropper and do not rub your eyes. Wear safety glasses. If acid comes in contact with any part of the body irrigate well with water. Turn gelatin blocks using a teaspoon; do not handle chemicals.

Procedure

■ Before starting the practical calculate the surface area to volume ratio of each of the cubes (see first three columns of Table 2.7). Assume that the cubes have exactly 3 cm, 2 cm or 1 cm sides. Due to the indicator (phenolphthalein) and the presence of the base (NaOH, sodium hydroxide) in the cubes they will appear pink in colour.
■ Place one of each size cube (1 cm^3, 2 cm^3 and 3 cm^3) into the 250 mL beaker, making sure that each cube is not touching the others.

■ Cover the cubes with the hydrochloric acid and leave for 10 minutes, using a plastic spoon to turn the cubes over every 2 minutes.

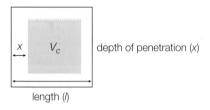

Figure 2.15 Gelatin cube showing depth of penetration

■ Remove the cubes from the acid after the 10 minutes, quickly blotting dry the surface of the cubes using paper towel. Cut each cube into two using the razor blade. Measure the depth of penetration of acid into each cube in millimetres and record it in Table 2.7.

■ Complete the remaining calculations in the table, determining the volume of each cube that remains coloured and calculating the percentage penetration in the last column as a measure of the proportion of the cube affected by diffusion, in cubes of different sizes.

Questions

1. Identify the trend in the change in surface area to volume ratio as the volume of the cube increases.
2. Assess the accuracy of any measurements and calculations made and their relative importance in this practical.
3. Describe the evidence that indicates that diffusion in the cube has occurred.
4. Identify which cube (cell) showed the greatest *depth* of penetration of acid and which showed the least. Was this what you would have expected? Explain.
5. Explain why it was necessary to calculate the *percentage* acid penetration for each cube.
6. When a cell divides, it produces two identical cells that, at first, are half the size of the original cell. Describe the change to the:
   (a) total surface area
   (b) total volume
   (c) surface area to volume ratio for each cell.
7. These cells will then grow, but to a limited size. Clarify the significance of the results of this investigation for multicellular living organisms.
8. Discuss the limitations of the jelly cubes representing cells.

Enrichment questions and 'Think about it'

Cube size (l) cm	Surface area, SA $(6 \times l^2)$ cm²	Volume, V (l^3) cm³	$\dfrac{SA}{V} = SA{:}V$	Depth of penetration (x) mm	Volume left coloured $V_c = (l - 2x)^3$	% left coloured $\left(\dfrac{V_c}{V}\right) \times 100$	% acid penetration = $100 - V_c\%$
1							
2							
3							

Table 2.7 Results of surface area to volume ratio investigation

REVISION QUESTIONS

Worksheet—
chemicals in cells

1. Draw a diagram of a plant/animal cell as seen under the electron microscope. Label each part of the cell. Identify what chemicals make up each part of the cell and, using a different-coloured pen, write the name of the chemicals next to each respective part that has been labelled.

2. Draw up a table to compare the organic chemicals found in cells. Your table should compare the monomers of each compound, the chemical elements (atoms) that comprise that compound, where in the cell each is found and two functions or uses of the chemicals in cells.

3. Identify the three main components of a cell membrane and describe how each is necessary for membrane functioning.

4. Complete the following table comparing the structure and function of a cell membrane with that of a cell wall.

	Cell membrane	Cell wall
Chemical composition		A simpler structure made mainly of strands of the carbohydrate cellulose; also contains pectin and may have additional thickening such as lignin and suberin.
Where it is found	Directly surrounds the cytoplasm, is the outer boundary in animal cells and forms part of the boundary in plant cells.	
Function		Provides support and shape, limits the expansion of cells and plays a role in cell turgidity. Allows water and most molecules to pass freely into and out of cells.
Access to molecules		It has pits with strands of cytoplasm passing through them to allow the passage of substances between cells.

Table 2.8
Comparison of the cell membrane and the cell wall

5. In the form of a table, compare the processes of diffusion, osmosis and active transport under the following headings:
   ■ Type of substances that move
   ■ Concentration gradient along which they move
   ■ Energy requirements
   ■ Only across selectively permeable membranes—yes or no?
   ■ Examples in living organisms.

6. If solution X contains more dissolved substances than solution Y, what process is involved in moving:
   (a) water from Y to X?
   (b) solutes from X to Y?
   (c) solutes from Y to X?

Answers to
revision questions

7. If we cut two identical cubes of potato and leave one to stand in water while the other stands in a 20 per cent glucose solution for 12 hours, predict which will have the greater mass and explain why. (*Hint*: In each case in which direction will water molecules move?)

8. The terms *turgid* and *flaccid* are used to refer to the condition of plant cells. Draw a labelled diagram to illustrate what is meant by each of these terms.

Obtaining nutrients

Plants and animals have specialised structures to obtain nutrients from their environment

Introduction

Organisms need to interact with their surroundings, taking up substances that they need for their functioning and getting rid of wastes which build up as a result of metabolic functions.

Unicellular organisms are so small that, in each organism, simple diffusion is adequate to supply the organism's requirements (e.g. oxygen for cellular respiration) and to remove waste products (e.g. carbon dioxide, urea and other metabolic waste substances). Water levels can also be maintained through the passive process of osmosis because the surface area to volume ratio of these organisms is large enough.

Multicellular organisms are larger in overall size and so their total surface area to volume ratio is smaller. As a result, passive transport would be insufficient to address their needs. This problem is overcome by the functional organisation of multicellular organisms:

- large organisms are made up of *numerous small cells*, so that each cell has its own large surface area to volume ratio. This leads to an increase in the efficiency of diffusion and osmosis in individual cells
- multicellular organisms are not simply thousands of similar cells lumped together. Cells have become *organised into groups* called **tissues** (e.g. blood tissue and skin tissue in humans, photosynthetic tissue and epidermal tissue in plants)
- some small multicellular organisms still rely on diffusion and osmosis for

exchange between their cells and the surroundings, but large multicellular organisms have their tissues further organised into organs and systems, such as those which have developed for the efficient uptake of nutrients (digestive system) and gases (respiratory system).

(a)

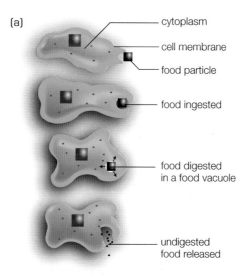

- cytoplasm
- cell membrane
- food particle
- food ingested
- food digested in a food vacuole
- undigested food released

(b)

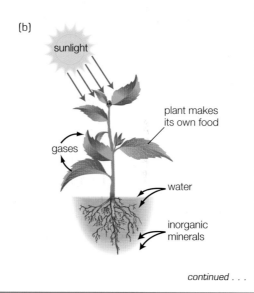

- sunlight
- plant makes its own food
- gases
- water
- inorganic minerals

Figure 3.1 Uptake of nutrients in unicellular and multicellular organisms: (a) a unicellular organism—amoeba (**heterotroph**); (b) multicellular—plant (**autotroph**) absorbs water, inorganic nutrients and light, but makes its own food

continued . . .

Figure 3.1 Uptake of nutrients in unicellular and multicellular organisms: (c) multicellular—human (heterotroph)

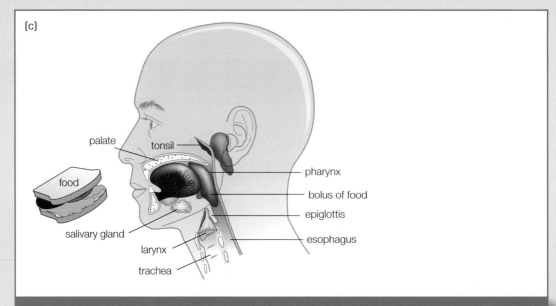

(c)

palate

tonsil

pharynx

food

tongue

bolus of food

epiglottis

salivary gland

esophagus

larynx

trachea

3.1

Functional organisation in multicellular organisms: cells to systems

- *identify some examples that demonstrate the structural and functional relationship between cells, tissues, organs and organ systems in multicellular organisms*

Figure 3.2 Cells that are structurally modified to increase surface area: (a) flattened, squamous epithelium of air sacs in lungs (light microscope view); (b) elongate palisade cells in a leaf (light microscope view); (c) an endothelial cell of the small intestine showing membrane folded to form **microvilli** (electron microscope view)

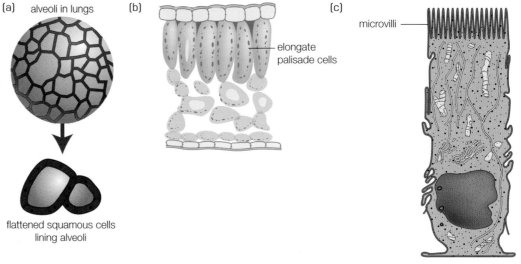

(a) alveoli in lungs

(b)

elongate palisade cells

(c)

microvilli

flattened squamous cells lining alveoli

Structure and function

In this chapter, we deal with the exchange of substances between cells and the environment, for example the uptake of nutrients by cells. The cells of those tissues involved in exchanging substances with the environment have special structural features to increase their surface area to volume ratio, allowing them to function more efficiently (see Fig. 3.2):

- cells may be flattened (e.g. in the tissue lining the air sacs in lungs) or elongate (e.g. photosynthetic cells in leaves), these shapes giving a greater surface area to volume ratio than cube-shaped cells
- the exposed edges of the cells may be extended into folds, for example root hair cells that absorb water and mineral salts in plants; or the cells lining the wall of the small intestine that absorb nutrients (see Fig 3.2c).

Cell differentiation and specialisation

In multicellular organisms a *division of labour* occurs—different cell types (tissues) become structurally suited to carry out different functions. This increases their effectiveness in carrying out their functions: some cells are involved in obtaining nutrients, whereas other tissues function in movement, growth and excreting. (See Fig. 3.3.)

A division of labour is quite a common phenomenon to increase efficiency—think of a community of people living together. Young babies and children need to be fed and nurtured by adults. Within a community, most adults are capable of raising their children and maintaining an efficient and safe household, but specialised things such as medical treatment, education, the building of homes, installing electricity, protecting the community and keeping law and order are carried out by adults who are suitably qualified to do the job (e.g. doctors, teachers, builders, electricians, soldiers, police, magistrates and lawyers). The division of labour within a community shows similarities to that amongst cells of a living organism.

Young cells (called **embryonic cells**) are similar to each other in structure and, in early life, their only function is to divide and give rise to new cells. Embryonic cells require protection and nutrients to grow, but it is only once they begin to mature that they develop suitable structural changes that allow them to carry out specialised functions: some cells fight infection, others store nutrients, some process and transmit information, some secrete substances like hormones and others have a protective function.

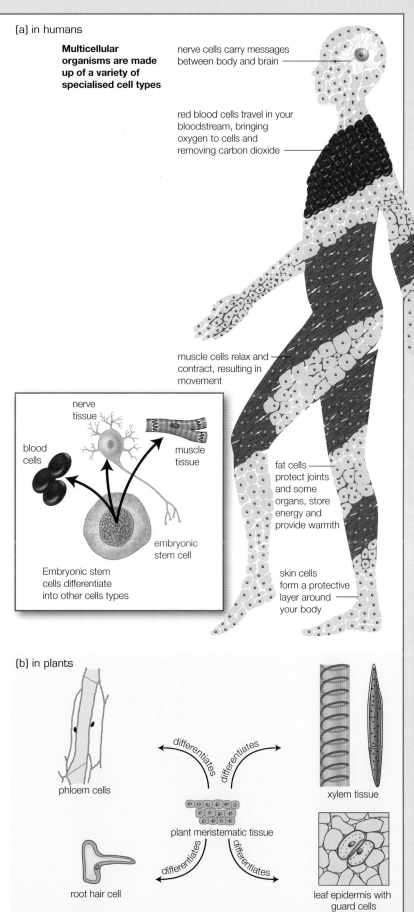

(a) in humans

Multicellular organisms are made up of a variety of specialised cell types

nerve cells carry messages between body and brain

red blood cells travel in your bloodstream, bringing oxygen to cells and removing carbon dioxide

muscle cells relax and contract, resulting in movement

fat cells protect joints and some organs, store energy and provide warmth

skin cells form a protective layer around your body

nerve tissue

blood cells

muscle tissue

embryonic stem cell

Embryonic stem cells differentiate into other cells types

(b) in plants

phloem cells

differentiates

differentiates

plant meristematic tissue

differentiates

differentiates

xylem tissue

leaf epidermis with guard cells

root hair cell

Figure 3.3 Cell differentiation and specialisation: (a) in humans; (b) in plants

127

Division of labour
analogies

When cells become specialised to perform a particular function, they are said to **differentiate**. They develop suitable structural features which allow them to carry out their functions and this makes them structurally *different* from other types of cells and from the embryonic cells from which they arose.

A group of cells that is similar in structure and works together to carry out a common function is called a *tissue*.

Once they have become specialised to form a particular type of tissue, *differentiated cells* lose their capacity to develop into other types of cells. Many even lose their ability to divide and give rise to the same type of cells. *Undifferentiated cells* that are able to divide and differentiate into other types of cells are known as **stem cells**.

Current interesting information

Stem cells may be *embryonic* (found in embryos) or *adult* stem cells (e.g. neural stem cells lining the ventricles of the adult brain; blood-producing stem cells in bone marrow). Stem cell research is a current area of contention: embryonic stem cells can give rise to all other cell types and may be useful for the treatment of injuries and diseases, but this involves harvesting cells from living embryos. In contrast, the use of adult stem cells is non-destructive, but adult stem cells are only able to give rise to their own particular types of tissue (i.e. neural stem cells give rise to cells of the nervous system or blood-producing stem cells to blood cells). Current researchers are trying to induce adult stem cells to transdifferentiate into cell types that they do not normally produce.

Functional organisation

Table 3.1 Functional organisation of living things: how living things are put together

	Functional organisation of body	Animals/humans	Plants
Largest	Multicellular organism		
	Systems	digestive system — stomach, small intestine, large intestine	transport system — xylem, transport tissue, phloem
	Organs	stomach and small intestine	roots and leaves — roots, leaves

Functional organisation of body	Animals/humans	Plants
Tissues	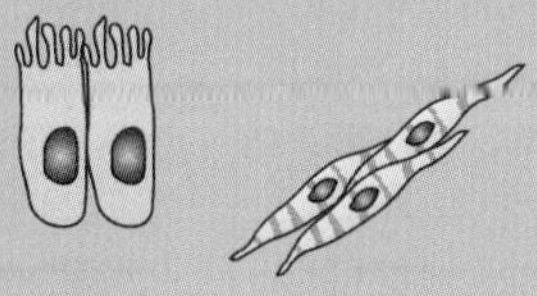 gland tissue muscle tissue gland tissue and muscle tissue	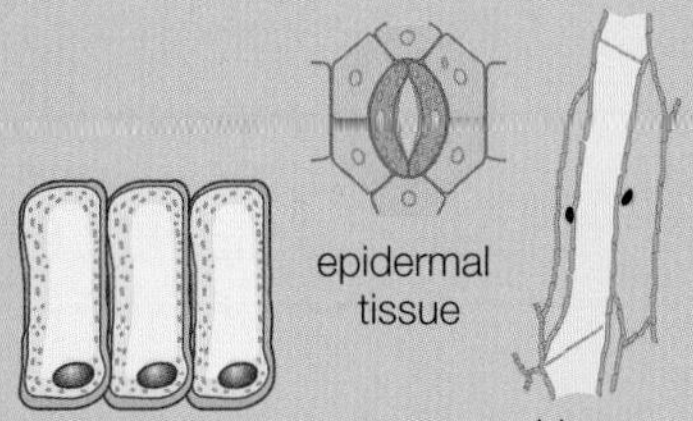 epidermal tissue photosynthetic tissue phloem tissue photosynthetic tissue, epidermial tissue and phloem tissue
Cells	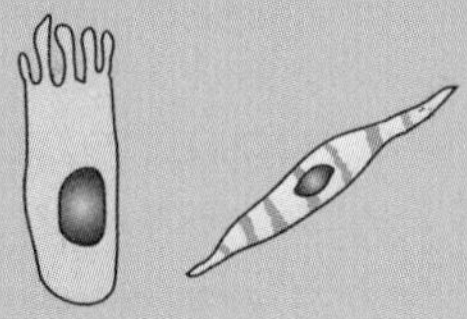 gland cell and muscle cell	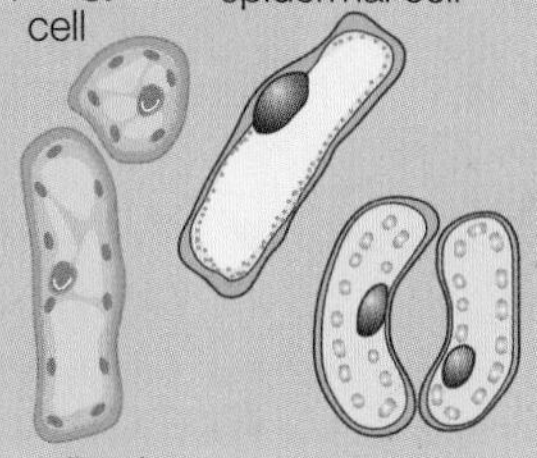 spongy cell epidermal cell palisade cell guard cells plant cells
Organelles	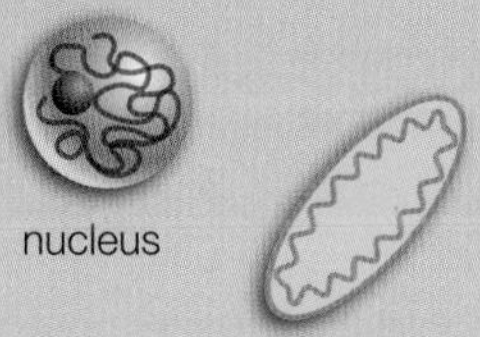 nucleus mitochondrion nucleus and mitochondrion	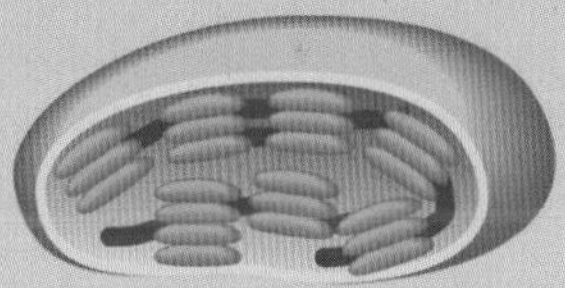 chloroplast chloroplast
Molecules	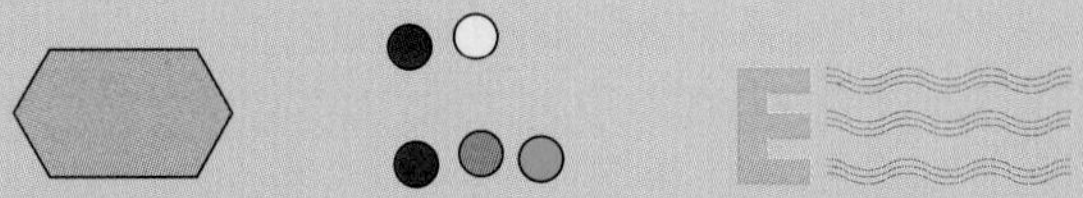 glucose amino acids lipids sugar, amino acids and lipids	
Atoms	 $-\overset{\mid}{\underset{\mid}{C}}-$ $\overset{}{\underset{\mid}{H}}$ $O=$ carbon (C), hydrogen (H) and oxygen (O)	

Smallest

129

Just as similar specialised cells that perform a common function are arranged together to form tissues, so groups of tissues collectively form organs. An **organ** is an arrangement of different types of tissues, grouped together for some special purpose; for example, the leaf is an organ for making food in a plant and the heart is an organ for pumping blood in animals. A **system** is a collection of organs that all work together to achieve an overall body function, for example the digestive system or nervous system in animals and the transport system in plants or animals. A multicellular organism is a living plant or animal composed of many systems which function together co-operatively to ensure its survival. When one or more of these systems malfunctions, the organism is no longer healthy and disease or even death may result.

3.2

Autotrophs and heterotrophs

- *distinguish between autotrophs and heterotrophs in terms of nutrient requirements*

Inorganic and organic nutrients

Living organisms need to obtain nutrients in the form of **organic** substances such as glucose, amino acids, fatty acids and glycerol, nucleotides and vitamins, as well as **inorganic** nutrients such as *minerals* (e.g. phosphates, sodium ions and chloride ions) and *water*. These nutrients, needed by living cells for their functioning, are used in two main ways:

1. as essential *building blocks* from which cells and living tissues are made
2. as a *source of stored energy* that can be converted to ATP, the form of chemical energy needed to power cell functioning.

Organic nutrients are the main supply of stored *energy* in living things, but they are also used extensively in the structure of cells. Inorganic nutrients are essential as structural parts of cells and tissues (e.g. iron in blood, calcium in bones and teeth) and they also play an essential part in assisting *enzymes*—the organic catalysts that control all chemical reactions within cells—but are not an energy source. Animals need to 'eat' food to obtain both organic and inorganic nutrients. In contrast, plants absorb inorganic nutrients from the soil, but they are able to manufacture (make) their own organic nutrients.

Autotrophs

Organisms such as plants that are able to make their own food are termed autotrophs (auto = self; troph = feeding). Most of these organisms produce their own food by the process known as **photosynthesis** ('photo' = light; 'synthesis' = manufacture); that is, they use the *energy of sunlight* to manufacture organic compounds, initially in the form of glucose, from water and carbon dioxide.

Although autotrophs are predominantly *green plants*, some bacteria and *unicells* are also able to make their own food. Most of them do this by the process of photosynthesis, but some bacteria are able to make organic nutrients by the process of *chemosynthesis*. This process relies on using *energy from breaking chemical bonds* to power their food-making, rather than using light energy as do photosynthetic organisms.

Heterotrophs

Photosynthesis does not occur in animals. Animals are termed heterotrophs because they are unable to make their own food ('hetero' = different; that is 'feeding on something different'). Heterotrophs need to take in organic and inorganic nutrients by eating other organisms (they are *consumers*) to obtain nutrients and energy for their cells to function. Most heterotrophs are animals, but some *bacteria*, *unicells* and *fungi* also use a heterotrophic mode of nutrition.

Nutrients that provide cellular energy

All living things require energy in order to survive. Machinery in a factory or the lights in your home need energy in the form of electricity to function. In a similar way, the cells in living organisms require energy so that they can function. The energy required by living cells is not electrical energy but a type of *chemical energy* called **ATP** (**A**denosine **Tri****P**hosphate). (See Fig. 3.4.) So we ask, what then is the role of organic nutrients (and glucose in particular) in providing living things with energy? The energy of ATP powers all cellular activities and can be produced by cells when they break down glucose—the energy trapped in glucose is released when it is broken down and can be captured and incorporated into the high energy compound, ATP. This process is known as *cellular respiration* and takes place in mitochondria in cells: glucose is combined with oxygen during this process, to produce ATP.

Summary

In summary, autotrophs make their own food, whereas heterotrophs rely on other organisms as a source of organic nutrients. Organic nutrients are a source of energy for cell functioning. Organic nutrients can be broken down to release energy that is captured in ATP, the energy essential for cell functioning.

electrical energy

light bulbs work with electrical energy

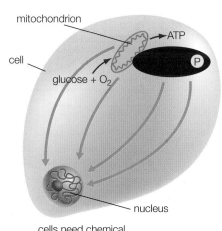

mitochondrion

cell

glucose + O$_2$

ATP

P

nucleus

cells need chemical energy from ATP so that they can function

Figure 3.4 Energy is required for things to work

3.3

Autotrophic nutrition

In this section we will deal with:
- photosynthesis
- absorption of minerals
- obtaining light and gases.

Autotrophic nutrition: photosynthesis

- ***identify the materials required for photosynthesis and its role in ecosystems***

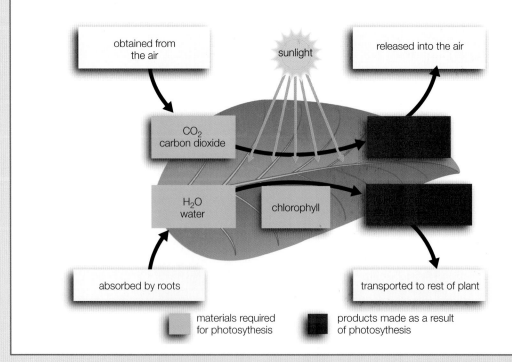

TR

Table outlining the requirements for photosynthesis

Requirements for photosynthesis

Carbon dioxide, water, chlorophyll and light are all essential for the chemical process of photosynthesis to occur in cells. Photosynthesis is the process by which all green plants and some unicellular organisms and bacteria make food: plants capture energy from **light** (radiant energy,) using **chlorophyll**, a green pigment present in plant cells. This energy is used to combine *carbon dioxide* (a gas obtained from the surrounding air) and *water* (which the plants absorb from the soil and then transport up to the leaves), to make sugars and oxygen. (See Fig. 3.5).

The water (H_2O) and carbon dioxide (CO_2) provide the basic chemical 'building blocks' (atoms) of

which the resulting sugar $(CH_2O)_n$ is made. Oxygen (O_2) is given out as a by-product of photosynthesis. The energy conversion involves a change from *radiant energy* (sunlight or artificial light) to *chemical energy* (stored in glucose). Chlorophyll is essential for this energy conversion to occur. This is an oversimplification of the process of photosynthesis, which involves a sequence or chain of many biochemical reactions, but these will be dealt with in more detail on pages 136–8.

The role of photosynthesis in ecosystems

Photosynthesis is the initial pathway by which energy enters all ecosystems. Organisms that photosynthesise are

Figure 3.5
Materials required for photosynthesis (and products of photosynthesis)

producers. Producers form the *basis of all food webs*, providing glucose (and other forms of stored food) for all living organisms, either directly (e.g. plants are eaten by herbivores) or indirectly (e.g. herbivores, which have eaten plants, are eaten by carnivores). Glucose, the product of photosynthesis, can be converted by plants into other organic compounds for storage. Glucose may be converted into and stored as:

- lipids (e.g. sunflowers and avocados which store their food as oils)
- proteins (e.g. legumes such as bean and pea plants)
- complex carbohydrates (e.g. potato plants).

These organic compounds, which rely on photosynthesis for their production, provide the structural basis of living cells and also provide a source of energy for all cellular processes (see Fig. 3.6).

Atmospheric *gases* essential to living organisms are *recycled* during photosynthesis—the process of photosynthesis provides oxygen for the respiration of all living things. It also removes carbon dioxide from the atmosphere and, since elevated carbon dioxide levels are harmful to living organisms and contribute to increased global warming, photosynthesis is beneficial to the environment.

Fossil fuels were formed from photosynthetic organisms approximately 300 million years ago. Huge tree ferns and other large leafy plants on land, as well as algae in the sea, sank to the bottom of the swampy oceans and were buried during the Carboniferous Period. Over many thousands of years, as a result of the pressure of the sand, clay and minerals that were deposited on top of them, they turned into coal, oil or petroleum and natural gas, which all serve as sources of fuel for us today.

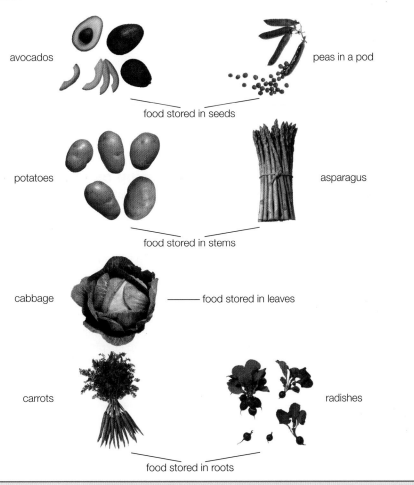

Figure 3.6 Plant organs that store organic compounds as food for consumers

avocados

peas in a pod

food stored in seeds

potatoes

asparagus

food stored in stems

cabbage —— food stored in leaves

carrots

radishes

food stored in roots

Investigating requirements for photosynthesis

FIRST-HAND
INVESTIGATION

BIOLOGY SKILLS

P11
P12
P13
P14

Students can be guided in their planning of these experiments using the scaffold 'Five steps of investigation'.

■ plan, choose equipment or resources and perform first-hand investigations to gather information and use available evidence to demonstrate the need for chlorophyll and light in photosynthesis

Background information

To find out what plants need in order to *photosynthesise*, we can test what factors allow the plant to produce *starch*—starch is one of the common end products of photosynthesis (glucose is stored as starch).

It has been suggested that plants need light, carbon dioxide, chlorophyll and water to produce starch by the process of photosynthesis. Factors such as these affect the ability of a plant to photosynthesise and so are termed the *limiting* factors of photosynthesis. The syllabus requires you to test whether light and chlorophyll are necessary for photosynthesis. You can do experiments to find out if chlorophyll and light are factors needed for starch production. You will need to do a separate experiment for each factor, since only one factor at a time can be varied in any experiment to make it scientifically valid.

Procedure

The following information will help you to design your experiments.

■ For each experiment you should use a control: in the *control* leave out one factor, for example, light or chlorophyll. In the *experimental*, provide this factor that you have left out in the control to validate your experiment and show that it is that factor which is essential for starch production.

■ To test a leaf for starch, the following method below should be followed:
—place a leaf into a beaker of boiling water for 1 or 2 minutes, until it is soft (the length of time depends on the thickness of the cuticle of the leaf). Placing it in water softens the leaf tissue and breaks down the cell wall to allow substances to enter and leave the cells more easily. *Turn off the Bunsen burner before proceeding*
—pour enough ethanol or methylated spirits into a test tube to cover the leaf. Place the leaf into the test tube of ethanol. Place the test tube into the hot water (with the flame beneath turned OFF) and leave it until the ethanol turns green and the leaf becomes pale. This step removes the chlorophyll from the leaf so that it does not mask any colour change that may be observed
—remove the leaf from the test tube and rinse it in the hot water in the beaker. This will re-soften the leaf tissue
—place the leaf on a watch glass and cover it with a few drops of dilute, fresh iodine solution. Leave it to stand for a 3 to 5 minutes so that the iodine can soak in. Give the leaf a quick rinse in water and examine it for dark (blue–black) patches which indicate the presence of starch.

Choosing equipment and resources

■ Read through the standard test for starch (above) and list the equipment you would need to safely carry out the investigation.

■ To test whether light is necessary for photosynthesis, start with a plant that has no starch. To remove starch, put the plant in the dark for a few days.

■ Expose the plant to light again, covering some leaves with aluminium foil or black paper to keep out the light. Alternatively, you could use an outdoor plant with some leaves covered or a black bag over part of the plant. Remember that it may take several days for these leaves to use up all the starch they already contain; however, you will need to perform the starch test before the exposed leaves begin to transport starch back to those lacking starch.

Figure 3.7 Extracting chlorophyll using a water bath

beaker — softened leaf
alcohol — water
turn off Bunsen burner
tripod

- It is not possible to remove the chlorophyll from a leaf without killing the plant, so you should make use of nature and use a plant that has *variegated* leaves; that is, leaves that occur in nature that are partly green (tissue contains chlorophyll) and partly white/yellow (tissue contains no chlorophyll). Some examples are geraniums (*Coleus*), certain types of ivy and 'hen and chicken' plants (Crassulaceae).
- Decide what plant you will use for each experiment, taking into consideration the fact that using softer leaves with a thin cuticle will make it easier to break down the cell walls and expose the starchy contents for the iodine test.
- Make sure the control plant has everything that it needs, excluding the one factor that you want to test.

Planning your investigation

(Refer to the biology skills table on pages x–xii.)
- Plan your investigation, preparing a written experimental procedure under the headings Aim, Hypothesis and Materials.
- Draw a diagram to illustrate the apparatus set up to use a water bath.
- Include a risk assessment and state what you will do to avoid risks and ensure a safe procedure.
- Plan and write out a method for your experiment (using a procedure text type).
- Remember to outline what you will do to make the test fair. For example:
  —identify dependent and independent variables
  —control all other variables
  —state how you will measure the independent variable
  —state any modifications made after a trial run
  —describe the controls.
- Prepare an appropriate way to record your results.
- Once your method has been checked by your teacher, conduct your experiment, record your results and use this evidence to state a valid conclusion from your results, looking at the aim of your experiment to ensure you 'answer' it. No inferences should appear in your conclusion.
- In your discussion, explain your results (any inferences should be made here), discuss whether your findings were what you expected and account for sources of any experimental errors. Describe how you could modify the experiment in the future to reduce error.
- Answer all discussion questions in the following list.

Discussion questions

To investigate whether chlorophyll is necessary for photosynthesis
1. Is chlorophyll soluble in ethanol? Justify your answer using evidence from your experiment.
2. State why it is necessary to remove chlorophyll before testing the leaf for starch.
3. Explain why removing chlorophyll at this stage of the experiment does not interfere with the aim of the experiment.
4. Identify which part of a variegated leaf contains starch.

To investigate whether light is necessary for photosynthesis
5. Explain why a leaf that has been left in the dark is 'destarched'. Propose a suggestion as to what may have happened to the starch that was originally present in the leaf.
6. Discuss why it is safer to use a hot water bath to heat ethanol, rather than using a Bunsen burner.
7. Explain why, if a destarched leaf is covered with foil and the remainder of the plant is left exposed to sunlight for three or four days, the results of your experiment may show that starch is present in the covered leaf.
8. Propose how you would investigate whether carbon dioxide is necessary for photosynthesis.
9. If only part of a leaf is covered with tin foil, a 'starch print' could be obtained. Predict the expected results if the following destarched leaf (see Fig. 3.8) was left in sunlight for a day and then tested for the presence of starch. Draw a diagram to illustrate the predicted result.

General questions
10. Discuss why it is necessary to use a control.
11. Explain what is meant by non-destructive testing and how this was achieved in your experiment.

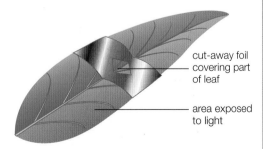

Figure 3.8
Destarched leaf exposed to sunlight for 24 hours

cut-away foil covering part of leaf

area exposed to light

3.4

Photosynthesis: biochemistry

■ *identify the general word equation for photosynthesis and outline this as a summary of a chain of biochemical reactions*

Photosynthesis: equations and reactions

The word equation for photosynthesis is:

| carbon dioxide + water | $\xrightarrow[\text{chlorophyll}]{\text{light energy}}$ | glucose + oxygen (+ water) |

The balanced overall chemical equation is:

| $6CO_2 + 12H_2O$ | $\xrightarrow[\text{chlorophyll}]{\text{light energy}}$ | $C_6H_{12}O_6 + 6O_2 + 6H_2O$ |

Although photosynthesis is often represented by the general equation shown above, it is not one chemical reaction but a series of many chemical reactions that take place in the chloroplasts of green plant cells and the cells of some photosynthesising bacteria.

Photosynthesis takes place in two stages. Each stage or phase is not a single chemical reaction, but consists of a series or chain of reactions:

1. the **light phase** (**photolysis**) involves the *splitting of water* using the energy of light (photo = light; lysis = splitting)
2. the **light-independent phase** (sometimes referred to as the **carbon fixation** stage) involves *using carbon dioxide to make sugar*. No light is used in this stage.

Each step of these reactions is controlled by a different enzyme (a chemical catalyst which speeds up reactions in living cells). These series of reactions have been outlined in a simplified form below.

The light phase—photolysis

Radiant energy from the sun (or an artificial source of light) is captured by chlorophyll in the thylakoids in the grana of chloroplasts. The energy is sufficient to excite an electron to a higher energy level, where it will break away from the chlorophyll molecule. This series of reactions is occurring in each chlorophyll molecule in every chloroplast exposed to light. There are thousands of chlorophyll molecules in one chloroplast and so thousands of excited electrons are produced. (See Fig. 3.11.)

An excited electron can follow one of two pathways:

1. it may be used to split water into a hydrogen component (H_2) and an oxygen atom (O). The oxygen atom combines with an oxygen atom from another chlorophyll molecule to form oxygen gas (O_2), which is released by the plant. The hydrogen atoms will go on to be used in the next phase, the light-independent reaction.
2. it may be used to form ATP, a high-energy compound that provides the cell with the energy it needs for functioning.

The light independent phase—carbon fixation

(See Fig. 3.11.) This phase uses carbon dioxide, but requires no chlorophyll and no light, so it is called the light-independent phase or carbon fixation phase:

■ hydrogen atoms from the light reaction are carried to the **stroma** to begin this phase

More detailed biochemistry

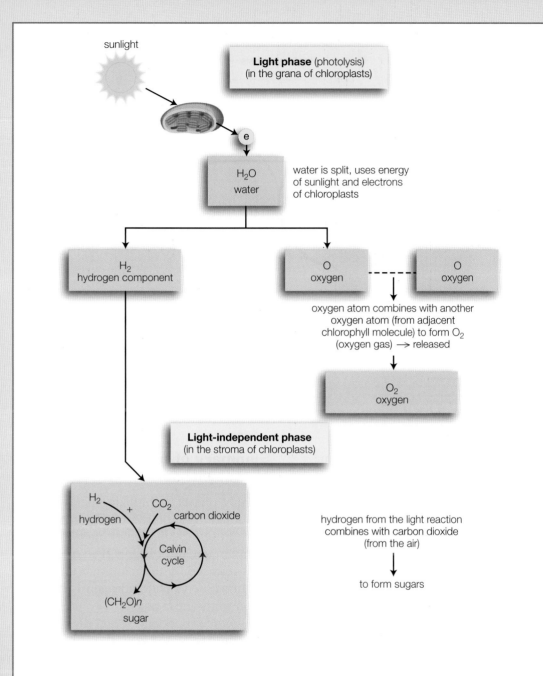

- carbon dioxide needed for this reaction was absorbed by the plant from the surrounding air. The *hydrogen* atoms go through a series of enzyme-controlled reactions, where they are *combined with carbon dioxide* to form a *sugar molecule* (glucose)
- this cyclical series of reactions is known as the Calvin cycle, named after Melvin Calvin, the Nobel Laureate who discovered it

- the *ATP* produced in the light reaction provides the necessary *energy* for the light-independent reaction to take place. The energy from ATP is incorporated into the new sugar compounds that are formed
- the *glucose* end-product of photosynthesis is converted to *starch*, the form in which food is most commonly stored in plants. Chloroplasts usually have large

Figure 3.10
Flowchart summarising a simplified overall reaction for photosynthesis, colour-coded for the light and light-independent phases

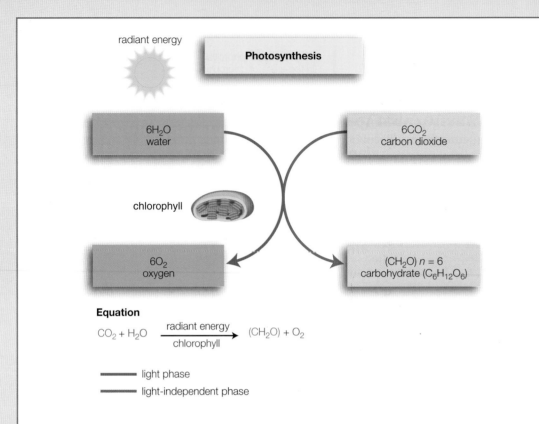

radiant energy

Photosynthesis

$6H_2O$
water

$6CO_2$
carbon dioxide

chlorophyll

$6O_2$
oxygen

(CH_2O) $n = 6$
carbohydrate ($C_6H_{12}O_6$)

Equation

$$CO_2 + H_2O \xrightarrow[\text{chlorophyll}]{\text{radiant energy}} (CH_2O) + O_2$$

——— light phase
——— light-independent phase

starch grains stored in the stroma, evidence that photosynthesis has occurred. To experimentally test whether photosynthesis has occurred in a plant, you need to simply test for the presence of starch
■ as mentioned previously, some plants convert starch into different organic storage compounds such as lipids or proteins, depending on the plant. The light-independent phase of photosynthesis takes place immediately following the light phase (since it relies on products of the light phase) and so both phases occur during daylight. At night, because there is no light and therefore no source of radiant energy, neither the light dependent nor the light- independent phases take place— there is no photosynthesis at night.

Summary of products of photosynthesis

■ Light phase: the radiant energy of sunlight is captured by chlorophyll and used to split water into hydrogen and oxygen. The oxygen is released by the plant into the atmosphere. The hydrogen is carried on to the next phase.
■ Light-independent phase: the hydrogen atoms from the light reaction are combined with carbon dioxide to form sugar molecules ($C_6H_{12}O_6$ = glucose):
—the carbon and oxygen atoms of the carbohydrate come from the carbon dioxide (CO_2)
—the hydrogen atoms of the sugar come from the water (H_2O)
—the oxygen that is released in the light stage also comes from the water.

Autotrophic nutrition: obtaining water and inorganic minerals

3.5

- *explain the relationship between the organisation of the structures used to obtain water and minerals in a range of plants and the need to increase the surface area available for absorption*

Roots are the structures in plants for absorbing water and inorganic minerals. These structures have an extensive surface area which allows water and inorganic mineral salts to be absorbed efficiently. The outermost layer of plant organs is the epidermis and it is through epidermal cells in the root that the absorption of water and minerals occurs (see Fig. 3.1).

Water uptake

Plants need to absorb large quantities of water and do so at a high rate to maintain their water balance. The uptake of water in roots occurs by the process of **osmosis**: when water in the soil is at a higher concentration than the cell sap of root cells, water will move from the soil into the root by osmosis. This is a *passive* form of transport and occurs *slowly*, so it is essential that the surface area of structures involved in absorption is increased.

Uptake of minerals

The uptake of inorganic mineral salts is mainly by the process of **diffusion**: if the minerals are in a higher concentration in the soil than they are in the cells of the roots they will move *passively* into the roots. Remember that salts dissociate into ions—charged particles—when they are dissolved in water and it is in this form that they are absorbed. For example, the *mineral salt* sodium chloride (NaCl) will *dissociate* into sodium ions (Na^+) and chloride ions (Cl^-). These *ions* are absorbed into the roots mainly by diffusion but, if diffusion alone

Fig 3.11 Roots of plants: (a) the structure of a root tip (external view); (b) longitudinal section through root tip showing the organisation of tissues (cross-sections are shown for three regions)

(a)

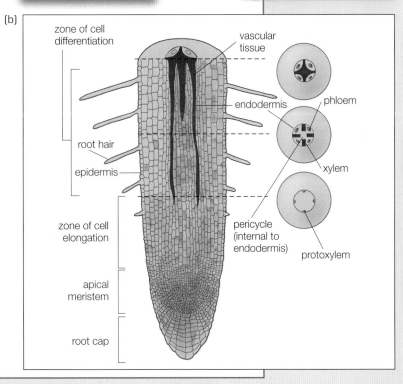

Mature region where branch roots form higher up

Specialisation zone
Cell differentiation

Elongation zone
Cell expansion

Meristematic zone
Cell division

(b)

zone of cell differentiation

vascular tissue

endodermis

phloem

root hair

epidermis

xylem

zone of cell elongation

pericycle (internal to endodermis)

protoxylem

apical meristem

root cap

is inadequate (e.g. if it is too slow or if the concentration gradient is not significant), **facilitated diffusion** and **active transport** may also be involved.

Increased surface area

The uptake of both water and mineral salts depends on a large area of contact between the roots and the soil water containing the dissolved minerals. An increased surface area in roots is achieved in the following ways:

■ the *root hair zone* is in the younger part of each root, near the tip. In this region, the epidermal cells protrude outwards into the surrounding soil, as microscopic extensions called root hairs. Their presence increases the surface area of a root up to 12 times. (See Fig. 3.11a.) The *growing regions* of the roots (the meristematic zone of cell division and the zone of elongation) are nearer to the root tip, protected by the root cap, while cells further away from the tip are older and more mature (see Fig. 3.11b). As a root grows, the older cells mature and differentiate. Root hairs which form in the young root hair zone should not be confused with much larger, macroscopic lateral or branch roots, which form in the mature region of the root where the cells have differentiated and are mature

■ *extensive branching* of root systems in the mature region increases the surface area of the root for absorption (and also provides good anchorage for the plant). Roots may branch in one of three ways, forming:

—a *tap root* system (one main root with many smaller branch roots which subdivide extensively)

—a *fibrous root* system (many main branches which subdivide) or

—an *adventitious root* system (the roots arise at regular intervals from the nodes of a stem, for example in creeping stems such as ivy or along horizontal stems such as grasses and onion bulbs). (See Fig. 3.12.)

■ water enters the root through the epidermal cells across the entire surface of the root system. The flattened nature of these cells increases their exposed surface, but the surface area of general epidermal cells is smaller than that of root hair cells and so less water is absorbed per cell than in the root hair zone (see Fig. 3.13a).

Figure 3.12
Types of root systems:
(a) a tap root system;
(b) a fibrous root system;
(c) an adventitious root system

(a)

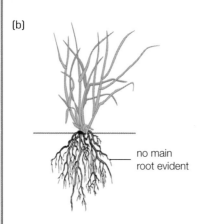

main taproot
branch root
root hairs
growing tip
new root hairs develop
growing tip

(b)

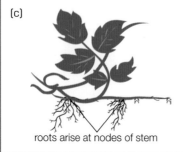

no main root evident

(c)

roots arise at nodes of stem

Movement of water and minerals across roots

Water and minerals that enter the root through the epidermal cells, particularly through the root hairs in the root hair zone, move across the root to the water-conducting tissue called **xylem**, found in the centre of the root. Water moves by osmosis through the cells (cytoplasm or vacuoles), but may also move through the cell walls. The water and minerals then move across the cortex, along a gradient, into the central vascular (transport) tissue. Xylem transports water and dissolved mineral salts upwards from the roots to the rest of the plant, where it is needed for normal plant functioning.

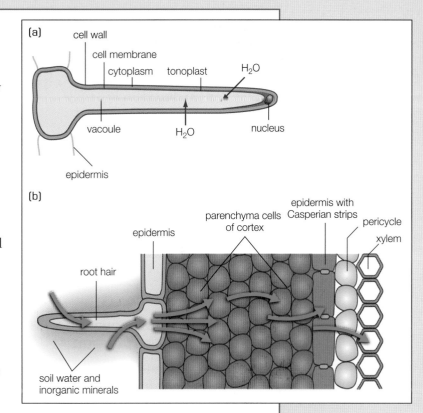

Figure 3.13 Water movement into a root:
(a) absorption of water by root hair;
(b) movement of water across the root

Autotrophic nutrition: obtaining light and gases

3.6

- *explain the relationship between the shape of leaves, the distribution of tissues in them and their role*

Leaves are the main photosynthetic structures in plants. Their primary function is to absorb sunlight and make food. They also carry out the function of transpiration.

The role of leaves

Plants need to *photosynthesise:*
- to absorb sunlight and carbon dioxide during the day
- to release oxygen
- to provide chlorophyll
- to make glucose and transport it to other parts of the plant where it can be stored as starch or other organic molecules.

Transpiration is needed to release water to cool the plant down and to create a suction pull to lift water from the roots to the top of a plant (similar to sucking water up a drinking straw—

the suction pull at the top causes water flowing in at the bottom to rise).

It is a common misconception that 'plants do not respire, they photosynthesise instead' or 'plants only respire at night'. Both of these statements are untrue. Respiration is a function of *all* living cells; in leaves it is simply 'masked' or hidden by photosynthesis and so the observed exchange of gases by plants during the day differs from that at night and it is this that causes confusion. In all plant cells, chemical respiration occurs during both the day and the night. During the day:
- the oxygen required for cellular respiration comes from the oxygen produced as a by-product of photosynthesis. Photosynthesis usually occurs at a greater rate than

that of respiration during the day, so any excess oxygen not used during respiration is released by the plant to the outside environment

■ the carbon dioxide released as a result of respiration during the day is used as a reactant for photosynthesis. With a high rate of photosynthesis, this carbon dioxide supply is usually insufficient so plants absorb more carbon dioxide from the air. The net gaseous exchange observed during the day is therefore that associated with photosynthesis, despite the fact that respiration is occurring.

The structure of leaves in relation to their function

Leaves are structurally adapted to enable them to effectively function. They are made up of a number of different tissues, arranged in a highly organised way to maximise their efficiency in carrying out their functions. Since all life depends on photosynthesis in plants as

a source of food and therefore energy, leaves must be extremely well suited to fulfill their role.

The structural features needed by a leaf for efficient photosynthesis are:

■ a large surface area, with an outer layer able to absorb light and carbon dioxide

■ pores in the leaf surface for the exchange of gases with the environment

■ cells inside that contain chloroplasts (and chlorophyll) to trap the energy of sunlight

■ a water transport system from the roots to the leaves

■ a food transport system from the leaves to other parts of the plant.

Absorption of light and photosynthesis

Leaves have an enormous diversity of shapes and sizes, but most are *flattened in shape* and relatively thin, resulting in a large surface area that is exposed to the sun. This allows the maximum

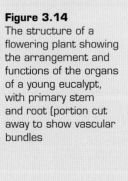

Figure 3.14
The structure of a flowering plant showing the arrangement and functions of the organs of a young eucalypt, with primary stem and root (portion cut away to show vascular bundles

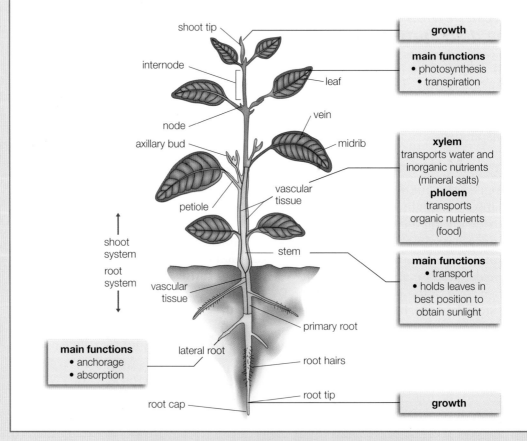

shoot tip

internode

leaf

growth

main functions
• photosynthesis
• transpiration

vein

node

axillary bud

midrib

xylem
transports water and inorganic nutrients (mineral salts)
phloem
transports organic nutrients (food)

vascular tissue

petiole

shoot system

root system

stem

main functions
• transport
• holds leaves in best position to obtain sunlight

vascular tissue

primary root

main functions
• anchorage
• absorption

lateral root

root hairs

root tip

growth

root cap

absorption of light for photosynthesis. Because of the thin nature of the leaf *no internal cell is too far from the surface* to receive light. In all leaves the outermost layer of cells, the epidermis, is *transparent*, allowing the sun to penetrate through to the photosynthetic cells inside the leaf.

Gaseous exchange

The surface of leaves is covered by a protective layer of cells, the **epidermis**. These are simple, flattened cells on the top surface (upper epidermis) and lower surface (lower epidermis) of leaves. Epidermal cells protect the delicate inner tissues and are able to secrete a waterproof cuticle to prevent the evaporation of water from the increased surface area of leaves. They are transparent to allow light to pass to the cell layers below.

Within the epidermis, there are specialised cells called **guard cells** that control both the exchange of gases (such as carbon dioxide and oxygen) and the loss of water through leaves. Guard cells are bean-shaped cells that occur in pairs, surrounding a pore (opening) known as a stoma (plural stomata). Stomata occur on both the upper and lower surfaces of leaves but are usually more numerous on the under-surface of leaves. (Because water is lost in the form of water vapour,

a gas, this is discussed in more detail in Chapter 4, which deals with gaseous exchange in leaves.)

Photosynthetic cells and tissue distribution

The cells that occur in the **mesophyll** or middle layers of the leaf are responsible for most of the plant's photosynthesis. Two main types of cell make up the mesophyll ('meso' = middle, 'phyll' = leaf)— **palisade cells** and **spongy cells**:

- *palisade cells* are elongate cells that contain numerous chloroplasts and they are the *main photosynthetic cells* in leaves. They are situated immediately below the upper epidermis (see Fig. 3.15a)
- *spongy cells* are the second most important *photosynthetic* cells. They have fewer chloroplasts than palisade cells and are irregular in shape. They have large intercellular

(a)

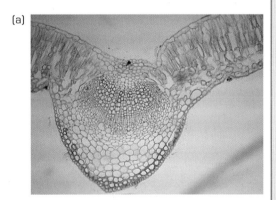

Figure 3.15
A dorsiventral leaf: (a) light micrograph cut in a transverse section; (b) plan diagram in a transverse section

(b)

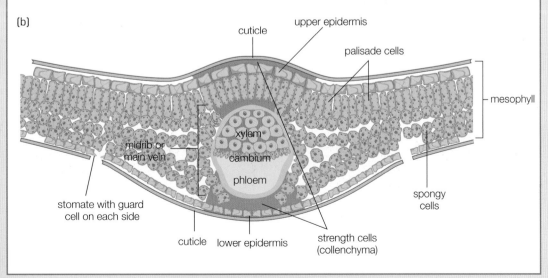

cuticle
upper epidermis
palisade cells
mesophyll
xylem
midrib or main vein
cambium
phloem
spongy cells
stomate with guard cell on each side
cuticle   lower epidermis
strength cells (collenchyma)

air spaces, particularly beneath the stomata, and their main function is *gaseous exchange*. They are situated lower in the leaf, beneath the palisade tissue and above the lower epidermis in **dorsiventral** leaves.

The distribution of photosynthetic tissue in leaves is directly linked to the orientation of leaves on the plant. Most plants have leaves that are *horizontal*—with the upper surface facing the sun and the lower surface shaded.

In these plants the upper surface faces the sun while the lower surface is shaded. The tissue distribution in these leaves is different on the upper and lower surfaces and so the plants are said to have dorsiventral leaves.

Some plants have leaves that hang vertically and so both surfaces are exposed to the sun. (See Fig. 3.16.)

In plants that have leaves that hang *vertically* (e.g. adult leaves in some eucalypts), both surfaces are exposed to the sun. In these leaves, tissue distribution on the upper surface is the same as that on the lower surface, so there is an additional layer of palisade tissue beneath the spongy mesophyll,

just above the lower epidermis. Stomata are also equally distributed in both the upper and lower epidermis. As a result, the upper and lower halves of the leaf have the same pattern of tissue distribution and so these leaves are termed **isobilateral** ('iso' = the same; 'bi' = two; and 'lateral' = sides). (See Fig. 3.16.)

Transport

The **vascular tissue** in the centre of the root is continuous, passing up the stem and into the leaves as 'veins' in the leaf, serving as the main transport tissue in the plant. (The term *vascular* means '*of vessels*', i.e. tissue arranged as vessels, as in transport vessels.) The main vein is called the **midrib** and many smaller veins branch out from it in certain types of plants (**dicotyledons**), such as eucalypts. In other plants (**monocotyledons**), such as grasses, all veins run parallel to each other. Veins are made of two types of tissue: xylem and phloem. *Xylem* transports *water* and *dissolved inorganic minerals* to the leaf for photosynthesis and *phloem* tissue transports the *sugars* produced by photosynthesis from the leaf to the rest of the plant.

The distribution of vascular tissue throughout the leaf ensures that no leaf cells are too far away from a source of transport. Vascular tissue also plays an important role in *supporting* the thin leaf blade (lamina). (Transport will be dealt with in more detail in Chapter 4.)

Figure 3.16
Isobilateral leaves:
(a) photomicrograph of isobilateral leaf in transverse section;
(b) diagram showing cell distribution

(a)

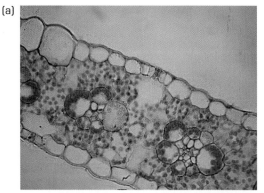

(b)

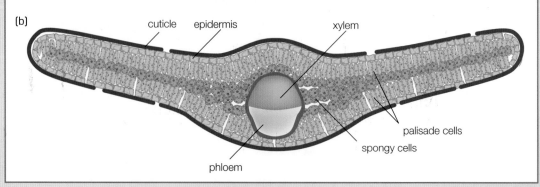

cuticle   epidermis   xylem

palisade cells

spongy cells

phloem

1. In the form of a table, compare the different types of tissue found in a leaf in terms of their structure, position within the leaf, functions, and structural features that suit them to their functions.

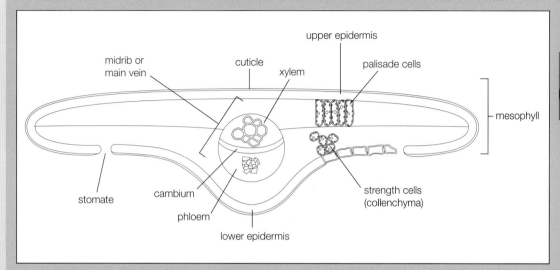

Table comparing tissues in leaves

Figure 3.17 Dorsiventral leaf plan diagram showing a few cells of each type

2. On Figure 3.17 shade each of the cell/tissue types listed in the table in a particular colour (refer to the Student Resource CD or the Teacher's Resource CD for a copy). Use a key to allow interpretation of the colours. Try to make the colours appropriate to the functions of the tissues (e.g. green for mesophyll, where dark green is palisade (more chloroplasts) and light green is spongy).

3. Draw and label a plan diagram of an isobilateral leaf (see Fig. 3.16), using the plan sketch of the dorsiventral leaf above as a guide. Shade the tissue types using the same key as you used for the dorsiventral leaf plan.

Heterotrophs obtaining nutrients

3.7

Introduction

Heterotrophs are living things that must feed on others because they cannot provide their own energy supply by photosynthesis. The **carbohydrates**, lipids and proteins, originally made by autotrophs, are consumed by heterotrophs and are then broken down—large chunks of food are broken down into smaller units (simple, single molecules which are usually soluble) so that they can be easily absorbed.

In heterotrophs, obtaining nutrients and energy from food involves several steps:
- **ingestion** is the intake of complex organic food ('eating')
- **digestion** is the breakdown of the chunks of food into smaller, simpler, soluble molecules that can be easily absorbed. Physical digestion is mainly as a result of food being chewed, but food may also be broken down by the churning action of the muscular stomach wall.

The aim of digestion is to obtain molecules small enough for absorption. These end products are: *amino acids* from proteins, *simple sugars* from carbohydrates and *fatty acids* and *glycerol* from lipids

- **absorption**: the basic units of food (glucose, amino acids, fatty acids and glycerol) are small enough to be absorbed across the wall of the digestive tract, into the bloodstream of the animal or, in simpler animals, directly into the cells

- **assimilation**: the end products of digestion can be built up by the body into useful substances, either as new biological material or as an energy source. For example, in mammals such as humans, blood transports the food molecules to where they are needed in the body and they can then be reassembled by the cells of the body into *structural parts* (e.g. lipids and proteins that form a structural part of cell membranes; protein fibres in muscle tissue) or into *energy storage* (e.g. fatty tissue or fat beneath the skin, glycogen a form of 'animal starch'; protein cannot be stored)

- **egestion**: the elimination of undigested food as waste. For example, fibre such as cellulose that cannot be digested by humans acts as 'roughage', passing through the digestive tract and out of the anus. (*Note*: The process of egestion should not be confused with *excretion*, which is the elimination of chemical waste products of metabolism—that is, excretion involves chemical wastes that are actually made by the body, for example urea excreted in urine.)

Figure 3.18
The human digestive system

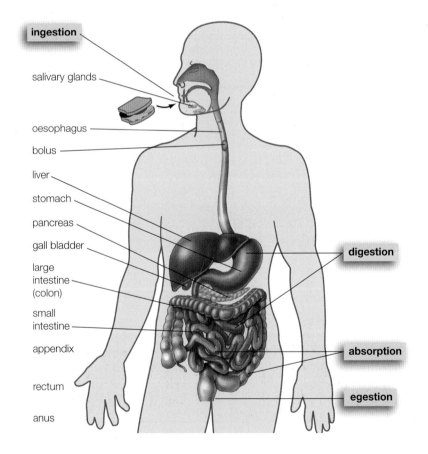

Teeth and surface area

■ *describe the role of teeth in increasing the surface area of complex foods for exposure to digestive chemicals*

Introduction

Most vertebrates have jaws and teeth that enable them to obtain and process foods. Fish, amphibians, reptiles and mammals all have teeth, but modern birds do not have teeth, only a beak. In this section, we will refer to vertebrates with teeth, mainly referring to mammals.

The digestion of foods involves two stages:

1. *mechanical* or *physical breakdown*, where food is chewed and large chunks are physically broken down into smaller bits
2. *chemical breakdown*, where digestive enzymes act on the food to chemically break down the large complex molecules into simpler, smaller molecules.

Digestive enzymes function more efficiently if the food to which they are exposed has a large surface area to volume ratio. Large chunks of food which are eaten are broken down by the teeth, exposing a greater surface area on which the chemicals can act. The volume of food is still the same, but the overall surface area to volume ratio is increased by chewing, increasing the rate of reaction with digestive enzymes. This can be demonstrated by performing a first-hand investigation to demonstrate the relationship between surface area and the rate of reaction (see page 149).

Teeth of mammals

Four main types of teeth, each with specific functions, are present in mammals:

1. **incisors** (front teeth): to grasp, hold and bite food
2. **canines** ('eye' teeth or 'fangs'): used for stabbing and gripping prey and for tearing flesh. Canines are long teeth that are conical in shape
3. **premolars** (cheek teeth): used for chewing and for cutting flesh and cracking hard body parts such as bones and shells. **Carnassial** teeth are modified premolars that shear and slice flesh in meat-eating animals
4. **molars** (back teeth): used for grinding and chewing. Molar teeth generally have ridges on them to help to crush and grind food.

Premolars and molars on the lower jaw work against those on the upper jaw to help break down the food and expose more surface area. Omnivores (animals that eat both plant and animal

Figure 3.19 Teeth of mammals: (a) different types of teeth in humans; (b) different types of teeth in mammals

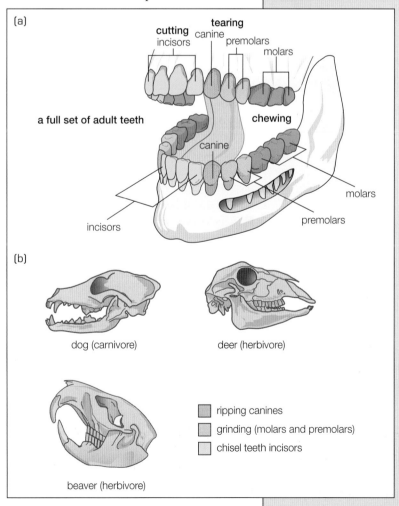

(a)

cutting incisors

tearing canine premolars molars

a full set of adult teeth

chewing

canine

molars

incisors

premolars

(b)

dog (carnivore)

deer (herbivore)

beaver (herbivore)

■ ripping canines
■ grinding (molars and premolars)
□ chisel teeth incisors

matter) have all four types of teeth present and the teeth do not show great variation in shape and size. The teeth of herbivores and carnivores differ in their structural detail and these adaptations equip each type of animal to best cope with its specialised diet. In some cases, particular types of teeth are absent altogether.

Herbivores

Herbivores are heterotrophs that consume *plant material*, and many fall into one of two groups: *grazers* who eat grass and *browsers* who eat leaves and vegetation on shrubs and trees. Herbivores have front teeth adapted to tearing off vegetation and back teeth adapted to chewing an abrasive, high-fibre diet. They also have powerful jaw muscles to cope with all the chewing required in their diet.

The incisors are used to bite off vegetation and a wide range of front teeth are typical across the herbivore group of mammals, including those adapted to nibbling, gnawing, tearing or biting. The molars are broad crushing teeth with relatively large surface areas, an adaptation to deal with an abrasive, fibrous diet. The molars are specially equipped with ridges to help break open the cellulose cell walls of plants.

Cellulose in the cell walls of plant tissue presents a challenge to animals which eat it because:
- it is very difficult to break cellulose down physically as well as chemically (cellulose is indigestible to most animals and therefore cannot be used by many as an efficient source of energy)
- cellulose cell walls enclose and surround cells, preventing enzymes from reaching the high energy, starchy contents of the cell which *can* be digested to release energy and nutrients.

In herbivores, an enormous amount of chewing is needed:
- to break down the high-fibre cellulose walls physically and expose the starchy contents

- because large quantities of vegetation must be eaten to provide a sufficient supply of energy
- to increase the surface area of cell walls exposed to microbes which can digest cellulose.

Many herbivores have developed a mutualistic feeding relationship with microbes that live inside their gut. These microbes are able to produce an enzyme which can chemically digest cellulose to release energy. Chewing the food physically breaks down the cellulose and this serves a second purpose—the damaged cell walls have a greater surface area exposed to the microbes within the digestive tract to chemically break down the cellulose.

Canine teeth are absent in herbivore jaws, leaving a gap called the *diastema,* which assists manipulation of food onto the molars, keeping chewed and unchewed food separate.

Carnivores

Large carnivores usually prey on other vertebrates. Their food is more difficult to obtain because they usually need to hunt and kill it. They have powerful jaws and well-developed canine teeth, conical in shape (coming to a point) and specialised for holding and killing prey and tearing meat from the bones. The meat is torn off in chunks, and many carnivores have molars with deep cusps to briefly chew the meat, increasing the surface area before swallowing it. Some carnivores such as cats have *carnassial* cheek teeth, adapted to for slicing and shearing meat, and they have lost their molars.

Small carnivores such as insectivores (where insects form the main part of their diet) have teeth adapted to piercing and penetrating the tough cuticle of their prey. They puncture and crush the exoskeleton with their premolars and molars and then use these teeth to shear the inner tissues.

Investigating surface area and rate of reaction

FIRST-HAND
INVESTIGATION
BIOLOGY SKILLS

P12.1; 12.2; 12.4
P13.1
P14.1; 14.2; 14.3

■ *perform a first-hand investigation to demonstrate the relationship between surface area and rate of reaction*

Background information

In this experiment using antacid tablets and hydrochloric acid, we simulate the conditions of digestion in mammals. The antacid tablet represents the 'food' and the hydrochloric acid represents a digestive chemical. When an antacid tablet comes into contact with acid, it reacts with it, producing carbon dioxide bubbles, water and salt. The stomach of mammals naturally secretes hydrochloric acid to provide the correct environment for stomach enzymes to work. If too much acid is secreted by the stomach, digestive discomfort (sometimes referred to as 'heartburn') results. Antacid tablets are usually taken to relieve digestive discomfort.

Aim

To demonstrate the relationship between surface area and rate of reaction.

Hypothesis

(Students should formulate a hypothesis.)

Equipment

■ Three test tubes in a test tube rack
■ Three antacid tablets
■ 30–50 mL dilute hydrochloric acid
■ Two spatulas or a pestle and mortar
■ Stopwatch

Safety

Avoid contact of skin or eyes with acid. If there is contact, rinse well with cool running water (tap for skin, water bottle for eyes).

Method

1. Place a whole antacid tablet, a tablet broken into quarters and a crushed tablet into three separate test tubes.
2. Pour the same amount of hydrochloric acid into each test tube. (Pour enough acid so that the whole tablet is covered—the exact amount depends on the size of the tablet—and then use this same amount in the other test tubes.)
3. Using a stopwatch, measure how long each test tube fizzes, until the tablet is completely dissolved.
4. Record your results in a table.
5. Write a conclusion for this experiment.

Discussion questions

1. Explain why it is necessary to avoid contact between hydrochloric acid and skin/eyes. Describe precautions that can be taken to prevent contact.
2. Identify the dependent and independent variables in this experiment.
3. Describe all other variables that were kept constant.
4. What is meant by the *rate* of a reaction? Explain why a stopwatch was needed.
5. What pattern or trend could you identify in the rate of reaction related to the surface area of the antacid tablets?
6. Refer to the background information about antacid tablets and then explain why you belch (burp) after taking an antacid tablet. Give a reason why taking an antacid tablet may relieve digestive discomfort.

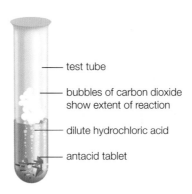

test tube

bubbles of carbon dioxide show extent of reaction

dilute hydrochloric acid

antacid tablet

whole tablet   tablet broken into quarters   tablet crushed into small pieces

Figure 3.20
Experiment to demonstrate the relationship between surface area and rate of reaction using antacid tablets

3.9

Digestive systems of vertebrates

■ *explain the relationship between the length and overall complexity of digestive systems of a vertebrate herbivore and a vertebrate carnivore with respect to:*
—the chemical composition of their diet
—the functions of the structures involved

Diet and the length and complexity of the digestive system

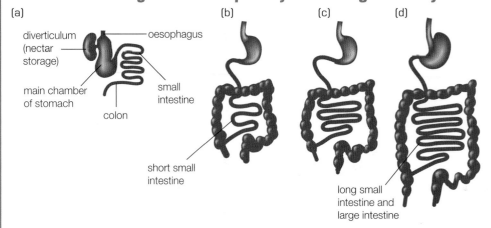

Figure 3.21
Diagrams comparing general length and complexity of animal digestive tracts:
(a) gut of a honey possum;
(b) gut of a carnivore;
(c) a human gut;
(d) gut of a herbivore

Herbivores

Diet

Plant material that is high in *fibre* (provided by the cellulose walls of plant cells) and *starch* (present in the plant cell contents) provides the main energy source in the diet of herbivores. Depending on what plants are eaten, sugars, proteins, oils and other nutrients also form part of a herbivore's diet, but usually in smaller amounts—the correct mix of nutrients, rather than the quantity, is important in herbivore diets.

Cellulose and other cell wall thickenings are difficult to digest and they create a barrier around the cell contents which are capable of being digested (e.g. starch, sugar and protein). Furthermore, cellulose cannot be used as an efficient energy source unless microbes are present. *Herbivores* have adaptations of their digestive tract that enable them to deal with this diet. (Some plants have evolved further defence mechanisms, such as thorns or the production of toxins to protect them against being eaten. Herbivores which eat these plants have even more specialised adaptations.)

Digestive tract

Most *large herbivores* rely on microbes in their gut to digest cellulose. To allow for the slow process of **microbial fermentation**, the gut of these herbivores is complex and very long, relative to their body size (see Fig. 3.21d). A close relationship is evident between the digestibility of food (how difficult it is to digest) and the length and complexity of the digestive tract. An increase in the length of the digestive tract provides space to hold the large quantity of food that must be eaten, gives the maximum opportunity for microbial action to take place and allows time for the nutrients to be absorbed. The increased complexity is evident by the presence of highly specialised digestive organs which are necessary to break down the high-fibre diet.

Small herbivores often eat plant tissue that has much less cell wall material and is energy-rich. These include fruit eaters (e.g. fruit bats) and small herbivores that live on a diet predominantly made up of nectar (e.g. the honey possum). The gut of these small herbivores is usually simple and short, in comparison to that of grazers and browsers, because the plant tissue in their diet is easier to digest, having a high sugar content and little or no fibre. (See Fig. 3.21a.)

Carnivores

Diet

Carnivores eat animal matter which is made up of cells the content of which is more accessible (animal cells have no cell wall, only a cell membrane, so their contents are more readily usable) and they are therefore more easily digested. Animal matter is high in protein, low in fibre and usually has a higher energy content than that of plants. The mineral content of animal matter is usually relatively constant. So even though the food of carnivores may be more difficult to obtain, it has high nutritional value and can be consumed in lesser quantities. The protein and fat in the diet are provided by the muscle, skin and internal organs of their prey; the small amount of fibre comes from cartilage and bones consumed. There is a small amount of fat and very little carbohydrates associated with their diet.

Digestive tract

In carnivores, the gut is relatively short and unspecialised, as protein and fat is relatively easy to digest (compared to complex carbohydrates or cellulose). Very little undigested material is egested, due to the low fibre content of their diet. (See fig. 3.21b.)

Functioning of the digestive system

The human digestive system

Understanding the human digestive system gives us a good overview of the structures involved and the necessary functions for heterotrophic nutrition. (See Fig. 3.20 and Table 3.2.) Humans

are omnivores, eating both plant and animal material, so the organs are suited to digesting both types of matter. The prescribed study of digestive tracts of herbivores, carnivores and nectar feeders which follows can then be easily compared.

Adaptations of parts of the digestive systems for specialised functions

Adaptations for feeding are thought to be one of the main forces behind the evolutionary process. If an adaptation in an organism makes it easier for them to obtain food or allows them to obtain nutrients from a type of food not sought by others, it is to their advantage because competition is reduced.

Herbivores

Very few animals have the ability to chemically digest cellulose and release energy from it, because they lack the enzyme *cellulase*. To overcome this problem, most herbivores have modified digestive tracts which contain symbiotic micro-organisms (bacteria and/or protozoans) that help digest the cell walls and make this energy available to the host animal. The process is known as microbial fermentation. The microbes exist in an expanded region of the gut, which protects them from being continually lost with the passage of food. The expanded part of the gut may be the stomach (fore-gut fermenters) or the caecum (hindgut fermenters).

In **fore-gut fermentation**, for example the eastern grey kangaroo, microbes use most of the dietary glucose and protein from the plant material that they digest, but they release short chain fatty acids as energy for the host and some of the microbes are digested by the host, providing a source of amino acids. Cattle are *ruminant* fore-gut fermenters—that is, they regurgitate the contents of the first part of the stomach and re-chew

Worksheet on human digestive system

Table 3.2 The digestive system of humans

	Part	Structure/position	Function
Mouth	Teeth	Four types, for biting, tearing, piercing and crushing food	To chew the food into smaller pieces that can be swallowed; increases the surface area exposed to digestive enzymes
	Salivary glands	Three pairs, emptying via ducts into the mouth cavity	Secrete saliva to: ■ moisten the food for easy swallowing ■ begin chemical break down of food, in particular starch digestion
	Pharynx (throat)	Muscular walls surrounding the opening to the **oesophagus**	To swallow food
	Oesophagus (gullet)	Long tube that lies behind the trachea (wind pipe) and their openings are separated at the top by a flap of cartilage, the epiglottis	Carries food from the mouth to the stomach, by waves of contraction called peristalsis. The epiglottis closes the windpipe during swallowing to prevent food entering the trachea
	Stomach (containing gastric juice)	Pouch-like organ with thick, strong muscular walls. Positioned in the upper abdomen, on the left-hand side of the body, immediately beneath the diaphragm	■ Churns the food by involuntary muscle contraction, mixing it with digestive juices ■ Stores the food and passes small amounts at a time into the small intestine ■ Contains enzymes for protein digestion ■ Contains hydrochloric acid which assists digestion and acts as an antiseptic, killing bacteria in food
Intestines	**Small intestine** (duodenum, jejunum, ileum)	Long narrow tube, greatly coiled, between the stomach to the large intestine. It is the longest part of the human digestive tract (6–7 m in length) but has a narrow diameter (less than 4 cm). Its inner walls are greatly folded to increase the surface area for absorption.	■ Receives digestive juices from the pancreas and liver, glands associated with the digestive tract: —bile from the liver emulsifies fats; —pancreatic juice contains digestive enzymes ■ Chemical digestion of foods—lipids, proteins, starch and complex sugars ■ Absorption of digested food
	Appendix (remnant of **caecum**)	Small pouch-like structure that forms the first part of the large intestine	Plays no role in human digestion (equivalent to caecum in other animals)
	Large intestine (ascending colon, transverse colon and descending colon)	A tube that is shorter than the small intestine, but wider (diameter about 6 cm, length about 1.5 m)	Absorption of water, vitamins and minerals from food. No chemical digestion
	Rectum	Last part of the large intestine	Stores undigested material (and any unabsorbed substances) as faeces, to be passed out of the body
	Anus	The opening of the digestive tract to the exterior	Egestion of undigested material to the exterior
Glands	Liver		■ Stores bile secreted by the gall bladder ■ Releases bile into the duodenum of the small intestine to emulsify fats and expose more surface area to enzymes ■ Centre of food metabolism: regulates the storage and release of the end products of digestion; changes excess amino acids into urea
	Pancreas	An elongated gland opening via a duct into the small intestine	■ Makes pancreatic juice with digestive enzymes and secretes it into the duodenum of the small intestine. ■ Secretes insulin, a hormone that controls the level of glucose in the blood.

it, further reducing the size of food particles. This is commonly termed 'chewing the cud' or they are said to *ruminate*. Ruminants have a complex stomach composed of four parts: two chambers for microbial fermentation, one for storage and one which functions as a true stomach. Food remains in the fermentation chambers for a long time, giving microbes plenty of time to digest the cellulose.

Hindgut fermentation, for example in the koala, takes place in the hindgut (caecum), but is otherwise similar to foregut fermentation. The main difference is that microbes cannot be digested by the host as a source of amino acids in hindgut fermenters, because they are in the large intestine where protein digestion no longer occurs and they are lost from the system by egestion. As a result, some hindgut feeders such as rabbits and ring-tail possums eat their egested faecal pellets to ingest and digest these nutrient-rich microbes.

Carnivores

Carnivores have a simple stomach (which may be enlarged to store food), a short intestine relative to their body size and the caecum may be absent or, if present, greatly reduced in size.

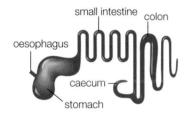
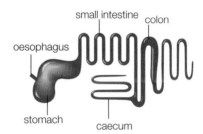

fore-gut digestion   **hindgut digestion**

Figure 3.22
Diagrams to compare fore-gut digestion and hindgut digestion in herbivores

Digestive systems of mammals

- *identify data sources, gather, process, analyse and present information from secondary sources and use available evidence to compare the digestive systems of mammals, including a grazing herbivore, carnivore and a predominantly nectar-feeding animal*

SECONDARY SOURCE INVESTIGATION

BIOLOGY SKILLS

P11.1; 11.3
P12.3; 12.4
P13.1
P14.1; 14.3

Identify data sources

- Remember to use a variety of sources and always assess the *reliability* and *accuracy* of your sources (see Biology skills table P12.4 e and f on page ix).
- Read the suggested background information on the previous pages (pages 150–3) and refer to other reference books as well.
- Identify some key scientific words that you might use in an Internet search for secondary sources. Here are some suggested words to start your search, keywords that you could type into a search engine such as 'Google': 'nectar-feeding

animals', 'digestive tracts of herbivores', 'digestive tracts of carnivores'.
- It is important to use scientific journals and publications to *validate* information.
- Acknowledge all sources appropriately in the form of a bibliography (see the Teacher's Resource CD and Student Resource CD for accepted formats for reference books, websites and journals).

Websites that may also assist as a starting point for your research are listed

Further information on identifying reliable and accurate sources and validating sources

Gather and process information

Identify relevant material from your source which addresses the aspects of the digestive tracts which are to be compared. Some sources may deal with information general to herbivores, carnivores and nectar feeders, whereas others will deal with specific animals.

Analyse and present information

Carefully examine the information you have gathered, identify the components that you need to compare and describe the relationships among them. Present your findings in the form of a table.

Research task

1. Select three animals to compare: one animal from each of the feeding groups stated below in italics. You may use the list to help you with your choice of animal, or you may select any other animal of interest to you, as long as it belongs to one of the listed feeding groups.
   Examples:
   - *grazing herbivores*
     —kangaroos, sheep and cattle (fore-gut fermenters)
     —koalas, possums and horses (hindgut fermenters)
   - *carnivores*: dogs; Tasmanian devil
   - *nectar-feeding animals*: honey possum.

2. Read information on teeth and the overall complexity of digestive systems and then research your three named examples as described below:
   - compare the digestive systems of one named *grazing herbivore*, one *carnivore* and one *predominantly nectar-feeding animal*
   - consider the structure and functions of the various parts of the digestive tract and how the structures are modified to achieve optimal functioning, in coping with the type of diet of each animal
   - present your findings in the form of a table
   - acknowledge your sources in an appropriate bibliography.

Suggested headings for table

- Main food source and its chemical composition
- Mechanical breakdown: teeth (shapes and explanations for this)
- Microbial action (fore-gut or hindgut fermenters and descriptions)
- Stomach (relative size and complexity; reasons why)
- Intestines (length relative to body size; how this relates to the type of food eaten)
- Caecum (relative size and complexity; reasons why)
- Chemical breakdown: time taken for food to digest
- Diagram
- Bibliography

REVISION QUESTIONS

1. Identify where the oxygen gas produced by photosynthesis comes from. Describe how it is produced.

2. List the chemical *products* of photosynthesis.

3. Write the overall word equation for photosynthesis.

4. Explain why the light-independent phase must occur during the day and not at night.

5. Describe two ways in which plants increase the surface area of their absorbing structures.

6. Explain why an increase in surface area is necessary for the normal functioning of plants.

7. Identify the processes necessary for the uptake of water and mineral salts by roots.

8. State the meaning of the terms: *heterotroph*, *herbivore*, *ingestion*, *absorption*, *assimilation* and *egestion*.

Answers to revision questions

9. Describe the role of both teeth and enzymes in the digestion of food (in relation of nutrition in animals) and give an example of each.

10. Compare hindgut fermentation with fore-gut fermentation, giving one similarity and two differences. For each, name an example of an animal that displays that mode of nutrition.

11. Carnivores obtain their food from other organisms. Describe two adaptations in carnivores which enable them to live on a diet of predominantly protein.

CHAPTER 4

Gaseous exchange and transport

Gaseous exchange and transport systems transfer chemicals through the internal and between the external environments of plants and animals

Movement of chemicals in plants and animals

4.1

For the normal functioning of plants and animals chemical substances needed by the organism must be transported into and around the body, while waste substances must be transported from where they are produced to the outside. The movement of these chemicals may be summarised as follows:

- from the external environment into the organism. For example:
  —oxygen from the air into an animal for respiration
  —carbon dioxide from the air into a plant for photosynthesis
- from inside the organism (internal environment) to the outside. For example:
  —wastes such as carbon dioxide or urea out of an animal
  —oxygen produced by photosynthesis carried out of a plant
- within the organism (internally), from the site where they have been produced within the organism to the site where they will be used or expelled. For example:
  —food carried from leaves in plants to storage organs
  —oxygen carried from the lungs of an animal to the muscle cells, where energy is required
  —carbon dioxide carried from the muscle cells where it is a waste, produced as a result of cellular respiration, to the lungs where it can be expelled

—chemical messengers such as hormones from glands where they are produced in animals, to organs where they act.

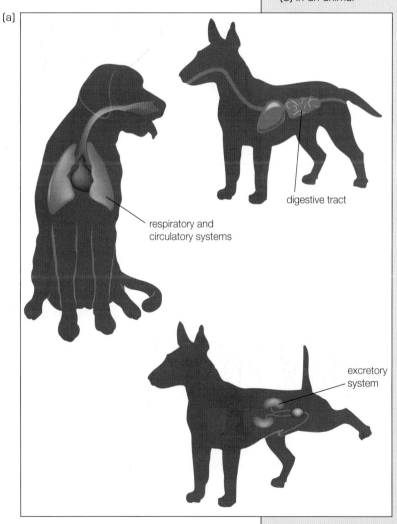

Figure 4.1 Gaseous exchange and transport: (a) in an animal

(a)

respiratory and circulatory systems

digestive tract

excretory system

continued . . .

(b)

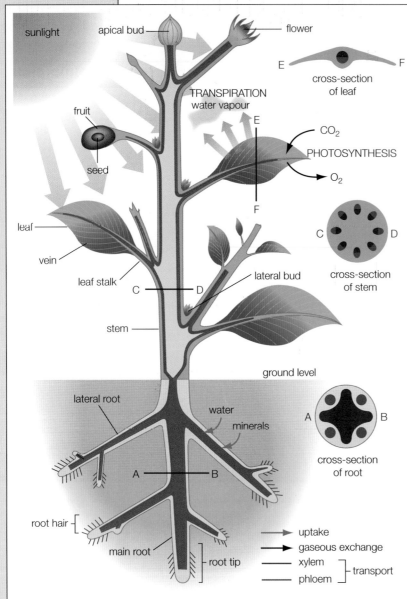

Figure 4.1
Gaseous exchange
and transport:
(b) in a plant

TR

Worksheet—diagrams
of a plant and a
human on which
transport can
be drawn

the **respiratory surface** or the *surface of gaseous exchange* and the gases move across this surface by the process of **diffusion**.

Once gases have crossed this surface, they need to be *transported* to the parts of the body that need them. Within unicellular organisms or small multicellular organisms, this movement of substances between the outside (external environment) and the inside (internal environment) of the organism occurs across the entire cell membrane or thin body wall. In large, multicellular organisms, however, special systems are needed to ensure the efficient movement of substances:

- a *gaseous exchange system* allows the exchange of oxygen and carbon dioxide with the external environment in plants and animals
- a *transport system* carries chemicals from where they enter the body or where they are produced, to where they are used. In the case of wastes, they are carried from where they were produced to where they leave the body (*excretion*).

The reason why special systems are needed in large multicellular organisms is dealt with in the following section.

If the chemicals that are being moved into the body, within the body or out of the body are gases, this movement is termed **gaseous exchange**.

When gases move into and out of an organism, they need to move across the *surface* of the body. In some organisms this could be a general movement across the *entire body surface* (e.g. across the moist skin of an earthworm), but in most, a *special surface area* has developed for this to occur (e.g. the gills in fish and tadpoles; lungs in land animals). The surface that these gases cross is called

The need for transport systems in multicellular organisms

4.2

- **explain the relationship between the requirements of cells and the need for transport systems in multicellular organisms**

As we already know, *unicellular organisms* are so small that their surface area to volume ratio is adequate to allow them to rely on simple *diffusion* to supply requirements such as oxygen for cellular respiration and to remove waste products such as carbon dioxide, urea and other metabolic wastes. Water levels can also be maintained simply, through the passive process of *osmosis* across the body surface, because the surface area to volume ratio of these organisms is large enough.

Multicellular organisms are bigger in size and so their total surface area to volume ratio is smaller. Cells near the centre of these organisms would be too far away from the surface for substances from the outside environment to reach them efficiently (remember, diffusion and osmosis are slow, passive processes). Large organisms that are active, such as complex animals, need more nutrients and oxygen to provide them with energy and they produce more wastes, so they have a greater need for transport. This problem is solved by the presence of a **transport system** within the bodies of large multicellular organisms.

The roles of respiratory, circulatory and excretory systems

4.3

- **compare the roles of respiratory, circulatory and excretory systems**

The respiratory system

Introduction to terminology

The terms **breathing**, **respiration** and gaseous exchange are all related, but do not mean the same thing. The following information should help you to distinguish between these terms:

- *Respiration* is a biochemical process occurring in *all* cells (inside the mitochondria). During this process, energy (in the form of ATP) is released from nutrient molecules (food) that had combined with oxygen. (Remember, the main purpose of **RE**spiration is to **R**elease **E**nergy from foods!) Details of this process were covered in Chapter 3 (see pages 20–22).

- *Breathing* is a mechanical (physical), rhythmic process involving muscles and the skeleton in animals, to allow an organism to inhale and exhale. It helps to increase the amount of gaseous exchange that can occur across a gas exchange surface.

- *Gaseous exchange* is a physical process in living organisms where gases move by diffusion, often across cell membrane(s). Gases required by organisms for their normal functioning move *into cells* and gases produced as a result of functioning are *expelled*. Gaseous exchange may take place either internally (between different parts of an organism's internal environment) or externally (between the external environment

Figure 4.2
The human respiratory system

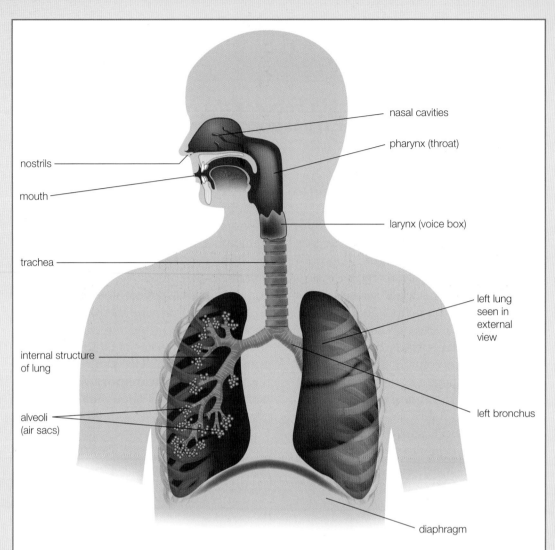

nasal cavities

pharynx (throat)

nostrils

mouth

larynx (voice box)

trachea

left lung seen in external view

internal structure of lung

alveoli (air sacs)

left bronchus

diaphragm

and the internal environment). In mammals for example (see Fig. 4.2) exchange of gases occurs:
—*externally*: oxygen required for respiration moves from the air into the lungs, across the surface of the air sacs, into the bloodstream; carbon dioxide moves out of the bloodstream into the air sacs and out of the lungs
—*internally*: oxygen moves from the bloodstream out of a blood vessel, into the surrounding cells; carbon dioxide moves from the cells into the blood vessels.

Gases move by diffusion across a surface which separates the internal and external environment or the parts of the internal environment. These gaseous exchange surfaces will be discussed in more detail on pages 161–2.

The role of the respiratory system

The respiratory system enables organisms to take in oxygen and to remove carbon dioxide from their bodies—it allows gaseous exchange between an organism and its external environment. Oxygen is essential for almost all living organisms as it is needed for the *release of energy* from food during cellular respiration. Carbon dioxide is a waste that must be removed because it is *toxic* in large quantities. The respiratory system is made up of tissues and organs that are specialised for gaseous exchange—in animals the respiratory organs are varied, such as lungs in mammals, gills in fish and tadpoles and a tracheal system in insects (see Fig. 4.5). In plants, respiratory tissues include **stomates** and **lenticels** (see Fig. 4.15).

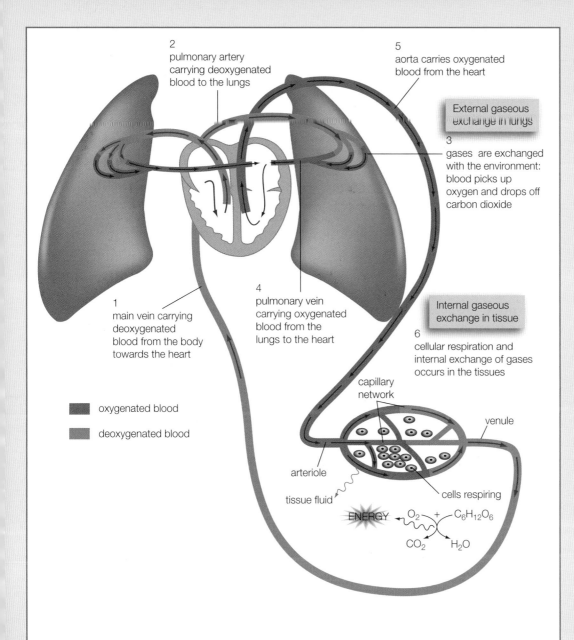

2
pulmonary artery
carrying deoxygenated
blood to the lungs

5
aorta carries oxygenated
blood from the heart

External gaseous
exchange in lungs

3
gases are exchanged
with the environment:
blood picks up
oxygen and drops off
carbon dioxide

1
main vein carrying
deoxygenated
blood from the body
towards the heart

4
pulmonary vein
carrying oxygenated
blood from the
lungs to the heart

Internal gaseous
exchange in tissue

6
cellular respiration and
internal exchange of gases
occurs in the tissues

oxygenated blood

deoxygenated blood

capillary
network

venule

arteriole

cells respiring

tissue fluid

ENERGY $\leftarrow O_2 + C_6H_{12}O_6$

$CO_2 \quad H_2O$

Figure 4.3 Scheme
showing the link
between external
and internal gaseous
exchange, cellular
respiration and the
transport of gases
in humans

Transport systems

A transport system is the link between all other systems in the body of an organism, ensuring that cells are supplied with the nutrients and gases that they need and that wastes are removed.

Effective transport systems have the following features:

■ a system of *vessels* in which substances are transported
■ some way of ensuring the materials *flow* in the correct direction
■ a *medium* in which the chemicals can be carried
■ a mechanism to ensure substances are *released* where they are needed

and *picked up* from where they are not needed.

The *circulatory system* is another name for the transport system in animals. It carries substances needed by the body from their point of entry into the body (or from their site of production) to the parts of the body where they will be stored or used. In animals, nutrients are carried in a fluid medium (most often blood) that *circulates* around the body, picking up and dropping off chemicals. This type of transport system is therefore termed a *circulatory* system.

159

The role of the circulatory system

The main functions of the circulatory system are:

- transport of gases (oxygen and carbon dioxide), nutrients, waste products, hormones and antibodies
- maintenance of a constant internal environment (pH, ions, blood gases, osmotic pressure)
- removal of toxins and pathogens
- distribution of heat.

Excretory systems

Introduction to terminology

Excretion involves expelling *metabolic wastes* from the body—wastes that have been made by cells as a by-product of metabolism (chemical reactions in cells). (This should not be confused with the term *egestion*, which is simply passing out food that could not be digested in the body. Undigested food or faeces are not a form of excretory waste because they are not a by-product of metabolic activity.)

Nitrogenous wastes are the main excretory wastes in vertebrates and are toxic if they accumulate in the body. As their name implies, they contain the chemical element *nitrogen* which cannot be stored in the body. The commonest examples of nitrogenous wastes that are excreted are the substances **urea**, **uric acid** and **ammonia**. These wastes are then transported from where they have been produced, to where they will be excreted.

The role of the excretory system

The main function of excretory systems is to remove metabolic wastes from the transport medium (e.g. blood) and to expel them to the outside:

- *nitrogenous wastes*, together with water and other substances such as salts are combined to form **urine** in mammals. The **kidneys** act as organs of excretion. Small amounts of urea are also lost in **sweat**
- *carbon dioxide* is an excretory waste which forms during cellular respiration (also a metabolic process). The lungs act as excretory organs, ridding the body of carbon dioxide.

Excretion is very closely linked with **water balance** in organisms—the more toxic the type of waste, the greater the amount of water required to dilute it for excretion. Different types of nitrogenous wastes excreted by vertebrates need different amounts of water to dilute them, depending on their toxicity. *Uric acid*, the least toxic form, is excreted with the least amount of water (e.g. white sludge excreted by birds). *Urea* requires more water to dilute it and is commonly found in the urine of mammals like us. *Ammonia*, the most toxic form of nitrogenous waste, must be diluted with the largest amount of water and so is usually only excreted by aquatic vertebrates such as freshwater fish, since water is freely available in their environment.

In some organisms, the role of the excretory system also includes the *elimination of excess salts*, *regulation of the pH* (acidity or alkalinity) of body fluids and their internal *pressure* (e.g. regulation of blood pressure in mammals).

Figure 4.4
The excretory system in humans

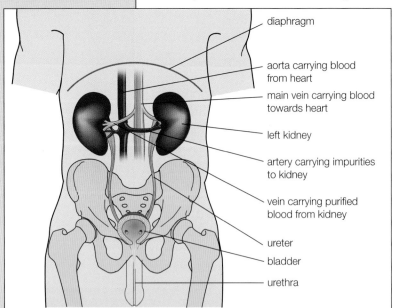

- diaphragm
- aorta carrying blood from heart
- main vein carrying blood towards heart
- left kidney
- artery carrying impurities to kidney
- vein carrying purified blood from kidney
- ureter
- bladder
- urethra

There is no recognised excretory system in plants (although some plants have specialised structures to eliminate excess salts). Most plants accumulate very few metabolic wastes because they are far less active than animals and do not rely on ingested food for their energy. Excretion of any excess materials in plants is by means of diffusion across the surface of organs such as roots and leaves.

Table 4.1 Excretory organs in mammals

Excretory organ	Excretory waste	Role of excretory organ	Excreted as:
Lungs	Carbon dioxide	Removes CO_2 from the blood	CO_2 gas in air
Kidney	Urea	Removes urea from the blood	Urine (fluid)
Skin	Urea	Removes small quantities of urea from the blood	Sweat
Liver	Bile salts	Removes pigment (haemoglobin) from worn out red blood cells	Bile salts (in faeces)

Summary of excretory system

STUDENT ACTIVITY

Comparing these systems in mammals

Complete the table of comparison between the respiratory, transport and excretory systems on the Student Resource CD.

Student activity table

Gaseous exchange in animals

4.4

- *identify and compare the gaseous exchange surfaces in an insect, a fish, a frog and a mammal*

Introduction

Animals have high energy demands because they actively move to search for food and escape predators. As a result, they need a large amount of oxygen and they release large amounts of carbon dioxide. Since their body surface area to volume ratio is not large enough to meet their high demand for gaseous exchange, specialised respiratory systems have evolved.

Respiratory surfaces

Respiratory surfaces are body surfaces that are in contact with the external environment and have become specialised for gaseous exchange. The types of respiratory organs vary from one type of organism to another but, to ensure efficient exchange of gases by diffusion, all gaseous exchange surfaces share certain common features:

Figure 4.5 Types of respiratory (gaseous exchange) surfaces in animals

(a) Gaseous exchange across the entire surface of body (e.g. unicells and earthworms)

(b) Gaseous exchange across the surface of a flattened body, flattening decreases the distance over which diffusion has to occur (e.g. flatworms)

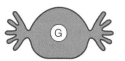

(c) External gills increase the surface area; gaseous exchange usually takes place across the rest of the body surface as well as the gills (e.g. young tadpoles)

(d) Highly vascularised internal gills (e.g. fish)

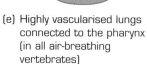

(e) Highly vascularised lungs connected to the pharynx (in all air-breathing vertebrates)

(f) Gaseous exchange at the terminal ends of fine tracheal tubes (e.g. insects and other arthropods)

- *a large surface area*—the surface area of respiratory organs may be increased by folding (e.g. lungs and gills), branching (e.g. tracheoles in insects) and/or flattening (e.g. the flat shape of cells lining air sacs in lungs) of tissue at the surface of gaseous exchange. An increased surface for gaseous exchange allows a faster rate of diffusion to supply oxygen and to remove carbon dioxide as required by the organism. It is also necessary to compensate for the small surface area to volume ratio of the animal's body

- *a moist, thin surface* creates the best possible conditions for efficient diffusion—moisture ensures that oxygen and carbon dioxide are able to be dissolved for easy diffusion across membranes of cells and the thin nature of the surface reduces the distance that gases need to travel

- they are *near to an efficient transport system* to allow gases to be carried to the cells where they are needed or from the cells where they have been produced. The continual movement of gases towards or away from the respiratory surface also ensures that an adequate concentration gradient is maintained: the steeper the gradient, the more rapid the overall diffusion that occurs.

Problems posed by habitiat

The habitat in which an animal lives poses its own set of problems in relation to respiration—terrestrial animals run the risk of dehydration as a result of the evaporation of water from the moist respiratory surface, while aquatic animals face the problem of a low oxygen content in water. The respiratory systems in animals have developed to overcome these difficulties.

Mammals: the human respiratory system

Overcoming the difficulty posed by habitat

Mammals that live on land have overcome the problem of dehydration which may arise from the evaporation of water at the respiratory surface, by having an internally-placed surface for gaseous exchange (i.e. lungs inside the chest cavity).

The respiratory surface

Gaseous exchange in mammals takes place in millions of **alveoli** (air sacs) in the lungs (see Fig. 4.6a). These alveoli (singular = alveolus) form the boundary between the *air* in the external environment and the *blood* **capillaries** in the body (internal environment). The lungs have all the typical features of an efficient gaseous exchange surface:

- *increased surface area:* contact between the alveolar air and the respiratory surface is increased by the folding of the thin lining of the air sacs making up the lungs. There are numerous examples of *folding* and *branching* to increase the respiratory surface of the lungs—the lungs are lobed, the trachea divides into bronchi which further sub-divide, eventually forming tiny tubules called bronchioles. Each bronchiole ends in a cluster of further folded air sacs, the alveoli (see Fig. 4.6b). An adult human has approximately 300 million alveoli, supplied by 280 million capillaries. If we were able to flatten the alveoli of the lungs out completely, they would cover the surface area of a tennis court!

- *thin:* alveoli have an extremely thin lining made up of a single layer of flattened cells called squamous epithelium. This facilitates diffusion as the thin layer which forms the respiratory surface reduces the distance that gases must travel to enter the body

- *moist:* the epithelial cells lining the respiratory tubules secrete *mucus* and moisten the air entering the lungs. The air inside the alveoli is saturated with *water vapour* and the mucus-lined epithelium reduces

Worksheet on human respiratory system

Figure 4.6 (a) Human respiratory system showing detail of alveoli in the lungs; (b) alveoli and blood supply; (c) one alveolus cut through in section, showing exchange of gases with an adjacent capillary

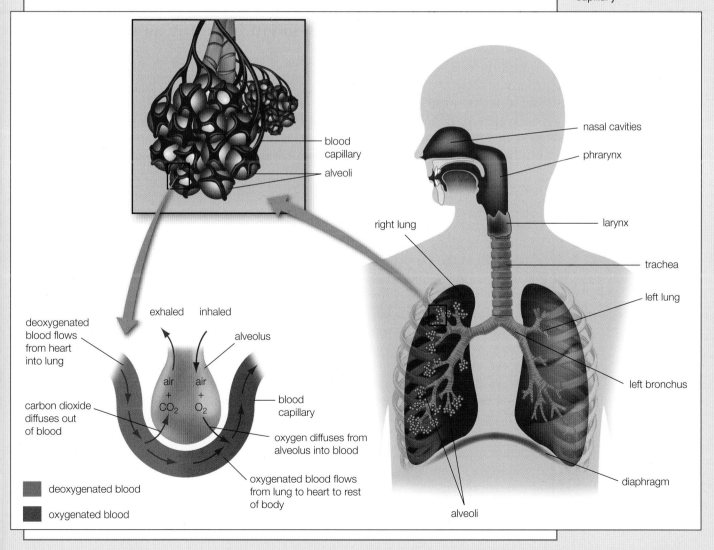

the evaporation of this water. The moisture ensures that the oxygen and carbon dioxide that diffuse across the gaseous exchange surface are in a dissolved form

■ *well supplied with blood*: numerous blood capillaries closely surround the outside of each alveolus. Capillary walls are also only one cell layer thick, keeping to a minimum the distance that gases need to travel between the respiratory surface and the bloodstream.

Gaseous exchange

When gases move between the air in the alveoli and the blood stream, they need to cross only a thin barrier—one layer of alveolar cells and one layer of capillary cells. Movement is by means of diffusion along a concentration gradient. Inhaled air contains approximately 20 per cent oxygen and 0.04 per cent carbon dioxide. Exhaled air contains approximately 15 per cent oxygen and 4 per cent carbon dioxide.

Oxygen in the incoming alveolar air is in a higher concentration than that in the bloodstream, so oxygen diffuses from the air sacs into the body; *carbon dioxide* is in higher concentrations in the bloodstream (a by-product of cell metabolism), so it diffuses along a concentration gradient from the capillaries, through the alveolar lining and into the alveolar air, where it will be breathed out. This movement of gases between the external environment (alveolar air) and internal environment (bloodstream) is known as *gaseous exchange*. There is always some air in the respiratory tract—it never empties completely. Breathing movements which cause ventilation (air to be drawn into and expelled from the respiratory organs) are brought about by the contraction of the diaphragm and muscles between the ribs, which influence the size and shape of the ribcage.

Fish

Overcoming the difficulty posed by habitat

The respiratory system of fish is adapted to use oxygen dissolved in water. The main problem that fish need to overcome is that water has a much lower oxygen content than air. Oxygen diffuses slowly from the surrounding air into a body of water across its surface. Water is much denser and more viscous than air and so it has an overall lower oxygen saturation. The exact percentage of oxygen saturation varies, depending on the depth of water, surface area exposed to air and temperature. The typical oxygen content of water would be approximately 4 to 6 per cent, as opposed to approximately 20 per cent oxygen saturation in air.

Gaseous exchange

The respiratory organs of fish are **internal gills**. (The scaly skin of fish cannot exchange gases with the surrounding water.) Water moves through the respiratory system in only one direction: as the fish swims, it opens its mouth so that water enters and flows over the gills and then it lifts its opercula (gill coverings) to let the water out (see Fig. 4.7a). Movements of the floor of the mouth (buccal cavity) and the opening and closing of the opercula assist with drawing water into the mouth and its movement over the gills and out of the body. As the water flows over the gills, gaseous exchange takes place. Water movement is slowed down because of the highly branched nature of the gill filaments and because the gills are closely stacked, relying on the flow of water to keep the tips of these filaments apart. There are four curved gill arches closely stacked together on each side of the pharynx (throat) of the fish, covered by an operculum (see Fig. 4.7b). Each gill has two rows of very delicate gill filaments, subdivided to increase the surface area for gaseous exchange with the surrounding water. The gill filaments

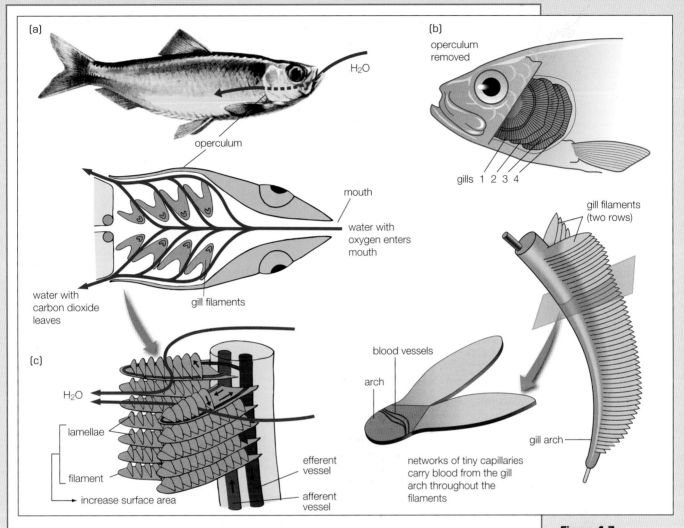

Figure 4.7
Respiratory system of fish: (a) water movement; (b) gill structure; (c) gill filaments

appear pink or red in colour, due to the presence of the rich supply of blood capillaries. Each filament is covered by an extremely thin layer of cells, so that gases can enter and leave easily (see Fig. 4.7c). As water flows over the gills, oxygen diffuses across the filaments into the blood capillaries and carbon dioxide diffuses out. The water leaving through the external gill slits has a higher concentration of carbon dioxide and a lower concentration of oxygen than the water which entered the mouth.

Frogs

Overcoming the difficulty posed by habitat

Frogs are partly aquatic (tadpole stages live in water) and partly terrestrial (most adult frogs live on land and are air-breathing, but must return to water for breeding). These changes of habitat within the life cycle present a challenge to the respiratory system:

- *adult frogs* have retained certain respiratory characteristics typical of simple aquatic organisms (e.g. a *naked, moist skin* that can be used for gaseous exchange) as well as developing terrestrial features (e.g. *simple lungs*)
- *tadpoles* (immature, larval stages of frogs' lifecycles) use their *thin, moist skin* and *gills* for respiration. Young tadpoles have external gills and these become internal gills as the tadpole matures. The surface area to volume ratio of tadpoles is large (because of their small body size and their long thin tail), so they can rely on diffusion to adequately meet their oxygen needs.

Extension activity: fish further increase their efficiency of gaseous exchange with a counter-current system

Gaseous exchange

Adult frogs use three surfaces for gaseous exchange:

1. the *skin* is the main site for respiration when the frog is in water or when it is relatively inactive on land. The skin is very well supplied with blood vessels

2. the *floor of the mouth* is large and well supplied with blood capillaries. It serves as a buccal pump, ventilating the lungs. Some gaseous exchange may also occur across the inner lining of the buccal cavity

3. frogs have two simple, sac-like *lungs* and, although internal (like those of terrestrial vertebrates), they are not greatly folded like those typical of mammals. Frogs only use their lungs for gaseous exchange when they are physically active (e.g. during hopping and when on land). The nostrils have valves which close to prevent the entry of water into the lungs during swimming.

All three respiratory surfaces are kept moist, are thin and are well supplied with blood vessels.

Figure 4.8
Respiratory organs of:
(a) tadpoles; (b) frogs

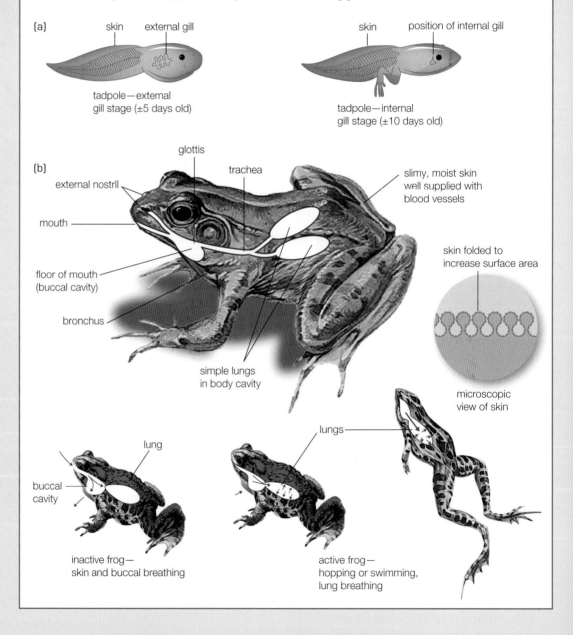

(a)

skin   external gill

tadpole—external
gill stage (±5 days old)

skin   position of internal gill

tadpole—internal
gill stage (±10 days old)

(b)

glottis

trachea

external nostril

mouth

floor of mouth
(buccal cavity)

bronchus

slimy, moist skin
well supplied with
blood vessels

skin folded to
increase surface area

simple lungs
in body cavity

microscopic
view of skin

lungs

lung

buccal
cavity

inactive frog—
skin and buccal breathing

active frog—
hopping or swimming,
lung breathing

In addition, frogs have an increased body surface area: the small size and flattened shape of the body of a frog means that its external surface area to volume ratio is quite high. Mucus on the skin and in the buccal cavity ensures that these surfaces remain moist and the air inside the sac-like lungs usually has high water vapour content. The fact that the lungs are internal reduces the chances of this water vapour from evaporating. The skin of some frogs shows microscopic folding on its surface, further increasing the surface area.

An interesting related fact is that, because of their permeable skins, frogs are extremely sensitive to changes in environmental chemicals such as air or water pollution. Frogs are therefore often used as a 'sentinel' species, similar to canaries in coal mines—their loss from an area is often a warning of harmful changes in the environment.

Insects

Overcoming the difficulties posed by habitat

Insects, being terrestrial, face the challenge of reducing the loss of water from their respiratory surfaces. Like other terrestrial animals, they have an internal respiratory surface. Insects take in and expel air through breathing pores called **spiracles**, which have *valves* to regulate their opening and closing, ensuring that they are not constantly exposed to the drying effects of the external environment. (See Fig. 4.9.) Furthermore, little or no gaseous exchange can occur through their body covering.

Although small in size, insects are very active during parts of their life cycle and many insects require large quantities of energy to sustain their flying. Therefore they need an efficient respiratory system.

Insects differ from terrestrial vertebrates in that insects do not have lungs or blood capillaries. Instead they have a system of branching air tubes called **tracheal tubes** which carry air directly to the cells of the body—blood is not involved in the transport of gases.

Gaseous exchange

Air enters the insect body through a row of small breathing pores or spiracles on each side of the abdomen of the insect's body (the locust has ten pairs of spiracles). Each spiracle has a valve which regulates its opening and closing. Air that enters the spiracles is drawn into **tracheae** (tracheal tubes) which are kept open by spiral rings of chitin (an insoluble chemical substance that functions in support) to prevent them from collapsing. Tracheae branch extensively into smaller tubules called **tracheoles**, creating a very large surface area for gaseous exchange. Tracheoles carry the air directly to and from the cells of the body.

The respiratory surface in insects differs from all other internal respiratory systems in that it has no blood or blood capillaries involved in the transport of gases. The ends of the tracheoles are filled with a watery fluid in which the gases dissolve. Oxygen from the air, dissolved in this fluid, diffuses directly into the cells and carbon dioxide diffuses directly out of the cells into the tracheoles. The rate of respiration in insects is generally controlled by the number of open and closed spiracles—more are open when the insect is active. Muscular movements of the thorax and abdomen during movement and general body movements when flying also help to ventilate the tracheal system.

Figure 4.9 Tracheal
system of insects

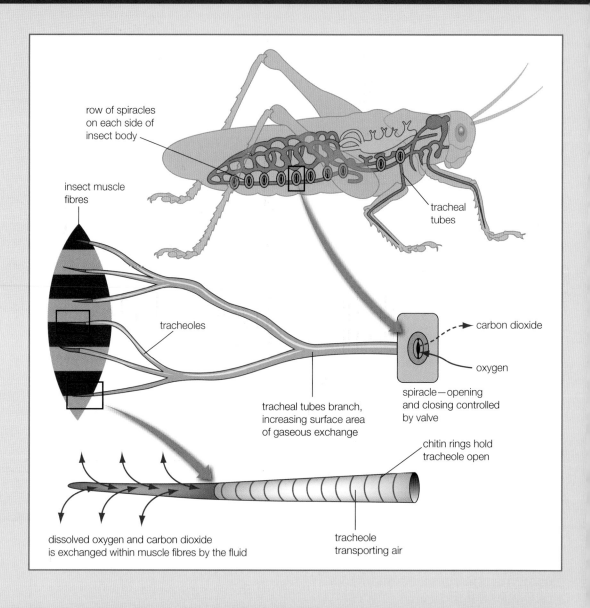

row of spiracles
on each side of
insect body

tracheal
tubes

insect muscle
fibres

tracheoles

carbon dioxide

oxygen

spiracle—opening
and closing controlled
by valve

tracheal tubes branch,
increasing surface area
of gaseous exchange

chitin rings hold
tracheole open

dissolved oxygen and carbon dioxide
is exchanged within muscle fibres by the fluid

tracheole
transporting air

STUDENT ACTIVITY

Draw up a table to compare the gaseous exchange surfaces of insects, fish, frogs and mammals.
The following headings are suggested:
- Respiratory organs
- Surface of gaseous exchange
- How it is kept moist
- How the surface area is increased
- Transport of gases to and from the respiratory surface.

Student activity

Transport in animals

4.5

■ *compare open and closed circulatory systems using one vertebrate and one invertebrate as examples*

Introduction: closed versus open circulatory systems

The cells of active, complex organisms require a large supply of nutrients and oxygen and the continual removal of waste products. A circulatory system is a more efficient means of achieving this than simply relying on diffusion, but not all circulatory systems are equally efficient. A circulatory system is classified as *closed* or *open*, depending on the flow of its transport fluid. In a **closed circulatory system** the transport fluid flows in *vessels* only, but in an **open circulatory system**, at some stage of circulation the transport fluid leaves the vessels and enters spaces or cavities in the body, bathing the organs directly.

Open circulatory system in an invertebrate

An open circulatory system is characteristic of invertebrate animals such as spiders, insects, crabs and snails. The transport fluid is pumped forward in the body by a long, pulsating vessel, the **heart,** into shorter vessels near the head end, which in turn *empty into large spaces* (called **sinuses**) *in the body cavity* (see Fig. 4.10). The transport fluid in an open circulatory system is called **haemolymph** (rather than blood) and it flows freely in the sinuses, directly bathing the cells. The exchange of nutrients and wastes relies on direct diffusion between the haemolymph and the cells. (Remember that gases in insects are not transported by the haemoplymph but by tracheoles.). Haemolymph returns to the heart by moving from the posterior (rear) sinuses back into the open end of the tubular heart, or it may enter the heart through tiny holes in the sides called **ostia**.

Open circulatory systems are not very efficient, as the *fluid pressure* is *low* and so the transport fluid *circulates slowly.* This type of transport system meets the needs of smaller animals such as insects.

Closed circulatory system in a vertebrate

A closed circulatory system (see Fig. 4.11) is characteristic of all vertebrates such as fish, frogs, reptiles, birds and mammals (including humans). The transport fluid is *blood* which is *contained in vessels at all times* and never flows through body cavities. The heart is a muscular organ that pumps the blood around the body. In mammals, the heart may be two-chambered (e.g. fish), three-chambered (e.g. frogs and some reptiles), or four-chambered as in other reptiles, all birds and mammals.

Blood flows through three types of **blood vessels**: **veins** which carry blood from body organs towards the heart, **arteries** which carry blood away from the heart to the organs and *capillaries* which form a link between arteries and veins. The *arteries* branch into smaller *arterioles* which subdivide

Figure 4.10
The open blood system of a locust

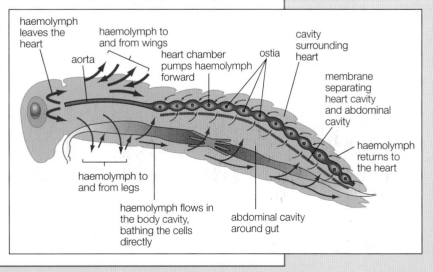

haemolymph leaves the heart

haemolymph to and from wings

aorta

heart chamber pumps haemolymph forward

ostia

cavity surrounding heart

membrane separating heart cavity and abdominal cavity

haemolymph returns to the heart

haemolymph to and from legs

haemolymph flows in the body cavity, bathing the cells directly

abdominal cavity around gut

further into a network of *capillaries*. These capillaries branch extensively throughout the tissues, so that no cell is very far from a capillary (see Fig. 4.11). The exchange of nutrients, wastes and gases takes place between blood in the capillaries and fluid surrounding the cells which the capillaries supply. Blood remains in the capillaries at all times, but any chemical substances required by cells leave the capillaries in a dissolved form—the fluid containing the nutrients, gases and wastes is called tissue fluid or **interstitial fluid**. (The tissue fluid makes internal organs appear 'wet'.). *Capillaries* join up to form *venules*, which in turn join up to from *veins*, returning blood to the heart.

In a closed circulatory system, the muscular heart pumps blood under *high pressure*, ensuring efficient transport, which suits large, active animals such as vertebrates. A four-chambered heart is the most efficient pumping mechanism, as it keeps oxygenated and deoxygenated blood separate. (Not all vertebrates have a four-chambered heart—fish have a heart with only two chambers.)

In both an open and closed circulatory system, the blood vessels are responsible for the transport of blood and its contents, but the capillary networks (closed system) or fluid in the body cavity (open system) carry out the other functions such as the exchange of nutrients and wastes, and maintaining a stable internal environment in the body of the organism.

Figure 4.11
The closed blood system of vertebrates: (a) turtles; (b) fish

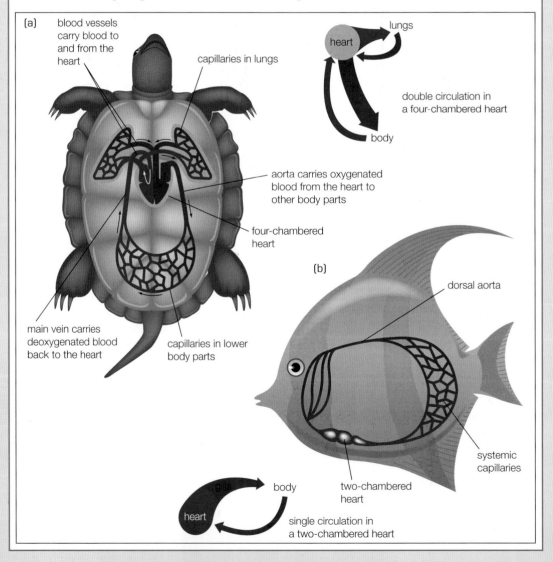

(a) blood vessels carry blood to and from the heart

capillaries in lungs

lungs

heart

double circulation in a four-chambered heart

body

aorta carries oxygenated blood from the heart to other body parts

four-chambered heart

main vein carries deoxygenated blood back to the heart

capillaries in lower body parts

(b)

dorsal aorta

systemic capillaries

gills

body

two-chambered heart

heart

single circulation in a two-chambered heart

STUDENT ACTIVITY

Draw a comparative diagram of each type of circulatory system and then complete the table below to compare open and closed circulatory systems.

Table 4.2 Comparing open and closed circulatory systems

Open circulatory system	Closed circulatory system
Similarities	
Differences	
Example: _____	Example: _____
Transport fluid is _____	Transport fluid is _____
Transport fluid flows through body cavities and bathes tissues directly	
	Vessels divide into capillaries, no cell is very far from a capillary
Tubular heart pumps fluid; pressure is low; suits smaller animals (e.g. insects)	
	Distributes and collects gases (O_2 and CO_2) as well as foods and wastes

Table 4.2

Gaseous exchange and transport in plants

4.6

- *outline the transport system in plants, including:*
  - *—root hair cells*
  - *—xylem*
  - *—phloem*
  - *—stomates and lenticels*

Introduction

In simple plants such as certain algae and moss, there are no specialised transport tissues and the movement of substances relies on diffusion and active transport through all cells. In more advanced plants such as ferns, conifers and flowering plants, special tissues have developed for transport. These transport tissues are the **vascular tissues** and there are two types—**xylem** and **phloem**.

The distribution of vascular tissue in flowering plants (see Fig. 4.12):
- in *roots*, xylem is found in the centre of roots, usually in a star or cross shape with phloem tissue between the arms of the xylem
- in the *stem*, the xylem and phloem tissue divides into vascular bundles
- bundles of vessels of xylem and phloem continue from the stem up leaf stalks, forming the veins in *leaves*.

Root hair cells

Transport in plants was been introduced in Chapter 3, root hairs were dealt with in detail and so only a general summary is given here. Revise the information on transport in plants on pages 140–1 and then read the additional details on xylem, phloem, stomates and lenticels in this chapter.

Water and dissolved nutrients move from the soil into roots, through the root epidermal cells, with most of the uptake being in the region of the root hairs. Water then moves across the root tissues from the outer epidermal layer to the vascular stele (transport tissue) in the centre of the root and into the xylem tissue (see Fig. 3.15b).

Gaseous exchange between roots and soil also takes place, relying on diffusion of gases. Cells of the root cannot photosynthesise (they are not exposed to sunlight and have no chlorophyll) but they do respire like all living cells.

Xylem

Xylem is specialised tissue for the transport of water and dissolved inorganic minerals from the roots to the leaves.

Structure of xylem

Xylem tissue consists of two main types of elements—**xylem tracheids** and **xylem vessels** (see Fig. 4.12), with other cells such as parenchyma and fibres in between. Tracheids are elongate with end walls that taper to a point. Most of the xylem in flowering plants occurs in the form of xylem vessels. Xylem vessels form continuous tubes for the transport of water. When cells specialise to become xylem vessels, their transverse walls break down, so the cells that are stacked on top of each other become continuous tubes. The cell contents die, leaving hollow vessels for the easy flow of water and dissolved mineral salts. The walls of xylem vessels and tracheids are reinforced with **lignin** thickenings laid down in rings, spirals or other regular patterns. These thickenings prevent the vessels from collapsing, and help the easy movement of water and dissolved substances.

Fibres give support to the xylem tissue and the parenchyma tissue conducts materials from one region of xylem to another and may function in storage.

Transport role of xylem

The function of xylem is to transport water and dissolved inorganic nutrients as *ascending sap*, from the roots up the plant to the leaves and the reproductive structures such as flowers. The movement of water up the xylem vessels occurs mainly as a result of a **transpiration stream** that develops: as water evaporates through the stomates of leaves, it sets up a concentration gradient across the leaf, creating a suction pull on the water and dissolved minerals in the xylem tissue. It is this suction force or *transpiration stream* that ensures the upward movement of ascending sap in the xylem (i.e. there is no pump mechanism such as the heart in animals).

Lateral movement of water in plants

Water moves laterally across plant organs (e.g. from root hairs to xylem, or from xylem tissue into mesophyll in leaves) via one of three pathways:
1. the apoplast pathway through the cellulose cell wall
2. the symplast pathway through the cytoplasm and palsmodesmata
3. the vacuolar pathway, from vacuole to vacuole.

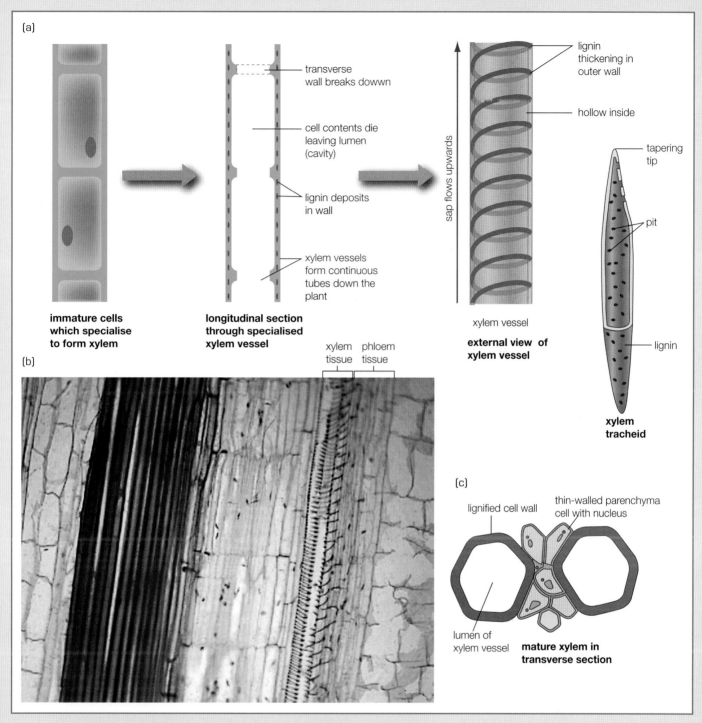

(a)

transverse
wall breaks dowwn

cell contents die
leaving lumen
(cavity)

lignin deposits
in wall

xylem vessels
form continuous
tubes down the
plant

sap flows upwards

lignin
thickening in
outer wall

hollow inside

tapering
tip

pit

lignin

**immature cells
which specialise
to form xylem**

**longitudinal section
through specialised
xylem vessel**

xylem vessel

**external view of
xylem vessel**

**xylem
tracheid**

(b)

xylem   phloem
tissue   tissue

(c)

lignified cell wall

thin-walled parenchyma
cell with nucleus

lumen of
xylem vessel

**mature xylem in
transverse section**

Figure 4.12 Xylem tissue: (a) scheme showing specialisation of xylem vessels; (b) photomicrograph of mature xylem tissue showing a variety of wall thickenings (light microscope view in longitudinal section); (c) mature xylem tissue seen in a transverse section (light microscope view)

173

Phloem

Phloem is specialised tissue that transports sugars (produced by photosynthesis) from the leaves to the rest of the plant.

Structure of phloem

(See Fig. 4.13.) There are two main types of phloem cells (sometimes referred to as phloem elements) in plants—**sieve tube elements** and **companion cells**. Unlike xylem vessels, sieve tube elements are cells with *living* contents (cytoplasm), but they do not have a nucleus or any other organelles besides mitochondria. Companion cells are small cells that are associated with each sieve tube element and they are responsible for keeping the sieve tubes alive.

Organic nutrients such as sugars, produced by photosynthesis, move down or up the plant through the sieve tube elements, a process known as **translocation**. The active loading of sugars (dissolved in water) into phloem requires energy. The sap flows from one element to the next, through perforated transverse walls. (Perforations are tiny holes that pierce the cell wall.) These perforated end walls are known as **sieve plates**.

Transport role of phloem

Movement in phloem occurs in both directions, upwards from the leaves to the upper parts of the plant and down towards the roots. Numerous experiments have been done to show this dual movement, but the current theories of how movement actually occurs in phloem will be studied in Year 12. It is sufficient for you to know that movement is as a result of flow along a concentration gradient, so once again there is no pumping mechanism. The flow of materials through phloem has been shown using radioactively-labelled carbon dioxide which is taken up during photosynthesis. Biologists can then trace the pathway taken by the radioactive carbon dioxide: from its entry into the leaf, its incorporation into sugar products and eventually the flow of sugars through phloem until they are used or converted into other organic compounds for storage.

Stomates and lenticels

Most gaseous exchange in plants takes place through stomates and lenticels.

Stomates

Structure of stomates

(The structure of stomates has been briefly dealt with in Chapter 3 on pages 142–3.) Stomates or *stomata* are pores in the epidermis of leaves, bordered by two bean-shaped guard cells. These guard cells are unlike other epidermal cells because guard cells contain chloroplasts and the inner wall of each is thicker than the outer wall.

Transport role of stomates

When stomates are open, gases are able to diffuse through them, but when they are closed, no gases are transported. Many theories have been put forward to account for the opening and closing of stomates, but at Preliminary Level it is probably sufficient for students to understand that—when the guard cells fill with water and become turgid, the thin outer walls (which are more elastic) stretch outwards, but the thick inner walls (fairly inelastic) do not bulge, so they are pulled apart and the pore between them widens. When stomates lose water, the outer walls no longer bulge, so the inner walls move together again, closing the pore (see Fig. 4.14c). What causes water to move into and out of the guard cells is still being researched, but current theories suggest it is linked to the movement of potassium ions.

Demonstration activity of stomate functioning using a long balloon and sticky tape

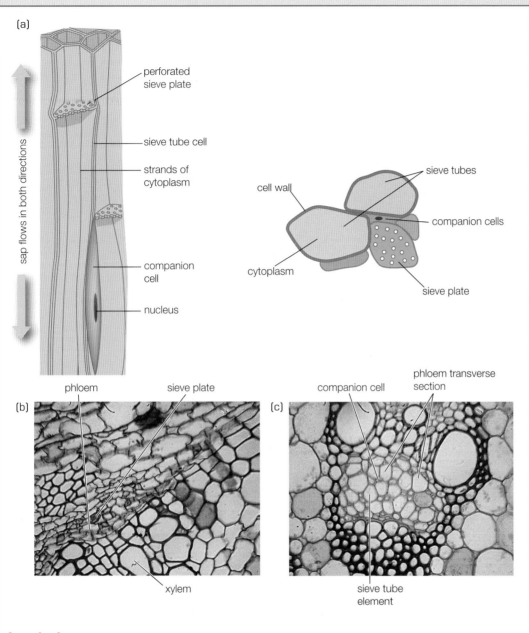

(a)

perforated sieve plate

sieve tube cell

strands of cytoplasm

companion cell

nucleus

sap flows in both directions

sieve tubes

cell wall

companion cells

cytoplasm

sieve plate

Figure 4.13 Phloem tissue: (a) longitudinal section (light microscope view); (b) light micrograph showing sieve plate (in transverse section); (c) transverse section (light microscope view)

(b)

phloem   sieve plate

xylem

(c)

companion cell   phloem transverse section

sieve tube element

Lenticels

Structure of lenticels

Lenticels are pores through which gaseous exchange occurs in the woody parts of plants such as the trunks and branches of trees and woody shrubs. They appear as small dots to the naked eye, but on microscopic examination, it can be seen that they are clusters of loose cells in the cork layer of bark.

Transport role of lenticels

The diffusion of oxygen, carbon dioxide and water vapour takes place through lenticels, relatively slowly.

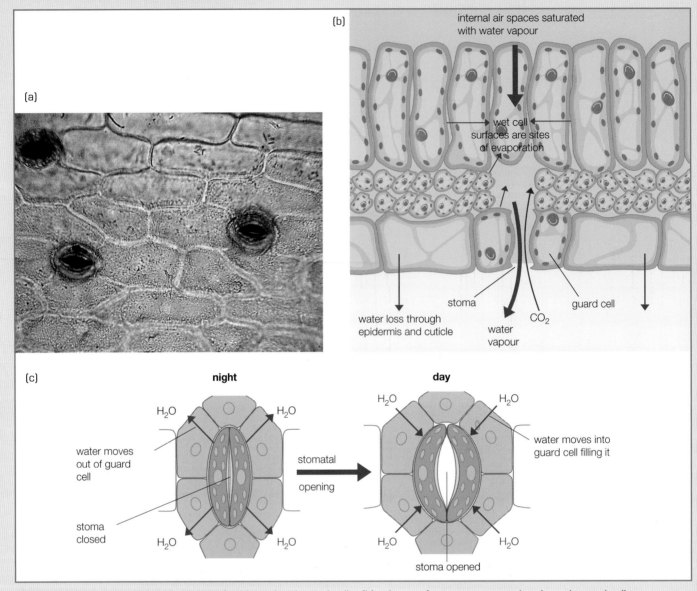

Figure 4.14 Stomates: (a) surface view of epidermal and guard cells; (b) scheme of a transverse section through guard cells (highly magnified) (c) the functioning of stomates (closed and open)

Investigating the movement of materials in xylem

FIRST-HAND INVESTIGATION

BIOLOGY SKILLS

P11.3
P12.1; 12.2; 12.4
P13.1
P14.1

■ *perform a first-hand investigation of the movement of materials in xylem or phloem*

Aim

To investigate the direction and rate of movement of water and a dissolved dye through xylem tissue.

Materials

- celery
- 1000 mL beaker
- water with eosin dye
- permanent marker pen
- stopwatch
- razor blade
- glass microscope slide and coverslip
- light microscope.

Safety

- Wear a laboratory coat to prevent eosin from staining clothes.
- Wear disposable gloves to keep hands stain-free.
- Handle razor blade with care.

Method

1. Place a stick of celery, with leaves attached, into a 1000 mL beaker containing 200 mL of water and eosin dye. (In the absence of eosin, food colouring may be used.)
2. Mark off three 1-centimetre intervals on the celery stem above the meniscus of the liquid. Time how long it takes for the dye to travel each centimetre and record your results. Calculate the average time taken for the dye to move 1 cm. Describe the direction of movement of the water and the dissolved dye.
3. Slice a very thin transverse section through a portion of the celery stem through which the dye has moved. Prepare a wet mount of this tissue to view under the light microscope.
4. On the diagram provided (see Student Resource CD) shade the tissue in the celery stem which has been stained orange by the dye and record the structural features which enabled you to identify the tissue. (Use Figs. 4.13 and 4.15 to help you to identify the tissue in the stems.)
5. Optional: Locate one vascular bundle and peel it longitudinally in the stained region. Make a wet mount of this tissue and observe it under the microscope. Draw a diagram to show its appearance in longitudinal section.
6. Replace the remaining celery stem, with leaves still attached, into the coloured water and leave overnight. Observe the celery after 24 hours and describe the distribution of the dye.

Results

Record your results in a suitable format.

Discussion questions

1. Explain the purpose of the eosin dye in this experiment.
2. Describe evidence from the experiment that supports the hypothesis that water and dissolved nutrients are transported by xylem tissue in stems.
3. Predict the rate of movement of a coloured solution that has been made using warm water.
4. In which direction(s) did water move in the stem?
5. Identify the force that you think made the water move. Is your answer to this question an inference or a conclusion? Explain why.
6. Describe another way (besides changing the temperature of the solution) that you could use to increase the rate of movement up the xylem.
7. Describe the appearance of the xylem tubes.
8. Relate the structure of the xylem tubes to their function.

Conclusion

(Students to write their own conclusion.)

Diagram of celery tissue

Figure 4.15
Movement of materials in xylem: (a) experiment using celery to show the movement of materials in xylem; (b) plan diagram of transverse section through celery stem showing tissue distribution (light microscope view)

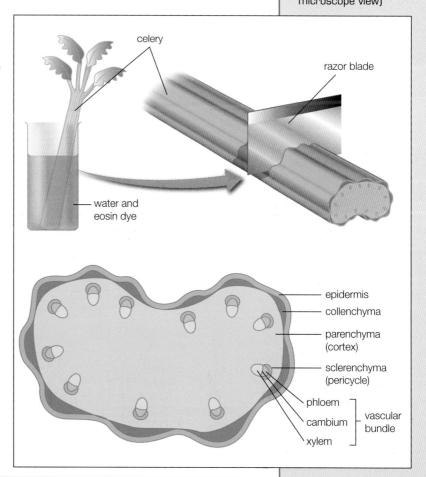

celery

razor blade

water and eosin dye

epidermis
collenchyma
parenchyma (cortex)
sclerenchyma (pericycle)
phloem
cambium
xylem
vascular bundle

Transpiration in plants

■ *use available evidence to perform a first-hand investigation and gather first-hand data to identify and describe factors that affect the rate of transpiration*

Background information

■ Transpiration is the evaporation of water through the stomates of leaves. To replace the water lost from a plant during transpiration, water is drawn in by the roots.

■ Water is needed by the plant:
   —for photosynthesis
   —to make the cells turgid and thus help support the plant
   —as a medium of transport and for the metabolic processes within the cells.

■ There are several factors which can affect the rate of transpiration: wind, temperature, light and humidity are the main factors which could increase or decrease the rate of transpiration.

■ A **potometer** is an instrument used to measure the rate of absorption by a shoot. This is assumed to be equal to the rate of transpiration from the leaves.
   The rate at which water moves along the potometer is a direct indication of the rate of transpiration.

Aim

To determine how wind, light and air movement affect the rate of transpiration.

Hypothesis

A plant exposed to hot windy conditions/bright light will transpire *more/less* than a similar plant exposed to cooler, still air/little light. (Cross out the italic word which you think does not apply.)

Materials

Read through the method and then list all equipment that you will need to conduct this investigation.

Safety

Read through the method and list safety precautions. Also record how you will ensure that your selection of plants complies with non-destructive testing.

Method

1. Set up the potometer as shown in Figure 4.16. Some factors that should be taken into account when selecting the plant to be used are listed below:
   ■ the diameter of the stem is important—have the rubber stopper that you will use on hand; try to match the diameter with that of the opening as closely as possible, as this needs to be airtight
   ■ you will need a set of experimental apparatus and a control. You may use the same plant twice (read the Method carefully and decide whether you should conduct the investigation under the experimental or control conditions first) or select two very similar plants. They should be of the same species, have a similar number of leaves on the stem and have a stem of similar diameter and length
   ■ cut the stem underwater with a sharp blade and insert it through the stopper underwater to prevent air bubbles from forming an air lock. Transfer the stem directly into the water in the potometer
   ■ ensure all seals (e.g. between the stem and the stopper and between the stopper and the photometer) are airtight and watertight by drying them well and placing vaseline where they meet

Figure 4.16
Potometer set up to measure the rate of absorption in relation to transpiration

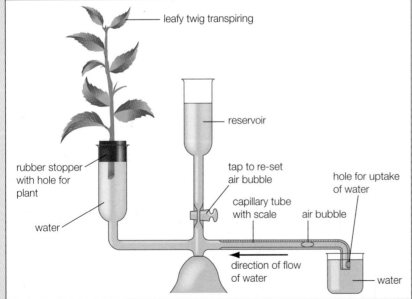

- create the air bubble in the capillary tube using a corner of a tissue or sponge to suck out some water from the hole in the side of the capillary tube and then place the end of the capillary tube into a small beaker of coloured water. The colour makes it easier to see the movement of the bubble along the capillary tube. The beaker should have enough water to create pressure to ensure the air bubble gets pushed upwards and across into the part of the capillary tube marked with a graded scale
- the air bubble can be reset to the end of the tube by opening the tap on the reservoir carefully.

2. Measure the rate at which water moves along the potometer under the following conditions:
   - plant exposed to air in the laboratory (describe the day's conditions). This will be the control
   - plant exposed to hot, windy conditions (simulated using a fan heater or a hair dryer). This will be the experiment.

Results

Present your results in a table with suitable headings.

Conclusion

Write your own conclusion, after reading the aim and hypothesis again.

Discussion questions

1. Explain the link between transpiration and the absorption of water in this experiment.
2. Discuss your choice of using either one plant or two separate plants for the experimental and the control in this investigation.
3. Predict whether the rate of transpiration would increase or decrease, as compared with your control plant, if you subjected a similar plant to each of the following conditions:

Environmental conditions	Rate of transpiration compared with control
1. Increased temperature	
2. Increased wind	
3. Increased humidity	
4. Brighter light	
5. Cool, windy conditions	

4. Explain how and why each of the following factors affects the rate of transpiration:
   (a) heat
   (b) light
   (c) wind
   (d) humidity.

Answers to investigation questions and the results table

The use of radioactive isotopes to trace transport in plants and animals

- *use available evidence to discuss, using examples, the role of technologies, such as the use of radioisotopes in tracing the path of elements through living plants and animals*

Introduction

Isotopes are atoms of the same element with different numbers of neutrons. Isotopes are usually named for their mass numbers (that is, the total number of neutrons and protons). For example, most carbon atoms have six neutrons and six protons, i.e. ^{12}C, but the isotopes of this are ^{13}C and ^{14}C. **Radioisotopes** are isotopes that emit radiation—they are said to be **radioactive** (they emit radioactive waves

or particles to try to achieve a stable state). The emission of these radioactive particles or waves can be measured using photographic film, a cloud chamber, a Geiger counter or a scintillation counter. Their presence can be detected and so they can be distinguished from normal elements. Therefore radioisotopes have become very useful as tracers— technologies can be used to trace their path in physical, chemical and biological systems.

SECONDARY SOURCE INVESTIGATION

PFAs

P5

BIOLOGY SKILLS

P11.1; 11.3
P12.3; 12.4
P13.1
P14.1; 14.3

Radioactive isotopes are produced in Australia by a company called ANSTO (Australian Nuclear and Science Technology Organisation). Radioisotopes are used as a diagnostic tool, giving information about the functioning of organs and tissues in living things. Radioisotopes can be chemically attached to molecules that will enter living tissue. The movement of these molecules can then be traced and their use in particular tissues and organs provides helpful information. Modern computer software can convert the information into three-dimensional images so that investigators can 'see' where the radioactively-labelled chemicals are moving or stored in living organisms.

The use of radioactive isotopes in animals and humans

In animals and humans, radioactively-labelled isotopes are used to study the circulatory system. Technetium-99 is the most commonly used medical isotope. It can be combined with a tin compound and injected into the bloodstream, where it readily attaches to red blood cells, so its path through the circulatory system can be traced. It is useful to determine abnormalities in heart functioning and in blood vessels.

Radioactive particles do carry some risk in that they may be mutagenic (cause mutations in the DNA of cells). Radioisotopes that have a short half-life (such as Technetium-99 which has a half-life of 6 hours) are used in humans, to reduce the person's exposure to radiation and any risks associated with it.

The use of radioactive isotopes in plants

Radioactive isotopes are also used to trace the path of nutrients and gases through plants.

Melvin Calvin was awarded a Nobel Prize in 1961 for his work using a carbon-14 tracer, to show that sunlight acts on the chlorophyll in a plant to begin the manufacturing of organic compounds, rather than on carbon dioxide as was previously believed.

Another experiment was conducted to trace whether the oxygen released during photosynthesis originated from the oxygen atom in water or that in carbon dioxide. Plants that were given water that contained radioactive atoms of oxygen showed that all of the radioactive oxygen atoms from the water molecules were released as oxygen gas, proving that *water* (and not carbon dioxide) was the source of oxygen gas released during photosynthesis:

$$CO_2 + H_2O^* \longrightarrow C(H_2O)n + O_2^* + H_2O$$

*Single oxygen atoms from the water in adjacent chloroplast molecules combine to form O_2, oxygen gas, which is then released.

Biologists can trace in which parts of plants the radioactive atoms have been taken up and how quickly they are absorbed. This helps them to develop fertiliser programs for commercial crops. ^{13}N is being used to trace nitrogen movement in plants.

STUDENT ACTIVITY

Investigate two more uses of radioisotopes—one in plants and one in animals. Use the search words: 'radioisotope tracers in . . .' (i.e. in plants, animals or humans); or 'technology to trace nutrients in plants' to refine your search on the Internet. You could also search for 'radioisotopes in medical imaging' or 'imaging the heart'.

REVISION QUESTIONS

1. Complete the Table 4.3 comparing examples of substances transported in plants with those transported in large multicellular animals.

Table 4.3 Comparing substances transported in plants and animals

	Direction of movement	Examples in plants	Examples in animals
Gaseous exchange system	Oxygen from external environment into the organism		Oxygen from the air into the lungs
	Carbon dioxide from organism to the external environment		Carbon dioxide from the lungs into the air
Transport system	From where the substances entered the body to where they are needed		Oxygen from the lungs to the cells of an animal's body
	From where the substances were produced in the body to where they will be used/excreted		Carbon dioxide from the cells where it was produced to the lungs
Excretory system	From where the substances were produced by metabolism in the body to outside the body		Urea from the kidneys to the outside

2. In the form of a table, compare the percentage content of oxygen and carbon dioxide in inhaled air with that of exhaled air in humans. Account for the difference.

3. Explain why unicellular organisms and small, simple multicellular organisms do not need respiratory structures.

4. (a) Distinguish between the terms *respiration* and *breathing*.
   (b) Write the word equation for cellular respiration.
   (c) Describe what is meant by a respiratory surface or surface of gaseous exchange.
   (d) Describe two differences and one similarity between the respiratory surface of a fish and that of an insect.
   (e) Explain how each respiratory surface compared in Question 4(d) above is suited to its particular environment.

5. Identify four gaseous exchange surfaces in frogs. Which of these surfaces is also used for gaseous exchange in:
   (a) fish
   (b) mammals
   (c) insects?

6. State two reasons why fish can survive in water, even though there is far less oxygen in water than there is in air.

7. Mammals have a closed circulatory system whereas insects have an open circulatory system. Discuss the advantages and disadvantages of these transport systems for the organisms in which they occur.

8. Two leafy shoots were taken, each one from a different species of plant, to determine which species had the greater rate of transpiration under set conditions.
   (a) What piece of apparatus would be used to measure the transpiration rate in each?
   (b) State two precautions that should be taken when selecting the leafy shoots to ensure that this is a valid experiment (fair test) and the results are comparable in the two species.
   (c) Describe the conditions under which you might measure the transpiration rate.

9. The following is a list of biological tools. Outline how each may be used to study transport in plants:
   (a) light microscope
   (b) radioisotopes
   (c) biological dyes.

10. Outline the path that water takes through plants: from when it enters the root from the soil, its passage through the plant and its final exit through the leaves. Specify the tissues that it travels through in each part of the plant and name the process(es) involved at each point.

Answers to revision questions

CHAPTER 5

Cell division

Maintenance of organisms requires growth and repair

How does a single fertilised egg cell become a large, complex, **multicellular** organism? How did we develop from the tiny embryo inside our mothers? This is the result of both growth and development.

Growth

Growth of an organism occurs as a result of increasing the *number* of cells in the body (**cell division**) and increasing the *size* of each cell already there (*cell enlargement* or *elongation*). Growth and development involves four phases:
1. cell division
2. cell enlargement
3. assimilation
4. differentiation.
(Some of these we have already addressed in earlier chapters. This chapter will concentrate mainly on cell division and tie it in with the other phases.)

Cell division

Cells can only arise from pre-existing cells. Cell division is essential for the *maintenance* of organisms, both for the continued *growth* of organisms and also for the *repair* of damaged or worn out cells within them. Cell division is the process by which cells give rise to other cells. One cell replicates its genetic nuclear content exactly and then divides to become two cells, distributing one full set of genetic material to each cell—it is essential for every body cell in an organism to contain the full complement of genetic material if it is to function properly.

Assimilation

These new cells now need to enlarge until they are equal in size to the original cell. *Assimilation* of material from the products of digestion in animal cells, or from photosynthesis in plant cells, supplies new cells with the material that they need for cell enlargement.

Cell enlargement

Most cells increase in size as a result of the assimilation of new materials into the boundaries and cytoplasm. In plant organs such as the root, this cell *enlargement* or *elongation* takes place in a particular area, the zone of elongation, just behind the dividing cells near the tip of a root. Cell elongation may also involve the intake of water, particularly in plant cells, resulting in an increase in the size of the vacuole. If the increase in size is due only to water absorption, it may be temporary, but if it is accompanied by assimilation, it is permanent. Growth refers to a permanent increase in the size of living things.

To scientifically measure growth in an organism, the most accurate method involves measuring an increase in its dry mass. Because it is unethical to maintain large colonies of animals and then kill and dry them, *dry mass* is used to measure growth in plants, and an increase in *height and wet mass* in animals is measured.

Differentiation

The differentiation of cells follows cell enlargement, resulting in an increase in the complexity of an organism, as cells with different structural features develop to enable them to perform their functions more effectively. As a result, a variety of different tissues form, each with a special function; that is, a *division of labour* occurs amongst the cells.

Repair

Cells in our bodies constantly need repair or replacement. If we fall and graze our skin, or the heat of the sun in summer burns the leaves of plants, or a tail is ripped from a lizard, the need for *repair* of these tissues is evident. Continuous *maintenance* of the body parts of an organism may not seem as obvious, but is easier to understand if we think of our own bodies. For example, in humans, red blood cells have a life span of only three months and so they need to be replaced on an ongoing basis; skin cells near the surface of the skin die and slough off continuously (each time we rub our hands together we lose thousands of dead skin cells), and these are only a few of the cell types that need to be constantly replaced. Therefore cell division is important not only for growth, but also for repair and maintenance to replace old, worn-out cells in living organisms.

Where mitosis occurs

5.1

■ *identify the sites of mitosis in plants, insects and mammals*

Although every cell in an organism has the potential to divide by **mitosis**, in reality, cell division occurs only in certain cells in mature multicellular organsisms. Living bodies have control mechanisms to ensure that only those cells required to divide do so. If cells divide in an uncontrolled manner, tumors or cancer may result.

Embryos

Most multicellular organisms start life as a single fertilised egg cell which grows into an embryo. It continues its development and eventually grows into an adult. During embryonic development, there is continual mitotic division and growth of all cells, starting with the division of a single fertilised egg into two identical cells which then divide into four and so on, until a ball of cells results. This ball of cells continues to divide by mitosis and eventually a variety of tissues begin to differentiate, until the young organism begins to closely resemble the adult it will eventually become. This growth and development or *life cycle* may occur in a continuous manner, or it may involve very noticeable changes in form—**metamorphosis** in some animals, such as insects, or an alternation of generation in plants. In all forms, *mitosis* followed by *differentiation* is the driving force behind embryo growth and development.

In mature organisms not all cells continue to divide. As cells differentiate and specialise, some lose the ability to divide by mitosis. Those cells that are able to divide occur at particular locations within the adult body. The tissue involved is known as **meristem** in plants, but is not given a generalised term in animals, although some cells are referred to as stem cells. Growth occurs in two directions: growth in length (height) is termed *primary growth*, whereas growth in girth (width or diameter) is termed *secondary growth*. Particular dividing cells give rise to these forms of growth in mature organisms. (See Fig. 5.2.)

Where mitosis occurs worksheet

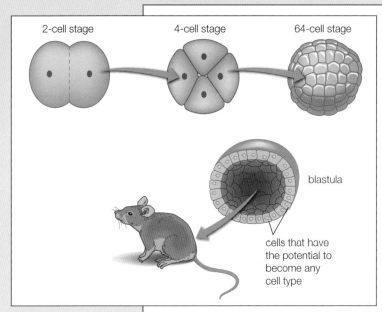

2-cell stage   4-cell stage   64-cell stage

blastula

cells that have
the potential to
become any
cell type

Figure 5.1
Early stages of
mammalian embryonic
development: Division
of embryonic stem
cells at early stages

Plants

Localised cells which are actively
involved in mitotic division in plants,
termed *meristematic tissue* or *meristems*
are found at the following locations
(see Fig. 5.2a):

- **apical meristems** are growing
  points near the tips of roots and
  stems (*apex* = top or tip). They result
  in *growth in length* (primary growth)
- *buds* near the tip contain
  meristematic tissues that divide
  by mitosis and give rise to *lateral
  branches*
- **cambium** is meristematic tissue
  that divides by mitosis, resulting
  in growth in diameter (secondary
  growth) in both stems and roots
  and allows young leaves to grow.
  There are two types:
  —*vascular cambium*, which occurs
  between xylem and phloem of
  vascular bundles in stems and gives
  rise to the woody tissue common in
  many trees and shrubs
  —*cork cambium*, which occurs just
  below the epidermis of stems and
  gives rise to cork cells and the bark
  of woody trees
- **pericycle** is a special layer of
  meristematic tissue in plant

roots and is responsible for the
development of *branch roots*.
(A tip for remembering the above
tissues—you simply need to remember
your 'abc': *a* is for apical mersitem,
b for bud meristem, *c* for cambium and
well, perhaps *d* could be an upside
down *p* for pericycle!)

Mammals

Some parts of mature mammalian
bodies require constant replacement,
while others require no replacement or
are unable to be replaced. For example,
skin cells are constantly being replaced,
but if a nerve cell is destroyed, it is not
replaced because nerve tissue cannot
divide once it has fully differentiated.
Active centres of cell division occur at
several locations in mature mammalian
bodies. The fastest cell replacement
rates occur in dividing tissue in the
following locations (see Fig. 5.2c):

- cells in the basal layer of the *skin*
  and *protective layers* (such as
  cells that give rise to hair, scales,
  fingernails, feathers, cornea of
  the eye)
- *bone marrow* which produces
  blood cells
- cells lining the *digestive tract*.
  Cells which divide, but have a
slower replacement rate, are:
- liver cells
- bone cells—growth plates in bones
  result in an increase in the length
  of long bones, and therefore height
  in upright mammals like humans,
  until they become sexually mature,
  at which stage primary growth shuts
  down and secondary bone growth
  occurs. Once adulthood is reached,
  living bone and cartilage cells found
  in clusters within bone tissue and
  cartilage can also divide to repair
  damaged or broken bones.

As cells age, their ability to divide is
impaired and cell division slows down
as a natural part of ageing.

Insects

Insects show a different type of development to that of plants and mammals. As mentioned before, their growth and development involves metamorphosis, a change in form (and a change in organs when one stage of the life cycle changes into another). (See Fig. 5.2b.) The transformation of an insect larva into an adult is illustrated by two typical examples outlined below:

- *complete metamorphosis*—e.g. the total transformation of a caterpillar into a pupa (cocoon) and then into a butterfly
- *incomplete metamorphosis*—e.g. the gradual transformation of a young grasshopper into an adult, through a series of stages or instars, each separated by a period of moulting (when the insect sheds its outer skin and forms a new one), with an increase in size at each stage until adulthood.

In both types of metamorphosis the change in form relies on the death of some cells and the division of other cells. During a *complete metamorphosis*, because the larval and adult forms are so different from each other, most of the larval tissues inside the pupa are destroyed by the action of lysosomes. Zones of embryonic tissue, made up of cells that have the potential to divide and differentiate occur in the larva. Growth hormones stimulate these larval cells to divide and begin to differentiate. Moulting hormones then act on the cells stimulating them to become new adult tissues, larval growth decreases and moulting and change occur. Just as the nucleus of a cell controls cell division, so it is thought to control cell death during the metamorphosis of pupae into adults. (See Fig. 5.2b.)

In insects which have an *incomplete metamorphosis*, less cell death occurs. A large proportion of the larval tissues remain and the cells simply enlarge to form the corresponding tissue in the adult. Mitosis is involved in the addition of certain adult body parts; for example, the growth of wings in mature locusts, grasshoppers and cockroaches from embryonic cells present in *wing buds*.

In insects mitosis also plays a role in the normal repair and maintenance of body tissues.

STUDENT ACTIVITY

The information in the remainder of this chapter deals with the process of mitosis. Once you have read through this information, answer these questions.

1. Draw a fully-labelled diagram to distinguish between the following terms:
   - *chromosome*
   - *chromatid*
   - *centromere*.
2. Observe the photographic images in Figure 5.5 and identify each of the stages shown.
3. Are the cells in Figure 5.5 plant or animal cells? Justify your answer using evidence visible in the images.
4. Draw a fully-labelled diagram of one of the images in Figure 5.5 showing cytokinesis and one image showing the chromosomes lined up in the centre of the cell. (Your diagram should resemble the images actually shown.)
5. Explain, giving examples, how mitosis plays a role in the addition of adult body parts to an immature (larval stage) insect during metamorphosis.

Word search gird

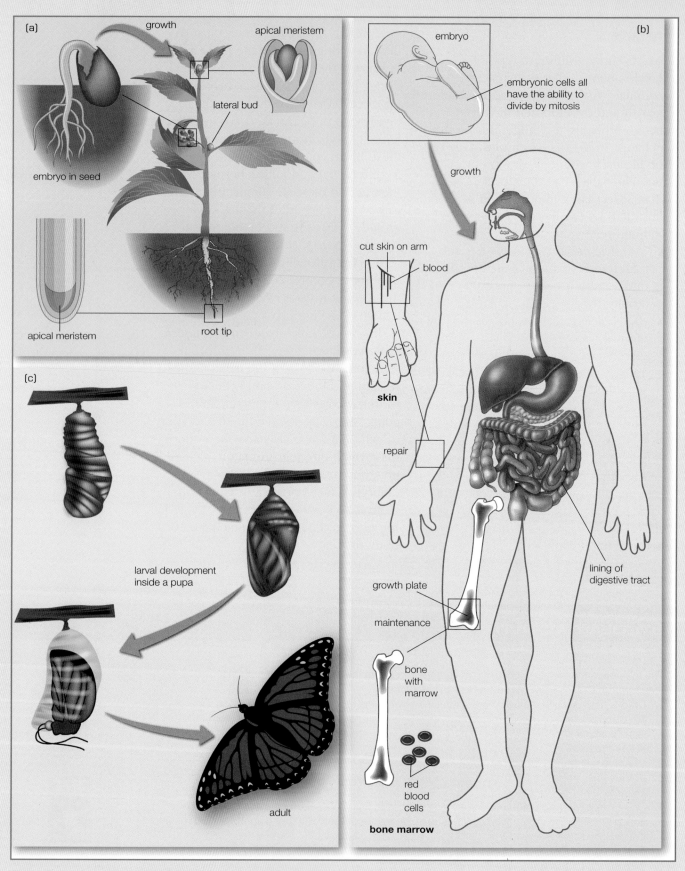

Figure 5.2 Where mitosis occurs in plants, humans and insects: (a) positions of meristems in plants; (b) position of meristems in humans (mammals); (c) complete metamorphosis in insects—transformation brought about by the division of some cells and the death of other cells

Mitosis: the process and its importance

5.2

- *identify mitosis as a process of nuclear division and explain its role*

The role of mitosis

If cell division occurs in **unicellular** organisms, it results in asexual reproduction—one organism becomes two, but no sex cells or gametes are involved. (This will be dealt with in more detail in the last module.) In multicellular organisms, cell division is a process that leads to the formation of new cells that form part of the organism and, as a result, contribute to the growth of the organism or the repair of damaged tissue.

The role of mitosis in multicellular organisms can be summarised as follows:
- *growth*
- *repair* of damaged tissue and replacement of worn out cells
- *genetic stability*: mitosis ensures the precise and equal distribution of chromosomes to each daughter nucleus, so that all resulting cells contain the same number and kind of chromosomes as each other and as the original parent cell
- *asexual reproduction*: e.g. *growing plants from cuttings* (more detail in the last module) and *cloning*—an artificially-induced form of asexual reproduction in multicellular organisms that relies on human intervention and genetic modification (more detail in the HSC course).

The process of mitosis

The growth or repair of tissue begins with the formation of new, identical cells by the process of cell division. Cell division occurs in a cycle and involves two main steps:
1. mitosis—the division of the nucleus
2. **cytokinesis**—the division of the cytoplasm.

The cell cycle

Cell division and enlargement occur in a repetitive sequence called the **cell cycle**, with one complete cell cycle taking about 18 to 22 hours in many species (see Fig. 5.3). Mitosis is only one part of this cell cycle and usually takes about an hour or two. Before a cell divides, it undergoes preparation for division and this preparation phase, called **interphase** takes much longer.

Interphase is subdivided into three stages, G_1, S and G_2:
- G_1 is a gap phase during which cell enlargement takes place before the DNA replicates
- S phase is a synthesis period when DNA replicates—that is, the DNA in the cell makes an identical copy of itself, so that each dividing cell has two copies at the start of mitosis. When the cell divides, one full copy ends up in each resulting daughter cell
- G_2 is a second gap phase after replication, when the cell prepares for division.

Mitosis (division of the nucleus) then occurs, followed by cytokinesis (division of the cytoplasm).

Mitosis is a highly co-ordinated process, ensuring that the replicated chromosomes separate and are equally distributed to the daughter cells. Mitosis is a gradual and continuous process, but is usually described in four phases to make it easier to understand. (It is not necessary for students to know the names of these phases, but learning them may make it easier to remember the process.)

Websites that show cell division and the cell cycle

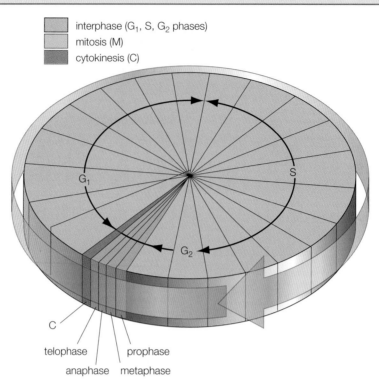

interphase (G$_1$, S, G$_2$ phases)
mitosis (M)
cytokinesis (C)

Figure 5.3 The cell cycle: each wedge represents 1 hour of the 22-hour cycle in human liver cells growing in culture

Mitosis: introduction to terminology

- **Chromosomes** (see page 90) contain linear sequences of genes, the units of heredity that code for the inherited characteristics of an organism; for example, in humans chromosomes code for our eye colour, hair colour and height.
- Each organism has a set number of chromosomes (e.g. humans have 46 chromsomes, a platypus has 52 and a lettuce has 18).
- In a non-dividing cell, the DNA and protein occurs as chromatin material, but once the DNA has replicated the chromatin material separates into short, thick individual rod-shaped structures called chromosomes.
- Each chromosome consists of two identical copies of a DNA sequence.
- Each chromosome splits longitudinally and the duplicated arms of the chromosome are held together by a structure called a **centromere.** These two identical arms are termed **daughter chromatids** (see Fig. 5.4) and their segregation to opposite poles occurs during mitosis. Following this, the cytoplasm divides, separating the two daughter nuclei from each other.

Mitosis: stages

Although mitosis is a continuous process, it is easier to understand it if we analyse the process and name identifiable stages within the process. There are four main stages of nuclear division in mitosis: following on *interphase* are the stages *prophase*, *metaphase*, *anaphase* and *telophase*. The late stages of nuclear division are accompanied by the start of cytokinesis. The process of cell division is summarised in Table 5.1 on page 189.

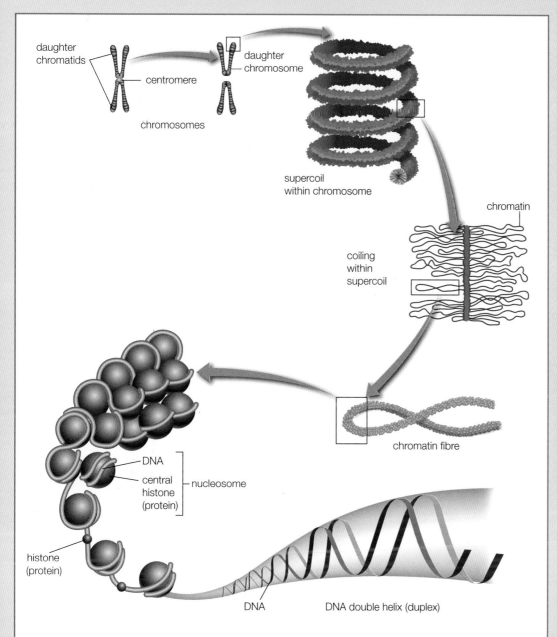

daughter chromatids

centromere

chromosomes

daughter chromosome

supercoil within chromosome

coiling within supercoil

chromatin

chromatin fibre

DNA

central histone (protein)

nucleosome

histone (protein)

DNA

DNA double helix (duplex)

Figure 5.4 Diagram representing the changes to chromatin material in dividing cells

Table 5.1 Mitosis in an animal cell

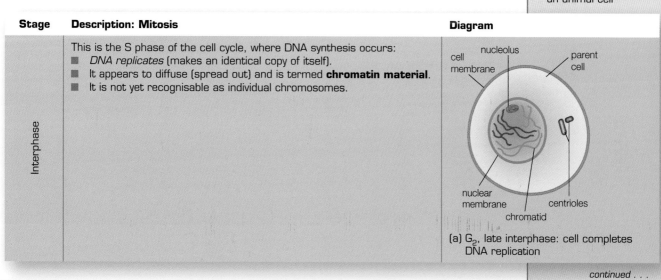

Stage	Description: Mitosis	Diagram
Interphase	This is the S phase of the cell cycle, where DNA synthesis occurs: ■ *DNA replicates* (makes an identical copy of itself). ■ It appears to diffuse (spread out) and is termed **chromatin material**. ■ It is not yet recognisable as individual chromosomes.	cell membrane; nucleolus; parent cell; nuclear membrane; chromatid; centrioles (a) G$_2$, late interphase: cell completes DNA replication

continued . . .

Stage		Description: Mitosis	Diagram
Mitosis	Prophase	■ Chromatin material shortens and thickens by spiralling and separates out into individual chromosomes. ■ Each chromosome contains two copies of the DNA. ■ The *chromosomes split longitudinally* into two arms, the daughter chromatids, held together by a centromere. Each chromatid contains one copy of the DNA. ■ The nuclear membrane begins to break down and is no longer visible; a spindle begins to form from the broken down material and extends across the cells (like the imaginary lines of longitude that are drawn on a globe of the world).	daughter chromatids — chromosome splits longitudinally — nuclear membrane breaks down — centromere — spindle fibres begin to form (b) Prophase: chromosomes condense, become visible and spindle apparatus forms
	Metaphase	■ The *chromosomes line up across the centre* or the 'equator' of the cell, each attached to the spindle fibres by a centromere. ■ The *centromere divides* horizontally.	spindle fibres — centromere — chromatid (c) Metaphase: chromosomes align along the equator of the cell
	Anaphase	■ The spindle fibres contract, and as the daughter chromatids begin to separate, they are now termed **daughter chromsomes**. They are *pulled towards opposite poles* of the cell and their movement is assisted by the centromere.	centromere — spindle fibres contract — daughter chromosome (d) Anaphase: sister chromatids separate to opposite poles of cell
	Telophase	■ The daughter chromatids gather at opposite poles of the cell. ■ The spindle breaks down. ■ The nuclear membrane and nucleolus reappear. ■ Nuclear division or mitosis is now complete. The result is two nuclei that have the identical chromosomes to each other and to the original nucleus in the parent cell.	nucleolus — daughter nuclei — nuclear membrane reforms — nucleolus — chromosome (e) Telophase: nuclear membranes assemble around two nuclei

Stage	Description: Cytokinesis	Diagram
Cytokinesis	■ Division of the cytoplasm, separating the two daughter nuclei so that each is in its own cell. Cytokinesis differs in plant and animal cells: —*animal cells*: the cytoplasm constricts in the centre of the cell, between the two daughter nuclei and it 'pinches off' (similar to squeezing a round balloon in the centre until the edges meet and then giving it a twist to separate it into two bubbles!) —*plant cells*: cytokinesis involves the formation of a cell plate while the nucleus is still in telophase; thickenings appear on the spindle fibres in the region of the equator and they join up to form a cell plate made of pectin compounds. Cellulose is then deposited on either side, forming a cell wall to separate the two daughter nuclei.	 (f) Cytokinesis: division of cytoplasm into two; G_1, early interphase of daughter cells (g) Cytokinesis in plant cells

Note: As we already know, the number of chromosomes varies from one species to another. To keep the diagrams simple, the diagrams show a progressive sequence of events in a cell with two pairs of chromosomes.

Cytokinesis

■ *explain the need for cytokinesis in cell division*

Cytokinesis is the final step in cell division. It is the division of the cytoplasm and begins while the nucleus is completing its division. Cytokinesis is important to separate the newly formed daughter nuclei, *to ensure that each cell has only one nucleus*. The outcome at the end of mitosis and cytokinesis is *two daughter cells* that have the *identical chromosomes* to each other and to the original parent cell. The daughter cells will then enlarge until they are the same size as the original adult cell (assimilation as well as cell enlargement occurs). The nucleus of each cell controls all the cell activities; it is interesting to note that the ratio between the proportion of nucleus and cytoplasm is kept constant. If the cytoplasm exceeds a certain proportion of the cell, the ability of the nucleus to control it decreases and this may be involved in triggering the cell to divide.

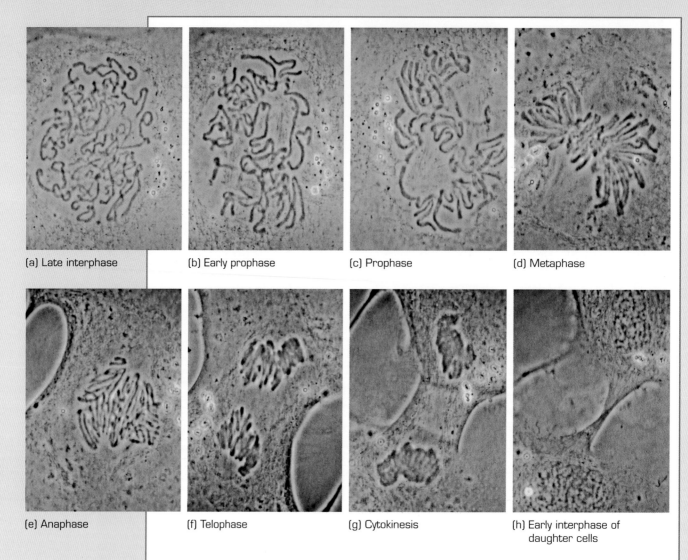

(a) Late interphase

(b) Early prophase

(c) Prophase

(d) Metaphase

(e) Anaphase

(f) Telophase

(g) Cytokinesis

(h) Early interphase of daughter cells

Figure 5.5 Images of a living cultured newt lung cell dividing, showing nuclear division followed by cytokenesis

5.3 DNA outside the nucleus

- *identify that nuclei, mitochondria and chloroplasts contain DNA*

When the cytoplasm divides, the organelles such as mitochondria and chloroplasts are distributed to the daughter cells in approximately equal numbers. It is then necessary for the organelles in the cytoplasm to replicate so that they are not reduced in quantity. Assimilation is important in the growth of many organelles, but mitochondria and chloroplasts contain their own small amounts of DNA and so they are able to *replicate* themselves. By the time the daughter cells have grown to the size of the original cell, they have a similar number of organelles as the original cell had.

Investigating mitosis using a microscope

■ *perform a first-hand investigation using a microscope to gather information from prepared slides to describe the sequence of changes in the nucleus of plant or animal cells undergoing mitosis*

Introduction

Mitosis can be observed relatively easily in cells in the growing region of a root tip. These cells have been specially dyed, using a stain (such as acetic orcein) which is specifically taken up by chromosomes. The prepared slides that you will look at show stained tissues that are no longer living, but with current, more advanced microscopy techniques, living tissue can be recorded today actively dividing. There are a number of videos available to demonstrate this process, some of which can be downloaded from the Internet.

Aim

To gather information on the sequence of changes in the nucleus of plant cells undergoing mitosis.

Background information

In this first-hand investigation, the first step is to identify the part of the plant organ where you would expect to find dividing cells.

If you are looking at root tips, this would be in the meristematic region immediately behind the root cap at the very tip of the root (see Fig. 5.2a).

Because the tissue is no longer living, you will see the cells in the stages of division that they were in when the root tip was killed with a fixative. The division of cells at any one time is not synchronised—different cells will be in various stages of division. The phase that is most commonly represented is probably the longest phase. Look at the cell cycle and predict which phase you would expect this to be.

Procedure

■ Using the correct microscope technique, observe a prepared slide of a root tip under the compound microscope.
■ Locate and identify the apical meristem (just behind the root cap) under low power and then look for cells with darkly-stained chromosomes, indicative of mitosis occurring.

For interactive websites

Figure 5.6
A photomicrograph of root tip cells undergoing mitosis

interphase

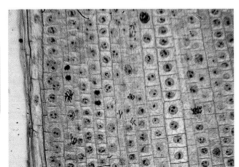

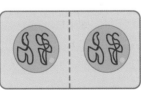
cytokinesis

metaphase

telophase

anaphase

■ Change to higher power and identify cells in at least four different stages of division.
■ Record your results in the form of fully-labelled diagrams. Remember to include a heading for each diagram and state the magnification used. Write a brief description of what is happening at each stage in the diagram.

■ Write out a practical report using a procedure text type, under the headings Aim, Materials, Method, Results and Conclusion.
■ Answer all questions at the end of this chapter.

REVISION QUESTIONS

1. Distinguish between the terms *cell division* and *mitosis.*

2. (a) Describe the main difference in the role of mitotic cell division in unicellular organisms, as opposed to multicellular organisms.
   (b) State two other important roles of mitosis in multicellular organisms.

3. Some meristems in plants result in growth in height (length), but other meristems result in growth in diameter. Give one example of each of these types of meristematic tissues and describe where it occurs in plants.

4. (a) Define *metamorphosis* and explain the role of both cell division and cell death in insects with a complete metamorphosis.
   (b) In the development of frogs from tadpoles, a tadpole loses its tail and grows legs. Apply what you have learnt from insect metamorphosis to explain how this change may come about.

5. Identify where, in mature mammals, dividing cells may be located.

6. Describe what is meant by each of the following terms:
   ■ centromere
   ■ chromosome
   ■ chromatin material.

Answers to
revision questions

7. Compare cytokinesis in plant cells with that in animal cells.

8. By means of a fully-labelled comparative diagram, compare early prophase with metaphase of mitotic division in an animal cell.

9. Explain the difference between a daughter chromatid and a daughter chromosome.

LIFE ON EARTH

Evidence for the origin of life

Analysis of the oldest sedimentary rocks provides evidence for the origin of life

1.1

Origin of organic molecules on early Earth

■ *identify the relationship between the conditions on early Earth and the origin of organic molecules*

The age of the Earth

The approximate age of the Earth, based on geological, magnetic, radiographic and palaeontological studies, is 4.5 billion years. During the **Hadean eon** (approximately 4.5 to 3.8 billion years ago) was the formation of the Earth as it was transformed from a gaseous cloud into a solid body. The heavier molten iron sank down to become the core, while the lighter rock rose to the surface, becoming the crust. As a result of the high temperatures at the centre of the Earth and due to volcanic activity, there was an emission of gases, or **outgassing**, of volatile molecules such as water (H_2O), methane (CH_4), ammonia (NH_3), hydrogen (H_2), nitrogen (N_2), and carbon dioxide (CO_2). The atmosphere was **anoxic** meaning it had no free oxygen (O_2), as all oxygen was bound within compounds such as water and carbon dioxide. This meant that there was no ozone (O_3) layer, exposing the Earth's surface to ultraviolet radiation. Most of the hydrogen gas escaped into space, as happens today, whereas the other gases accumulated and remained in the atmosphere.

Forming the first organic molecules

Early Earth, with an atmosphere of water vapour, hydrogen, methane and ammonia, provided an environment in which the production of **organic** carbon-containing molecules would be fairly easy. The energy for driving these reactions could have come from a number of sources, in particular the sun. Ultraviolet light would easily have reached the Earth's surface because no ozone layer existed. Other possible energy sources could have been lightning, hot springs and volcanoes, radioactivity in the crust, and impact from meteorites. At this stage, organic molecules would have most likely formed in the lower atmosphere or the Earth's surface. The stages of change thought to have occurred in early Earth are listed below:

■ Dense clouds formed in the steamy atmosphere. These clouds were formed of water from meteorites and hydrated minerals. These clouds then formed a reflective shield from the sun's penetrating heat.

■ Eventually, **meteorite** impacts declined (approximately 500 million years later) due to the protective layer of the atmosphere and the friction caused on entry, the Earth cooled and the temperature fell below 1000°C, forming a stable rocky crust.

■ The release of gases from volcanoes in turn increased air pressure in the atmosphere. When the temperatures cooled, this assisted in causing the immense clouds of water vapour to condense into liquid and fall as rain.

■ Rain would have washed organic molecules into lakes and ponds rich in dissolved minerals and created an environment for reactions to occur producing new organic molecules. In turn this would have created an environment for further reactions forming other molecules and compounds and so on (see Table 1.1).

■ Carbon dioxide dissolves readily in water to form carbonic acid (H_2CO_3). It would have been flushed out of the atmosphere by the rain and into the oceans where it reacted with calcium to form calcium carbonate ($CaCO_3$).

■ At first, the rain evaporated as it fell on the hot rock surface but the evaporation gradually cooled the crust until the water could accumulate in the lower regions of the Earth's surface forming oceans. The first rivers were created on the continents where the water, dissolving and transporting minerals along the way, eventually ran back into the oceans.

■ The dissipation of heat into space cooled the Earth, causing crust fragments to become numerous enough to form a first thin, solid cover.

■ Over the next 3.5 billion years, the amount of carbon dioxide in the atmosphere was reduced as it became incorporated in rocks (e.g. limestone). The main gas remaining in the Earth's atmosphere was nitrogen. The composition of Earth's atmosphere today is somewhat different to that proposed on early Earth: 78.1 per cent nitrogen, 20.9 per cent oxygen, and 1 per cent consisting of small traces of different gases such as carbon dioxide, water vapour, methane, hydrogen and ozone.

There is much known about the composition and conditions of early Earth; however, there are a lot of questions remaining unanswered. Scientists continue to search for more evidence reflecting the conditions of early Earth that may have existed when life began. If these conditions are known then we may perhaps discover more about the building blocks from which life began.

Table 1.1

Bioelements	Biomolecules
Water: hydrogen (H), oxygen (O)	Monosaccharides (e.g. glucose), polysaccharides (e.g. starch), glycoproteins and proteoglycans
Organic: hydrogen (H), carbon (C), nitrogen (N), oxygen (O), phosphorus (P), sulfur (S)	Triacyglycerols (e.g. animal fat, seed reserves), phospholipids, glycolipids, polyisoprenoids
Ionic: sodium (Na), potassium (K), magnesium (Mg), calcium (Ca), chlorine (Cl)	Amino acids (20) Proteins, glycoproteins and proteoglycans Nitrogenous bases (4); ribose or deoxyribose phosphate Nucleic acids (RNA and DNA)

1.2

Implications of the existence of organic molecules

- *discuss the implications of the existence of organic molecules in the cosmos for the origin of life on Earth*

There is very little evidence towards the existence of organic molecules in the universe or **cosmos**. The finding of the 100 kg Murchison meteorite in Victoria, Australia in 1969 was significant in providing evidence that organic molecules from outside of Earth are similar to those we have today. Nineteen of the 92 amino acids identified in the meteorite were found on Earth. This suggests that the source of organic molecules needed for the origin of life on Earth may in fact have originated from outside of the Earth. Theories presented by Haldane and Oparin (1923) did not provide much evidence for the existence of organic molecules until the supporting experiments of Urey and Miller (1953). Urey and Miller created the conditions thought to be those of early Earth and these conditions resulted in the change from **inorganic** molecules to organic molecules. Although this theory is commonly supported, there is no evidence to support the mechanism for complex organic compounds changing to early life forms. The question still remains unanswered and quite controversial.

1.3

Evolution of chemicals of life: theories and their significance to the origin of life

- *describe two scientific theories relating to the evolution of the chemicals of life and discuss their significance in understanding the origin of life*

The major theories accounting for the origin of life on Earth are:
- steady state—life has no origin
- spontaneous generation—life arose from non-living matter on numerous occasions
- special creation—life was created by a god or supernatural event at a particular time
- cosmozoan/panspermia—life arrived on Earth from elsewhere
- biochemical evolution or chemosynthetic theory—organic compounds are produced from inorganic molecules, leading to early life forms.

Steady state theory

This theory suggests that the Earth and its species had no origin; they always existed. The Earth has always been able to support life and has changed very little over time.

Spontaneous generation theory

This theory by Aristotle (384–322 BC) suggests that life arose spontaneously, assuming that certain 'particles' of matter contained an 'active principle' which could produce a living organism when conditions were suitable. He was correct in assuming that the active principle was in a fertilised egg, but incorrect to extrapolate this to the belief that sunlight, mud and decaying meat also had the active principle. Louis Pasteur (1861) later demonstrated the theory of biogenesis and finally disproved the theory of spontaneous generation.

Special creation

This theory is upheld by most of the world's major religions and civilisations and attributes the origin of life to a god or supernatural event at a particular time in the past. Since the process of special creation occurred only once and therefore cannot be observed, this is sufficient to put the concept outside the framework of scientific investigation. Science concerns itself only with observable phenomena and as such will never be able to prove or disprove special creation.

Cosmozoan/panspermia theory

This theory suggests that life could have arisen once or several times, at various times and in various parts of the universe. Russian and American space probes have provided evidence that the likelihood of finding life within our solar system is remote but cannot look outside the solar system. Materials found in meteorites and comets have revealed the presence of organic molecules which may have acted as 'seeds' falling onto early Earth. There is as yet no compelling evidence to support or contradict it, particularly due to the challenges of survival and transport in space.

Biochemical evolution theory

This theory suggests that certain conditions of early Earth (see Section 1.1) generated the organic compounds and the right environment for the first production of a living organism.

Evidence leading to the support of the biochemical evolution theory

Oparin (1923)

Aleksandr Oparin suggested that organic compounds could have formed in the early Earth oceans from more simple compounds, the energy for these reactions probably being supplied from the sun's strong ultraviolet radiation before the formation of the ozone layer which now blocks out most of it. Oparin argued that, considering the amount of simple molecules in the oceans, the energy available and the time scale, it was conceivable that the oceans would gradually accumulate organic molecules to produce the 'primeval soup' in which life could have arisen. J. B. S. Haldane independently arrived at the same idea as Oparin in 1929. Oparin's hypothesis was not tested until the 1950s by Urey and Miller.

Oparin's theory has been widely accepted; however, major problems remain in explaining the transition from complex organic molecules to living organisms. This is where the theory of a process of biochemical evolution offers a broad scheme which is acceptable; however, there is no agreement as to the precise mechanism by which it may have occurred.

Oparin considered that protein molecules were crucial to the transformation to living things and through a possible sequence of events it would have produced a primitive self-replicating organism feeding on the organic-rich primeval soup. While this is supported by some scientists, others such as Sir Fred Hoyle argue that the probability of random molecular interactions giving rise to life is 'as ridiculous and improbable as the proposition that a tornado blowing through a junk yard may assemble a Boeing 747'.

Urey and Miller (1953)

In a series of experiments, Stanley Miller (a student of Harold Urey) simulated the proposed conditions of early Earth. He successfully made many substances including amino acids and simple sugars. More recently, Leslie Orgel succeeded in making a simple nucleic acid molecule in a similar experiment. Recent experiments using Miller's equipment but using mixtures of carbon dioxide and water and only traces of other gases have produced similar results to those of Miller.

Urey and Miller's experiments

■ *gather information from secondary sources to describe the experiments of Urey and Miller and use the available evidence to analyse the:*
  —*reason for their experiments*
  —*result of their experiments*
  —*importance of their experiments in illustrating the nature and practice of science*
  —*contribution to hypotheses about the origin of life*

Aim

■ To describe the experiments of Urey and Miller.
■ To analyse the result, reason for, importance and contribution of Urey and Miller's experiments.

Background information

In the early 1950s, Harold Urey and his student Stanley Miller carried out the first experiment simulating hypothetical conditions present on early Earth in order to look at the chemical reactions that may have occurred. Using the equipment set up in Figure 1.1, Urey and Miller placed water, methane, ammonia and hydrogen into sealed glass tubes and flasks connected by a loop. One flask was half-filled with liquid water and the other contained a pair of electrodes. The water was heated to cause its evaporation into steam. Sparks were created between the electrodes in the other flask to simulate lightning storms in a steamy atmosphere. The steam then cooled and condensed back into liquid trickling back into the first flask, simulating an atmosphere that was cooled. This was continuously repeated simulating a cycle in the early atmosphere.

After a week, they observed that as much as 10 to 15 per cent of the carbon was now in the form of simple organic compounds, with two per cent forming amino acids. Of the amino acids formed, 13 of the 22 were those used to make proteins in living cells.

This experiment tested Oparin and Haldane's hypothesis that conditions on early Earth were favourable to chemical reactions producing organic compounds from inorganic precursors. Urey and Miller demonstrated through these experiments that organic compounds such as amino acids, which are essential to cellular life, could be made from inorganic substances under conditions hypothesised as being present on early Earth.

Method

Read the background information provided and gather information from a variety of secondary sources on the experiments of Urey and Miller. Select the relevant information that is needed to address each aim and complete the two parts below.

Description of Urey and Miller's experiments

1. Write a one-page description (including a simplified diagram of the equipment used) of Urey and Miller's experiments.

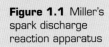

Figure 1.1 Miller's spark discharge reaction apparatus

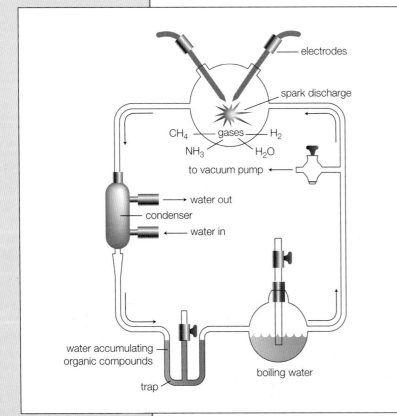

Analysis of Urey and Miller's experiments

2. Write a paragraph for each of the following:
   (a) the reason for their experiments
   (b) the result of their experiments
   (c) the importance of their experiments in illustrating the nature and practice of science
   (d) the contribution of Urey and Miller's experiments to hypotheses about the origin of life.

Results

Once you have answered and addressed the two parts above, now summarise that information. You may choose to do this in table form and use Table 1.2 below to assist in structuring your answer, or you may prefer another method such as using point-form under subheadings.

Discussion/conclusion

1. Why were the experiments of Urey and Miller important?
2. How did they contribute to the idea of the origin of life?
3. What other scientists and their research were affected by Urey and Miller's experiments?

Answers to Table 1.2

Table 1.2
The experiments of Urey and Miller

Description	Reason	Result	Importance in illustrating nature and the practice of science	Contribution to hypotheses about the origin of life

Urey and Miller's experiments

1.4

■ *discuss the significance of the Urey and Miller experiments in the debate on the composition of the primitive atmosphere*

Support for Urey and Miller's experiments

Urey and Miller's experiments provided the first experimental evidence that it is possible for inorganic substances to produce living (organic) substances. This has been called the theory of biochemical evolution (or the chemosynthetic origin of life). The experiments have been replicated successfully to provide a similar outcome each time. Although some scientists have argued that electrical energy might not have efficiently produced organic molecules in the atmosphere of primitive or early Earth, other energy sources such as cosmic radiation, high temperature impact events and even the action of waves on a beach could have been quite effective. Some scientists replicated experiments that have been modified using ultraviolet light instead of electricity to produce nitrogen bases and nucleotides (genetic material) as well as amino acids. In 1961, Juan Oro found that amino acids could be produced from ammonia and hydrogen cyanide solution. His experiments produced a significant amount of the nucleotide base, adenine, which is an organic compound found in **DNA** (deoxyribose nucleic acid). It is also a component of ATP (adenosine triphosphate). The results of the experiments indicate that the building blocks of life could have originated on the primitive Earth, changing from inorganic molecules to organic molecules. Scientists concluded in the 1950s that this was the significant first step in the evolution of life on Earth, and they were optimistic that the origin of life would be solved in a few decades. However, these discoveries have created a stir in the scientific community as the origin of life has not been solved.

Recent debate

There has been recent doubt concerning Urey and Miller's experiments as it is now believed that the atmosphere of early Earth did not contain free hydrogen and was not a '**reducing**' **atmosphere**. There is geological evidence for the existence of an '**oxidising**' (not free hydrogen) **atmosphere** in the precipitation of limestone (calcium carbonate) in great quantities, the oxidation of ferrous iron in early rocks and the distribution of minerals in early sedimentary rocks. There is also evidence to suggest the existence of an oxidising, (not free hydrogen) atmosphere in the composition of volcanic gases and the destruction of molecules by UV radiation. To the contrary, however, there is strong evidence to support an oxygen-free primitive Earth atmosphere in fluvial uranium sand deposits (1999) and banded iron formations documented in 1998 and 2000. In 1994, Holland documented the paleosols (ancient soils) as a source to determine atmospheric composition suggesting very low oxygen levels

2.1 billion years ago. There is also 2001 data from mantle chemistry suggesting oxygen was essentially absent from the earliest atmosphere. In 2005, simulations conducted by the University of Colorado indicated that the early atmosphere of Earth could have contained up to 40 per cent hydrogen, implying a more favourable environment for the production of organic molecules, and supporting Urey and Miller's experiments.

Another objection is that these experiments required a significant amount of energy. It is argued that although lightning storms were common on primitive Earth, they did not occur continuously as portrayed in Urey and Miller's experiment. This means that amino acids and organic compounds may have only formed in smaller amounts.

Other sources for organic compounds

Many of the organic compounds made in the Urey and Miller experiments are now known to exist in outer space. There are other sources of organic 'building blocks of life', such as meteorites, comets, and **hydrothermal vents**. The Murchison meteorite found in Victoria in 1969 was found to contain over 90 amino acids, of which 19 are found on Earth. The primitive Earth is believed to be similar to many of the comets and asteroids found in our galaxy. In 1997, Douglas C. B. Whittet published an article in *The Astrophysical Journal* on the conditions favourable to the formation of organic compounds that exist in interstellar dust clouds. If amino acids are able to survive the extreme conditions of outer space then this might suggest that amino acids were present when the Earth was first formed. More importantly, the Murchison meteorite has demonstrated that the Earth may have received organic compounds and amino acids from outside the planet.

In 2000, some scientists argued that organic compounds could have formed in areas other than the atmosphere, such as hydrothermal vents and **volcanic aquifers**.

Even though we continue to obtain more evidence towards the composition of the atmosphere of primitive Earth, each piece of evidence may support different ideas and theories. Scientists may interpret the evidence in different ways and continue to oppose each other's theories and models. Hence, the controversy and debate continues.

The experiments of Urey and Miller remain significant in the advancement of ideas surrounding the composition of the primitive atmosphere. They supported Oparin and Haldane's proposed theory and led to further experimental testing of variations in conditions favourable for the production of organic compounds.

Extension activity— class debate

Technology has increased our understanding of the origin and evolution of life

1.5

- *identify changes in technology that have assisted in the development of an increased understanding of the origin of life and evolution of living things*

Improved technology over the years has increased our understanding of the origin and evolution of living things. In particular, biochemical and molecular technologies have significantly improved in recent times, therefore having a profound impact on our understanding of the evolution of life.

Early technologies

(See 'Patterns in Nature' for revision.)

Early technologies included:
- glass jars and cotton—used by Francesco Redi for a spontaneous generation experiment with flies and meat, testing the idea that organisms originate directly from non-living matter
- swan-necked flasks designed and used by Louis Pasteur in his experiment for disproving the spontaneous generation theory
- the light microscope (Leeuwenhoek, 1676)—allowed us to see organisms that cannot be seen with the naked eye.

Recent technologies

Recent technologies have included:
- electron microscope development— this led to the understanding of structures at the molecular level, the remains of micro-organisms and the mineral nature of early rocks
- radiometric dating (the principle of superposition, stratigraphic correlation)—developed for dating the relative ages of fossils and surrounding rock material
- seismology—provided knowledge of the structure of the Earth and the characteristics of earthquakes
- geology—determined the composition of meteorites and volcanoes, the fossil record and geological history of the Earth
- geophysics—used the concept of continental drift and sea floor spreading (magnetic surveys) to indicate properties of the Earth's structure and age
- atomic absorption spectrophotometry —used to measure the concentration of metal elements in a rock materials

- relative proportions of stable isotopes —used to determine the absolute age of fossils
- X-ray crystallography—used to determine the structures of an immense variety of molecules and compounds
- gas and liquid chromatography (chemical separation technique)— used to isolate molecules for further study
- radioactive tracing—used to measure the speed of chemical processes
- developments in engineering have enabled both space and deep sea exploration
- amino acid and nucleotide sequencing—comparisons with ancient organic material and biological compounds today
- biochemical analysis (DNA)— comparative studies of different organisms
- genetic engineering—used to increase the understanding of relatedness between organisms and possible evolutionary pathways.

REVISION QUESTIONS

1. List the atmospheric gases believed to have existed on early Earth.

2. Describe the hypothesised environment and conditions on early Earth.

3. Identify the composition of Earth's present-day atmosphere.

4. Describe the contribution of the Murchison meteorite finding to the understanding of the origin of life.

5. Discuss the implications, to the existence of organic molecules in the cosmos, for the origin of life on Earth.

6. Describe two scientific theories relating to the evolution of the chemicals of life and their significance in understanding the origin of life.

7. Describe Urey and Miller's experiments (reason, method and result), including a simple diagram of the apparatus used.

8. Discuss the significance of Urey and Miller's experiments in the debate on the composition of the primitive atmosphere.

Answers to revision questions

9. Identify three examples of different types of technology that have assisted in increasing the understanding of the origin of life and the evolution of living things.

CHAPTER 2

Evolution of life and the fossil record

The fossil record provides information about the subsequent evolution of living things

Evolution of living things

- *identify the major stages in the evolution of living things, including the formation of:*
  —*organic molecules*
  —*membranes*
  —*procaryotic heterotrophic cells*
  —*procaryotic autotrophic cells*
  —*eucaryotic cells*
  —*colonial organisms*
  —*multicellular organisms*

It is believed that the environment on early Earth provided conditions for inorganic molecules to form organic molecules, and then for organic molecules to react with each other to form more complex organic compounds. To metabolise effectively, complex organic compounds needed to separate from their surroundings and form membranes, which would result in the first primitive living cells. To advance, cells over time needed to develop specialised compartments to carry out different chemical reactions. These would have become the first **eucaryotic cells**. Co-operation between these cells would have resulted in **colonial organisms**, which in turn with higher organisation develop into **multicellular** organisms. Of course, these changes would have extended over very long periods of time and involved far more complex changes than those summarised above. The changes that were believed to occur from one organism type to another did not mean that all organisms of that group changed, in fact some members of these groups are known to have continued in the original forms to present-day (e.g. cyanobacteria— **procaryotic autotrophic organisms**).

The major stages in the evolution of living things can be best described and illustrated in Table 2.1.

Table 2.1 The major stages in the evolution of living things

Stage of evolution	Number of million years ago	Oxic or anoxic environment	Outline changes that occurred	Examples of life and evidence
Organic molecules	4500	Anoxic	■ Formation of organic molecules (e.g. amino acids) from inorganic molecules, as suggested by the chemosynthetic theory (Haldane, Oparin, and Urey and Miller) ■ Atmosphere consists of N_2, NH_3, H_2O, CO_2, CO and CH_4; water vapour condenses to form seas; there are high levels of ultraviolet radiation, lightning and volcanic activity	N/A
Membranes	4000–3500	Anoxic	■ Membranes may have enclosed chemicals within a microstructure to control metabolic reactions ■ Proteins or nucleic acids may have been able to replicate	N/A
Procaryotic heterotrophic cells	3500–2500	Anoxic	■ Cells may have obtained their energy from the organic molecules in their environment ■ Cells may not have had membrane bound organelles	■ Evidence of microfossils ■ Bacteria
Procaryotic autotrophic cells	2500–2000	Anoxic and oxic	■ Ozone layer forms ■ Cells may have developed the pathway to make their own food (photosynthesise)	■ Evidence of stromatolites (cyanobacteria)
Eucaryotic cells	2000–1500	Oxic	■ Single cells may have developed from procaryotes to increase metabolic efficiency ■ First eucaryotic cells may have appeared with a nucleus, using cyanobacteria as chloroplasts and bacteria as mitochondria ■ Some of the simple procaryotic cells may have engulfed other cells which became internal structures or organelles and evolved into first eucaryotic cells ■ Eucaryotic cells may have advanced to form membrane bound organelles such as mitochondria	■ Protozoans and algae ■ Evidence in deposited banded iron formations (oxic environment) ■ Evidence to support this comes from mitochondria having their own nucleic acids
Colonial organisms	1500–1000	Oxic	■ Many cells may have worked in a cooperative group	■ *Volvox* ■ Slime moulds (can exist at times in colonies of up to 50000 cells and search for their food supply) ■ Sponges and corals
Multicellular organisms	1000–500	Oxic	■ Cells may have been more organised to become specialised and work together as a multicellular organism (forming into tissues, then from tissues into organs)	■ Evidence of multicellular animals in Ediacaran fauna around 640–680 million years ago ■ Simple organisms such as sponges, jellyfish and coral ■ More complex organisms such as worms, echinoderms (starfish) and algae

Evolution of life timeline

SECONDARY SOURCE
INVESTIGATION

BIOLOGY SKILLS

P12
P13
P14

■ *process and analyse information to construct a timeline of the main events that occurred during the evolution of life on Earth*

Aim

1. To process and analyse information.
2. To construct a timeline of the main events that occurred during the evolution of life on Earth.

Background information

Scientists have gathered information about life in the past through varying types of fossil evidence and placed it into a **geological timescale** (see Table 2.2). The subdivision of the Earth into eons, eras and periods is related to the fossil assemblages found in each time zone. Unique fossil groups are found in each period of time.

Method

Process and analyse information

Use a variety of different secondary sources in your search such as scientific journals, Internet sites, textbooks, library reference books, CD-ROMs, newspaper articles and experts in the field. Using a variety of sources increases the reliability of your information. Using recent sources, such as scientific journals, improves the accuracy and validity of your information. Start your search with the use of keywords. These are used when looking up the index of a textbook or using a search engine to find relevant Internet sites. After you have sourced the information, you next need to analyse and select the information relevant to this task.

Construct a timeline illustrating the evolution of life

Once you have processed and analysed the information in order to obtain the dates for the evolution of different life forms over time, draw a timeline to scale (e.g. 1 million years = 10 cm) that covers all eras and periods of time up to the present day. You will need to use a large A3 or a number of A4 sheets of blank or graph paper joined together so that your scale can fit each organism type. Once you have plotted the timeline, including the period, era and eon names, start adding in the names of each organism type at each time interval. You may wish to draw diagrams of selected organisms to help assist in presenting your evolution of life timeline, illustrating key moments in time.

Results

After you have completed your timeline, swap with another member of the class to ensure that you have not forgotten any important points along the timeline.

Discussion/conclusion

1. Describe the distribution or spread of organisms through time.
2. List five of the most significant changes in organisms over time.
3. Discuss why there may be slight differences in your timeline compared to other members of the class.
4. When gathering your information for this task, how did you ensure that the information you collected was reliable and valid?

Table 2.2 The geological timescale *Note*: Diagram is not to scale.

Eon	Eras of time	Periods of time	Age	Millions of years	Major biological events
Phanerozoic	Cenozoic	Quaternary	Age of mammals	2	Humans expand in range Major ice ages and extinction of large animals in northern hemisphere
		(Tertiary)		12	Extensive radiation of flowering plants and mammals Dominance of gastropods
	Mesozoic	Cretaceous	Age of reptiles	135	First flowering plants Extinction of ammonites, marine and aerial reptiles
		Jurassic		181	Cycads, conifers, ginkgoes, dinosaurs dominant. First birds, flying reptiles, marine reptiles
		Triassic		230	Dominance of mammal-like reptiles Dominance of ammonites
	Palaeozoic	Permian		250	Extinction of trilobites and many invertebrates Reptiles more abundant as amphibians decline
		Carboniferous	Age of amphibians	345	Coal swamp forests Amphibians on land First reptiles Algal-sponge reefs Echinoderms and bryozoans dominant
		Devonian	Age of fishes	405	Oldest land vertebrates Radiation of land plants and fishes Corals, brachiopods and echinoderms
		Silurian		425	Oldest life on land: plants, scorpions First jawed fishes
		Ordovician	Age of invertebrates	560	Diverse marine communities: brachiopods, bryozoans, corals, graptolites, nautiloids First jawless fishes
		Cambrian		600	Evolution of invertebrates with hard skeletons Dominance of trilobites
Proterozoic	Precambrian		Simple organisms diversify in the sea	2500	Eucaryotes evolve
Archaen			Origin of primitive life	3800	Origin of procaryotes
Hadeon				4500	

Evidence suggesting when life on Earth originated

2.2

■ *describe some of the palaeontological and geological evidence that suggests when life originated on Earth*

Palaeontology is the scientific study of **fossils** and all aspects of extinct life. **Geology** is the scientific study of the origin, history and structure of the Earth as recorded in rocks. These two studies are valuable in combination to produce evidence from the past.

Palaeontological evidence

The discovery of two 3400–3500 million year-old Precambrian fossils from Western Australia provided some of the first evidence towards the origin of life on Earth:

■ fossils of single-celled anaerobic procaryotes, or **microfossils**, were found in this discovery. These were very similar to those found living today.

■ the fossilised remains of **stromatolites** found provided valuable information about the structure of early organisms. Simple bacteria existed in structures called stromatolites which were very similar to present-day living stromatolites. In water, colonies of the simple bacteria, or cyanobacteria, trap layers of calcium carbonate and 'grow' upwards in columns towards the sun. Deposits of living stromatolites can be found in Western Australia at Shark Bay (see Fig. 2.1), growing at a rate of about 1 mm per year with individual domes reaching a diameter of about 200 cm and a height of 50 cm.

The 1999 discovery of what could be the remains of **nanobacteria**, or **nanobes**, in a meteorite from Mars (found in Antarctica) indicated similarities to nanobes discovered in Queensland in 1996. Nanobes are filament-type structures found in rocks. The nanobes found on Mars

were able to withstand radiation, cold and acidic conditions, a time when the environment on Mars may have paralleled that existing on Earth for a few hundred million years. Given the presumed sharing of debris generated from meteorite impacts amongst the early planets, the origins of nanobes on Mars and Earth may be the same. Some scientists hypothesise that nanobes are the smallest form of life (they are ten times smaller than our smallest bacteria). However, some researchers believe nanobes to be merely crystal growths. The debate continues.

In general, palaeontologists using the fossil evidence from different rock layers have found that the more primitive cells and marine organisms are found in the lower layers of rock compared to the more complex and land-dwelling organisms. This trend

Figure 2.1 Living modern stromatolites in Western Australia

suggests that simple organisms evolved into more complex organisms and marine organisms preceded land-dwelling organisms. We can make inferences about extinct organisms by studying their closest modern living relatives.

Geological evidence

2500 million year-old Archaean rocks from north Western Australia were examined by scientists in 1999. They found **biomarkers**, or chemical evidence, for the existence of cyanobacteria. Biomarkers are chemicals that are produced by only one group of organisms providing evidence of their existence in the past.

Oxidised rocks such as banded iron and red bedrock formations provide geological evidence towards the origin of photosynthetic life. Oxygen produced by photosynthetic organisms accumulated in the rocks until fully saturated, before building up as a gas in the atmosphere.

Plant and animal fossils

FIRST-HAND AND SECONDARY SOURCE INVESTIGATION

BIOLOGY SKILLS

P13

■ *gather first-hand or secondary information to make observations of a range of plant and animal fossils*

Aim

1. To gather first-hand or secondary information.
2. To make observations of a range of plant and animal fossils.

Background information

Fossils are any remains, traces or imprints of an organism preserved over a long period of time. It is an extremely rare occurrence as it requires very specific conditions at the time of death of the organism.

There are four main conditions under which fossils may form:

1. *quick burial*—rapid covering of dead organism or evidence such as footprints (see Fig. 2.2) or coprolites (faecal remains)
2. *prevention of decay*—needs conditions such as lack of oxygen, high acidity, very low temperatures and low moisture to prevent decay by bacteria and fungi
3. *organism lies undisturbed*—completely covered by sediments to prevent scavenger organisms from breaking up and scattering body parts
4. *presence of hard body parts*—for fossilisation to occur organisms need hard parts such as bone, exoskeleton, teeth or shells.

Occasionally the conditions for fossil formation are perfect and impressions for soft-bodied organisms can be preserved, for example, the jellyfish and worms found at Ediacara Hills in South Australia (see Fig. 2.3).

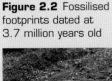

Figure 2.2 Fossilised footprints dated at 3.7 million years old

Figure 2.3 *Dickinsonia costata*—almost a metre long, resembling an annelid worm

Method—gather first-hand or secondary information

If you do not have fossil samples to observe, gather secondary information on a variety of plant and animal fossils. In your search, collect photographs or diagrams of at least four named plant and four named animal fossil specimens. List as many observations as you can for each specimen. Some fossil examples are provided in Figure 2.4. Observations can be listed (as follows for Fig. 2.4c) as comments such as:

- fern-like appearance
- only a sample of one fern frond or branch (not entire plant)
- similar to present-day fern appearance
- leaflets are opposite each other and contain vein-like structures.

Other fossil suggestions to research:
- Wollemi pine
- *Gangamopteris* and *Glossopteris*
- *Archaeopteryx*
- coelacanth
- ammonite.

Results

List your observations next to each fossil photograph or diagram (or description of the real fossil sample you observed).

Discussion/conclusion

1. Identify the four main conditions under which fossils may form.
2. List three examples of the types of things that can be classified as fossils.
3. Describe the types of structures in your observed plants and animals that were able to be preserved.
4. List the plant and animal fossils you observed that are still around today?
5. Of those fossils that are not plants or animals from the present day, try to estimate their age using your previously constructed *evolution of life* timeline (see page 207). Name the plant or animal fossils and your estimate of their age.

Plant and animal fossils—extension

Figure 2.4 Fossils: (a) trilobite—extinct organism found only in rocks dated around 500 million years ago; (b) fossil bones of baby diprotodont (large, herbivorous marsupial) found at Riversleigh, Queensland (one of the richest fossil deposits of the world); (c) fossil impression of an extinct seed fern; (d) *Baragwanathia longifolia*—plant fossil found in Victoria, preserved with planktonic marine organisms (graptolites)

(a)

(b)

(c)

(d)

2.3

Change from an anoxic to an oxic atmosphere

■ *explain why the change from an anoxic to an oxic atmosphere was significant in the evolution of living things*

An **anoxic** atmosphere is one defined as being deficient, or lacking, in oxygen. An **oxic** atmosphere is one where oxygen is available. A change from an anoxic atmosphere free of oxygen to an oxic atmosphere with plenty of available oxygen had a significant influence on the conditions of early Earth and hence the evolution of living things.

This major change to the atmosphere and increase in oxygen inhibited the growth of **anaerobic** organisms and caused them to decline, while photosynthetic organisms became more abundant. Today, anaerobic organisms only survive in environments with a very low oxygen concentration, such as in swamps, bogs, deep underground or in deep ocean hydrothermal vents.

The next significant change as a result of the increase in oxygen in the atmosphere was that aerobic organisms became more efficient in energy production (respiration), providing a large energy source for increased activity and eventually this led to an increased complexity and size of these organisms.

As the oxygen began to accumulate in the atmosphere, it reacted with the sun's ultraviolet radiation to produce ozone (O_3). As the amount of oxygen increased, so did the amount of ozone produced, until an ozone layer formed. This ozone layer reduced the amount of ultraviolet radiation reaching the Earth's surface. This had a significant influence on future organisms as it protected them from dangerous ultraviolet radiation. Even more so, organisms were able to succeed in inhabiting the land.

Impact of increased understanding of the fossil record

SECONDARY SOURCE INVESTIGATION

BIOLOGY SKILLS

P13
P14

■ *identify data sources, gather, process, analyse and present information from secondary sources to evaluate the impact of increased understanding of the fossil record on the development of ideas about the history of life on Earth*

Aim

1. Identify data sources, gather, process, analyse and present information from secondary sources.
2. Evaluate the impact of increased understanding of the fossil record on the development of ideas about the history of life on Earth.

Method

Identify data sources, gather and process information from secondary sources relating to the increased understanding of the fossil record and the development of ideas about the history of life on Earth. Once you have refined and summarised your gathered information, analyse the information so that you may evaluate the impact of this increased understanding. Use Table 2.3 to help create your answer.

(Remember to revise your definition of the verbs in this dot point, in particular: analyse—*identify components and the relationship among them; draw out and relate implications*, and evaluate—*make a judgement based on criteria; determine the value of*. Other verbs are defined at the beginning of this textbook on page viii.)

Increased understanding of fossil record	Development of ideas about the history of life on Earth	Impact of increased understanding of fossil record on development of ideas about the history of life on Earth	Evaluation of the impact

Table 2.3
Evaluating the impact of increased understanding of the fossil record on the development of ideas about the history of life on Earth

Results:

Once you have completed your evaluation, present it in one of the following formats:
- a letter
- a scientific journal article
- a flowchart or mind map
- a PowerPoint or overhead presentation to the class.

Discussion/conclusion

Make a conclusive brief statement of your final evaluation of the impact of increased understanding of the fossil record on the development of ideas about the history of life on Earth.

A blank copy of Table 2.3

The effect of scientific developments on ideas about the origins of life

2.4

- *discuss the ways in which developments in scientific knowledge may conflict with the ideas about the origins of life developed by different cultures*

Developments in scientific knowledge about the origins of life are constantly occurring as discoveries are made and new technologies provide more advanced approaches to unanswered questions. These changes may conflict with the ideas about the origins of life held by different cultures. Some of these different ideas for the origin of life have developed over thousands of years in certain cultures.

The difference between science and religion has been the basis of conflict for a long time between the knowledge of scientists and the beliefs of different cultures. It appears that most of these conflicts have arisen as a result of overlooking the distinguishable differences between science and religion:
- *science* can be defined as the pursuit of knowledge through observations that produce testable hypotheses and models
- *religion* can be defined as involving a god or superior being and is based on a set of beliefs which do not need to be tested. They do not produce testable hypotheses and models.

Extension activity—scientific developments

This means that logically there can be no intellectual conflict. For many people, cultural beliefs about life's origin form part of their religion, where the belief is that God is the creator of the universe. For example, in Australia, Aboriginal beliefs involve an ancestral being that created the Earth a long time ago during a time called alcheringa or Dreamtime. These beliefs have a long tradition in the culture they belong to.

Charles Darwin received much opposition surrounding his 19th century theory of evolution (*On the Origin of Species*, 1859). Heated debate occurred between biologists, religious leaders and people from his culture, particularly in response to his allegations that God was not creator of the universe. Science should not make statements about God and religious beliefs.

Scientific theories and belief systems can co-exist without conflict. Scientific theory is based on rational analysis and is subject to change, particularly with recent technological developments. Belief systems are based on unchanging views. Many scientists with religious beliefs investigate concepts surrounding the origin and evolution of life using scientific experimentation.

Classification systems and modern genetic technologies create issues which also may be potential sources of conflict.

When attempting to resolve or avoid such conflicts, we must consider the beliefs of the individuals, and carefully balance culture, religion and science. This balance may be achieved by recognising the distinctions between each group and understanding what each one means, aiming to be fair to all groups involved.

REVISION QUESTIONS

1. Construct a flowchart summarising the major stages in the evolution of living things (including organic molecules, membranes, procaryotic heterotrophic cells, procaryotic autotrophic cells, eucaryotic cells, colonial cells and multicellular organisms), labelling the time and whether the environment was oxic or anoxic when each stage occurred.

2. Identify three major periods in the geological time scale and *state* the organisms that existed and dominated in each period.

3. Distinguish between the studies of palaeontology and geology.

4. Briefly describe three pieces of palaeontological and geological evidence suggesting when life on Earth originated.

5. Describe the four main conditions under which fossils may form.

6. Identify an example of fossil evidence showing the existence of soft-bodied organisms.

7. *Name* one plant and one animal fossil that you have observed. Give a brief description of each.

8. Distinguish between an oxic and an anoxic atmosphere.

9. Explain why the change from an anoxic to an oxic atmosphere was significant in the evolution of living things.

Answers to revision questions

10. Discuss the ways in which developments in scientific knowledge may conflict with ideas about the origins of life developed by different cultures.

CHAPTER 3

Understanding of present-day organisms and their environments

Further developments in our knowledge of present-day organisms and the discovery of new organisms allows for better understanding of the origins of life and the processes involved in the evolution of living things

Technology and increased knowledge of procaryotic organisms

3.1

- *describe technological advances that have increased knowledge of procaryotic organisms*

Structural methods of classifying **procaryotic organisms** in the past have been very valuable (see Fig. 3.1); however, these methods did not always reflect the organisms' possible evolution. New technological advances have changed the way procaryotic organisms are now classified, and increased our understanding and knowledge of biological structures, chemical composition and biochemical (genetic) characteristics of procaryotic organisms, as summarised below in Table 3.1.

(a)

(b)

Figure 3.1 Electron micrographs of procaryotic cells: (a) *Streptococcus thermophilus*; (b) *Escherichia coli*

Table 3.1 Technological advances that have increased knowledge of procaryotic organisms

Technological advances	Increased knowledge of procaryotic organisms
Light microscope	■ Ability to identify cells as being unicellular and small (up to 1 micrometre) with a cell membrane and cell wall
Electron microscope	■ Ability to see fine details such as the lack of a nuclear membrane and membrane-bound organelles, the presence of a single strand of DNA and small ribosomes, and that cell division is not by mitosis
Chemical analysis	■ Ability to determine the chemical composition of cytoplasm and membranes ■ Enzymes and photosynthetic pigments are attached to cell membrane ■ Respiratory coenzymes are unique ■ Metabolism of some carbon compounds is different in Archaea than other organisms
Genetic sequencing	■ Determine number of chromosomes
Amino acid sequence	■ Amino acid sequence in proteins and DNA/RNA nucleotides vary ■ Nucleotide sequences of the Archaea RNA are different to bacteria and eucaryotic organisms.

3.2

Environment and the role of organisms in Archaea and Bacteria groups

■ *describe the main features of the environment of an organism from one of the following groups and identify its role in that environment:*
—*Archaea*
—*Bacteria*

Bacteria have evolved along two main evolutionary lines, classified as super kingdoms—Bacteria and Archaea.

Bacteria

Bacteria are an enormously diverse group that share many environments with, or live on and in humans and other animals and plants. These are habitats of moderate temperature with water freely available, low in salt or other solutes, and where sunlight or organic compounds are plentiful. Oxygen is not so important since many of the Bacteria have powerful fermentation capabilities, producing ATP (adenosine triphosphate) under anaerobic conditions. In fact, the group Bacteria contains almost every variety and combination of biochemical energy extraction and carbon-fixation that is thought to be feasible on the basis of the molecular composition of the biosphere. However, this group also contains some of the most specialised and sensitive cellular organisms known.

These are often pathogens that have become highly adapted to particular environments within animal or plant hosts.

Cyanobacteria

Cyanobacteria resemble algae and plants in that they contain chlorophyll and generate oxygen during photosynthesis. They occur as individual cells or as filamentous aggregates of many individual cells joined end to end (see Fig. 3.2). Cyanobacteria contains pigments called phycobilins which give them a blue-green appearance (see Fig. 3.3). Cyanobacteria often form dense mats of growth in shallow marine or estuarine environments.

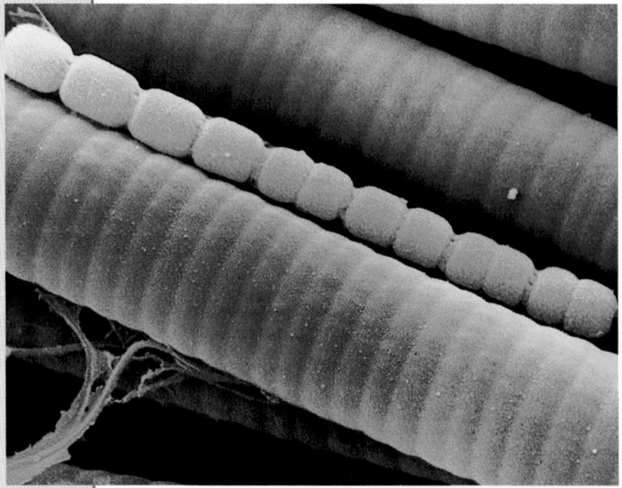

Figure 3.3 Cyanobacterial bloom in Lake Burley Griffin, Canberra

Figure 3.2 Two types of cyanobacteria common in lakes and rivers—*Microcoleus* (large cylindrical form) and *Anabaena* (smaller bead-like filament of cells)

Endospore-forming bacteria

These are bacteria that produce endospores, the most resistant form of life known (see Fig. 3.4). Endospores form inside the mother cell rather than by budding from it. They form under conditions of nutrient depletion or as a result of other environmental signals forewarning difficult times ahead for the bacterium. Endospores appear to have evolved as a resistant, virtually metabolically inactive cell type, increasing the probability of survival under conditions of excessive cold, heat or desiccation. Bacterial endospores are remarkably resistant to high temperature, high radiation and many chemicals (e.g. disinfectants and detergents) which would rapidly destroy all other living cells.

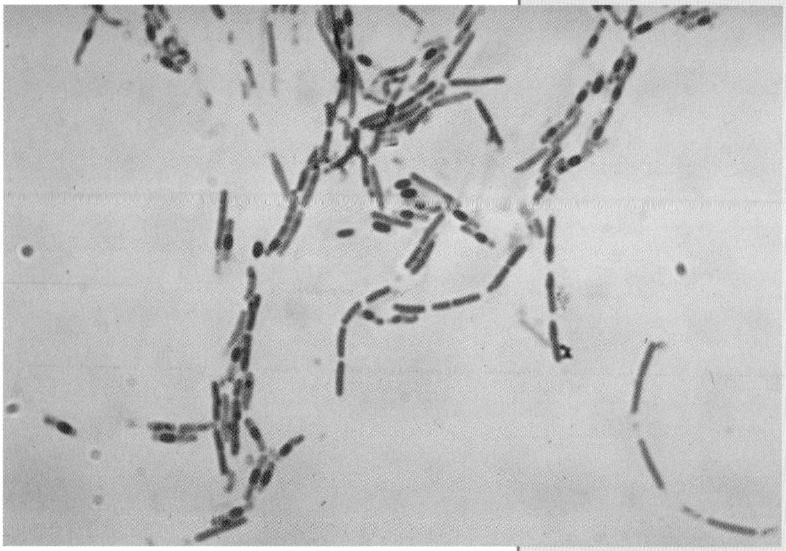

Figure 3.4 Spores of the *Bacillus* species stained red

Archaea

Archaea are still commonly referred to as archaebacteria in scientific literature; however, this term has now been discarded as they are not bacteria. Archaea is a more specialised group than Bacteria and its members are therefore more restricted to the environments they inhabit. Archaea are single-celled, microscopic organisms which do not contain any membrane-bound organelles. They do not require sunlight for photosynthesis, nor do they require oxygen. Most live in extreme environments and are called **extremophiles**; however, some still live in ordinary temperatures, salinities and acid levels. Types of extremophiles are described in Table 3.2.

Figure 3.5 Thermophiles grow in thermal reserves near Rotorua in New Zealand

Table 3.2 Environments of different types of extremophiles

Type of Archaea extremophile	Environment prefers to live in
Thermophiles	Water temperatures greater than 50°C (up to 110°C) or in volcanoes because they prefer the high temperature environments (see Fig. 3.5)
Halophiles	High saline (hypersaline) environments greater than 9% salt concentration (up to 32% salt concentration)
Acidophiles	Acidic environments lower than pH 2 (as low as pH 0.9)
Thermoacidophiles	Hot, acidic environments

Table 3.3
Environmental features and roles for organisms from the Archaea and Bacteria super kingdoms

Methanogens

Methanogens produce methane as a metabolic end product. They do not necessarily occupy extreme environments and are strictly anaerobic.

Examples of an organism from each of the two super kingdoms above are described in Table 3.3 below, focusing on the main features of their environment and their role within that specific environment.

Super kingdom	Organism	Main features of environment	Role in that environment
Archaea	*Methanobrevibacter* (methanogen)	■ Anaerobic environments such as deep soils or bogs, the digestive system of herbivores, and in sediments of marine and freshwater ecosystems	■ Converts carbon dioxide and hydrogen produced from the fermentation into methane ■ Releases methane into the atmosphere contributing to the carbon cycle ■ Plays an essential role in the final steps in the decomposition of organic matter and recycling of carbon in anaerobic environments ■ Those in the digestive system of cattle assist the breakdown of cellulose and aid in digestion
Bacteria	*Oscillatoria brevis* (cyanobacterium)	■ Warm temperature ■ Water freely available (e.g. ponds, streams and soil) ■ Low in salt or other solutes ■ Sunlight or organic compounds are plentiful ■ Oxygen is not so important.	■ Evolve oxygen into the atmosphere ■ Primary producers in the food chain (photosynthetic) ■ Powerful fermentation capabilities, producing ATP under anaerobic conditions ■ Some replace oxygen with nitrate and sulfate ■ Some are also capable of substituting, as a source of energy, reducing inorganic compounds in place of organic carbon.

Similarities in environments past and present

SECONDARY SOURCE INVESTIGATION

BIOLOGY SKILLS

P13
P14

Assessment task

■ *use the available evidence to outline similarities in the environments past and present for a group of organisms within one of the following:*
—Archaea
—Bacteria

Aim

To outline similarities in past and present environments for a group of organisms within Archaea or Bacteria.

Method

Read the information provided in Section 3.2 describing the main features of the environment of organisms from the super kingdoms Archaea and Bacteria. Select your category of Archaea or Bacteria and choose a group of organisms within this category. You may choose the same as presented in Section 3.2. If not, you may select one of your own. A Google search on the Internet of examples of Archaea and Bacteria groups will assist in making your decision. Expand your search of secondary sources by looking at available evidence regarding the *environment* of your chosen group of organisms. Search in particular for information outlining features of both the *past* and *present* *environments*.

Results

Address the following tasks:
1. Select the group: Archaea or Bacteria.
2. Name of the group of organisms.
3. Indicate the main features that are similar between past and present environments for your chosen group of organisms by completing Table 3.4.

Discussion/conclusion

1. List the main features similar in both past and present environments for your chosen group of organisms.
2. Describe the possible characteristics of the group of organisms that make them suited to these same features of their environment.

Table 3.4 Similarities in past and present environments

Kingdom	Group of organisms	Main features of past environment	Main features of present environment	Similarities between past and present environments

Alternative environments in which life may have originated

SECONDARY SOURCE INVESTIGATION

BIOLOGY SKILLS

P14

■ *analyse information from secondary sources to discuss the diverse environments that living things occupy today and use available evidence to describe possible alternative environments in which life may have originated*

Aim

1. To analyse information from secondary sources.
2. To discuss the diverse environments that living things occupy today.
3. To use available evidence to describe possible alternative environments in which life may have originated.

Background information

So far we have looked at the extreme and diverse environments that Archaea and Bacteria live in today and also looked at the similarities in their past and present environments. This may provide support in suggesting possible alternative environments in which life may have originated. The large diversity in present-day environments provides varying habitats for a significant proportion of the Earth's advanced multicellular organisms. However, present-day environments are not as extreme as those proposed to exist on early Earth in which life may have originated, and so would not promote the beginnings of early life forms. This suggests a possible reason why we do not see any new primitive life spontaneously forming on Earth today. The diversity of environments may be only available to complex organisms that are suited to them, as opposed to simple organisms that may only really have a small variety of environments to exist in.

However, extreme environments do still remain on Earth to the present day. Archaea still exist in these extreme environments (e.g. hydrothermal vents or volcanoes) and Bacteria in diverse environments (e.g. environments with extreme temperatures or estuarine environments). With this in mind we can describe possible alternative environments in which life may have originated. Some of these extreme environments are thought to have existed on early Earth and so may be alternative places where life may have originated and primitive life forms began.

Method

Analyse information from secondary sources

Revise information you have already covered in this chapter and read the above 'Background information' before sourcing a variety of other secondary sources. Gather and analyse information to address Parts 1 and 2.

Part 1 Discuss the diverse environments that living things occupy today

Use Table 3.5 as a guide in attempting to *discuss* or *give points for and/or against* the diversity of environments that present-day organisms live in. Once you have analysed information gathered for the task, you must decide the appropriate method of approaching the verb discuss, that is whether you provide *both points for and against*, or you just provide *one or the other*. Modify your final table accordingly.

Part 2 Describe possible alternative environments in which life may have originated

Name and give a brief description (provide any specific conditions) of three alternative environments in which life may have originated. Start by listing the possible alternative environments for the origin of life using your understanding of the conditions of early Earth and your knowledge of environments that simple organisms can live and survive in. Once you have your list, expand on this by describing each environment. Use Table 3.6 as a guide.

Results

Ensure that your findings are presented in a clear and succinct manner for each of the two parts. You may choose to summarise and simplify your tables by using point form, making it easier to remember throughout the course.

Discussion/conclusion

1. Discuss the difference in diversity of available environments for organisms between the present day and early, primitive Earth.
2. Describe the characteristics of two possible alternative environments in which life may have originated.

Table 3.5 Diverse environments that living things occupy today

Diverse environments that living things occupy today	
For	**Against**

Table 3.6 Possible alternative environments in which life may have originated

Alternative environment in which life may have originated	Description of alternative environment
Hydrothermal vent	Water temperatures sometimes above 350°C, high pressure, no sunlight, high amounts of dissolved minerals (toxic to animals on land) such as hydrogen sulfide

REVISION QUESTIONS

1. Describe three examples of different types of technological advances that have increased our understanding of procaryotic organisms.

2. Describe the main features of the environment of a named organism from either the Archaea or Bacteria group and identify the role of your named organism in that environment.

3. List three types of extremophiles from the Archaea group and identify the environment they prefer to live in.

4. Discuss the diverse environments that living things occupy today and the possibility of alternative environments in which life may have originated.

Answers to
revision questions

CHAPTER 4

Classification of past and present life on Earth

The study of present-day organisms increases our understanding of past organisms and environments

4.1

The need to classify organisms

■ *explain the need for scientists to classify organisms*

Biological classifications

The discovery of how different organisms are related is summarised in biological classifications. The classification of life is a hierarchy of names (taxonomic ranks) that includes all known organisms on Earth.

Scientific classification is an efficient, accurate way of communicating about organisms between scientists. When based on relationships, this kind of classification is predictive and useful for all areas of biology. Current knowledge of how organisms are related through evolutionary descent is summarised in biological classification.

The word 'taxonomy' is used to refer to the methods and principles of classification and rules that govern the naming of all organisms.

Scientists need to classify organisms to assist in the following:

■ *simplification*—a classification system makes it easier to simplify the description of groups of organisms and bring a sense of order to the vast range of groups on Earth

■ *communication*—by using a name for a group of organisms, such as *Eucalyptus*, scientists can communicate with one another about the group without having to list all the properties known to occur in every member species

■ *predictions*—classification serves also to predict information that we do not yet have. For example, if a set of organisms share characteristics A, B, C and D that no other organisms have, and we find another organism of which all we know is it has characteristics A and B, we can predict that it will also have C and D. Predicting information is useful. For example, the Morton Bay chestnut (see Fig. 4.1) contains a chemical that controls cancer cells and inhibits the AIDS virus. In search for other sources of this medically important chemical, researchers turn to taxonomy to reveal related plants, finding the *Alexa* from South America.

■ *relationships*—current knowledge of how organisms are related through evolutionary descent is summarised in biological classification. It assists in interpreting relationships between groups of organisms, which has become a rapidly changing and evolving area due to advances in biochemical technologies

■ *conservation*—classification provides us with information about the relationship of groups of organisms with their environment. The classification of groups of organisms

(a)

provides valuable information about the functioning of the organism and their interactions with their environment, and those that are rare or endangered may be protected from further decline using this vital information.

(b)

Figure 4.1 Morton Bay chestnut (*Castanospermum australe*)—native to coastal rainforests and beaches in Australia from around Lismore, NSW to Cape York Peninsula, Queensland: (a) tree can reach heights of up to 30m; (b) seeds inside seed pod may contain valuable chemicals

Construction and use of dichotomous keys

■ *perform a first-hand investigation and gather information to construct and use simple dichotomous keys and show how they can be used to identify a range of plants and animals using live and preserved specimens, photographs or diagrams of plants and animals*

FIRST-HAND INVESTIGATION

BIOLOGY SKILLS

P13
P14
P15

Aim

1. To gather information.
2. To construct simple dichotomous keys.
3. To show how dichotomous keys can be used to identify a range of plants and animals.

Background information

A dichotomous key is used to make the identification of organisms simple, quick and easy. This type of key uses two alternatives

at each level of a flowchart or diagram in order to classify organisms. Each key level requires the observation of structural or visual characteristics of the unknown organism and a decision of one description out of two alternatives as you move through the key. Dichotomous keys may be represented in two ways: one is represented as a basic flowchart; the other is represented by numbers and letters. They both work exactly the same; however, one is more simplified than the other (see Fig. 4.2).

Extension activity

Figure 4.2 Flowchart design for easy construction of a dichotomous key

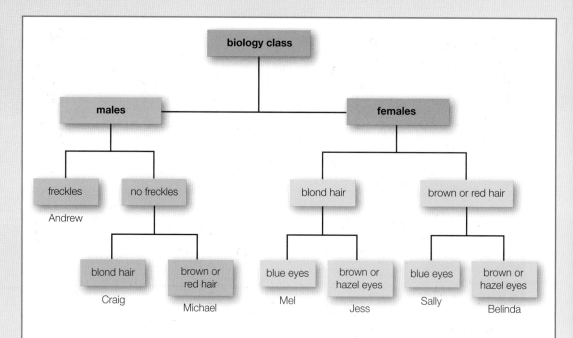

Figure 4.3 Key to plant groups

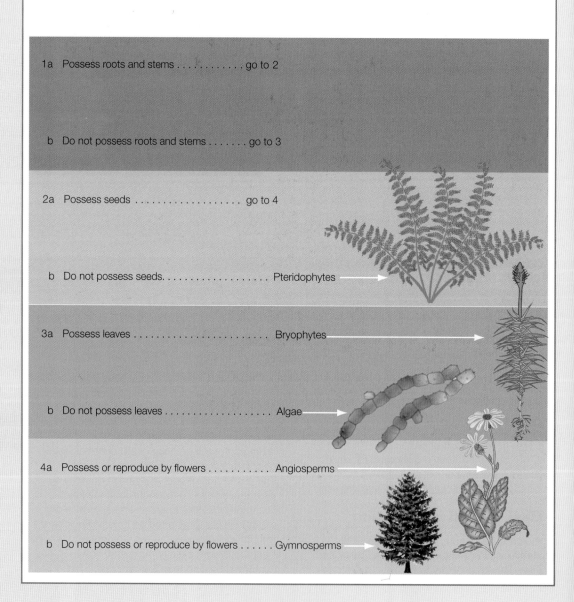

1a Possess roots and stems . . . . go to 2

b Do not possess roots and stems . . . . go to 3

2a Possess seeds . . . . go to 4

b Do not possess seeds. . . . . Pteridophytes

3a Possess leaves . . . . Bryophytes

b Do not possess leaves . . . . Algae

4a Possess or reproduce by flowers . . . . Angiosperms

b Do not possess or reproduce by flowers . . . . . . Gymnosperms

1a   More than three pairs of legs .... **not an insect**

1b   Three pair of legs only .... go to **2**

2a   With wings .... go to **6**

2b   Without wings .... go to **3**

3a   Ant-like with a narrow waist; ranges in size from
     1 to 25 mm (ants) .... **Hymenoptera**

3b   Not as above .... go to **4**

4a   Ant-like with a wide waist; ranges in size from
     5 to 20 mm .... **Isoptera**

4b   Not as above .... go to **5**

5a   With two antennae-like appendages located at the
     end of the abdomen which is used as a spring;
     may be held under the abdomen; ranges in size
     from 3 to 6 mm (springtails) .... **Collembola**

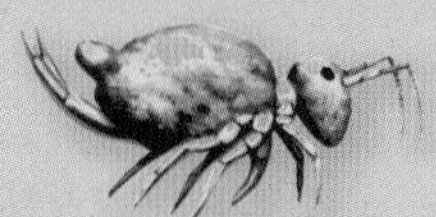

5b   Has three pairs of legs, but no wings, and does not
     fit any of the above descriptions .... **Immature insect**

6a   With only a single pair of wings; the second
     pair of wings modified into a pair of knob-like
     gyroscopic organs (attached to the end of stalks)
     known as halteres; ranges in size from 6 to 65 mm (flies)   **Diptera**

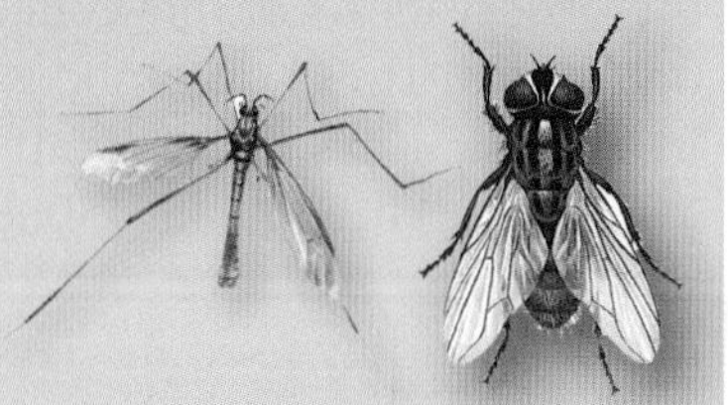

6b   With two pairs of wings .... go to **7**

7a   The wings are equal or nearly equal in size with a
     long slender abdomen; dragonflies range in size
     from 28 to 150 mm, damselflies range in size
     from 25 to 65 mm (dragonflies and damselflies) .... **Odonata**

7b   Not as above .... go to **8**

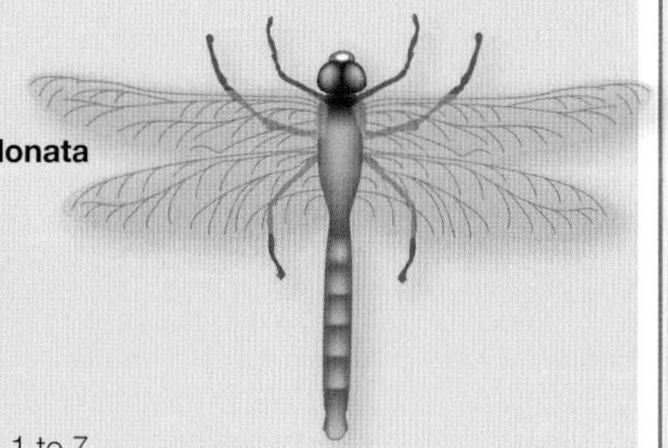

Note: Key only includes selected insect orders and numbering from 1 to 7.

Figure 4.4 Key to selected insect orders (only includes a small portion of all insect orders using numbering from 1 to 7)

Method

Gathering information

Read the 'Background Information' provided and collect information on a variety of different plant and animal dichotomous keys. You may be provided with keys by your teacher or you may need to source the keys from secondary sources. Make sure that you use valid sources for your keys, such as universities or government environmental agencies.

Part 1 Constructing dichotomous keys

Use the students in your biology class today to construct a dichotomous key for identifying the names of each person in the class. You may start off which a large dichotomous division such as male or female, or freckles or no freckles. Depending on the visual characteristics of the class members, continue on from there dividing the group into two until each individual has been identified. Figure 4.2 illustrates an easy way of constructing a dichotomous key. Other more advanced versions may be set up similar to those in Figures 4.3 and 4.4.

Part 2 Using dichotomous keys to identify plants and animals

Select a number of live and preserved plant and insect specimens (at least four of each). Using the keys provided in Figures 4.3 and 4.4 (or those provided by your teacher), determine the scientific name used to classify each specimen into its correct group.

Results

List the pathway selected through your key in order to determine the name of each specimen. Write the name of the group you classified each organism into. For example, 1b, 2b, 3a, Hymenoptera (ant).

Discussion/conclusion

1. Describe what is meant by a dichotomous key and what it is used for.
2. List the groups of plants and animals that you identified using keys.
3. Discuss any possible disadvantages you found in using the keys for identifying specimens.

4.2 Classification systems

■ *describe the selection criteria used in different classification systems and discuss the advantages and disadvantages of each system*

Classification systems are based on the observations of differences between organisms. These differences may occur in an organism's anatomical structure, physiology (functioning), behaviour or biochemistry. Anatomical structure is most commonly used in classification systems; however, more recently biochemistry has been favoured by scientists. As a result, different systems of classification have been used over time; however, with advances in technology, systems have become more refined and use different selection criteria on which to base groups of organisms. There are advantages and disadvantages to all systems of classification. Examples of both modern and older classification systems are compared in Table 4.1. These examples are represented diagrammatically in Figure 4.5.

Table 4.1 Comparison of advantages and disadvantages in three selected classification systems

Classification system	Kingdom names	Selection criteria	Advantages	Disadvantages
6 kingdoms (most recent system: Woese 1990)	■ Bacteria ■ Archaea ■ Protista ■ Plantae ■ Fungi ■ Animalia	■ Molecular criteria ■ Order of bases in particular genes (e.g. ribosomal RNA genes and gene coding for specific proteins)	■ Provides genetic similarities between organisms ■ Biochemical information provides more information leading to possible evolutionary relationships between organisms ■ Can infer when past evolutionary diversion from a common ancestor occurred	■ Requires costly, time-consuming procedures ■ Requires the use of experts in molecular techniques ■ The use of both biochemical and anatomical information may lead to differing interpretations ■ Upsets the traditional method of classification

Classification system	Kingdom names	Selection criteria	Advantages	Disadvantages
5 kingdoms (Whittaker 1969)	■ Monera ■ Animalia ■ Plantae ■ Fungi ■ Protista	■ Structural features ■ Procaryotic or eucaryotic	■ Easily observed in an organism ■ More constant in an organism's lifetime (no seasonal change or change with maturity) ■ Can infer reproductive methods	■ Structures may vary between males and females of a species ■ Internal biochemistry (genetic similarities) is not available for inferring evolutionary links
3 kingdoms (Linnaeus 1735)	■ Mineral ■ Vegetable ■ Animal	■ Structural features ■ Organised body and life (to separate mineral from vegetables and animals) ■ Power of locomotion (to separate vegetables and animals)	■ Easily observed in an organism ■ More constant in an organism's lifetime (no seasonal change or change with maturity) ■ Can infer reproductive methods	■ Structures may vary between males and females of a species ■ Internal biochemistry (genetic similarities) is not available for inferring evolutionary links

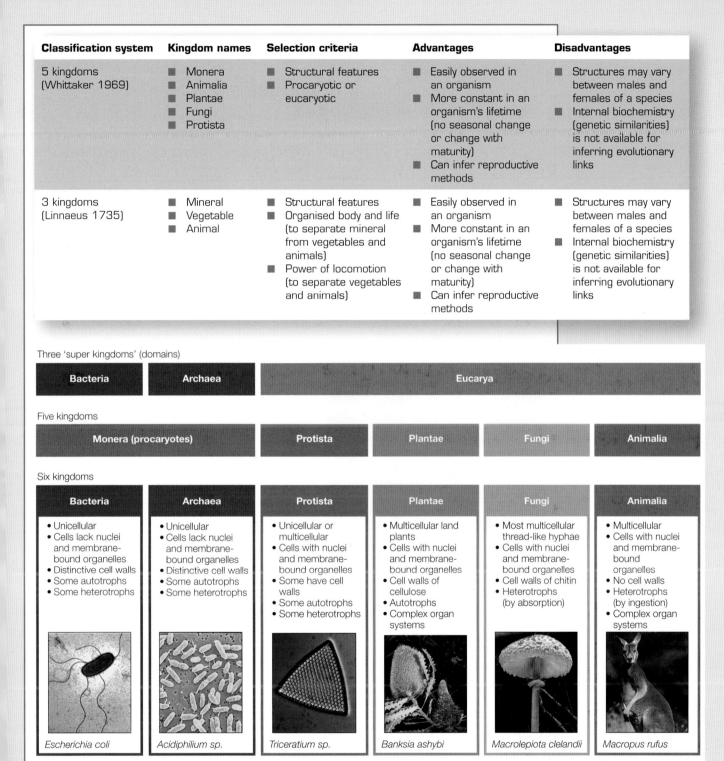

Three 'super kingdoms' (domains)

Bacteria	Archaea	Eucarya

Five kingdoms

Monera (procaryotes)	Protista	Plantae	Fungi	Animalia

Six kingdoms

Bacteria	Archaea	Protista	Plantae	Fungi	Animalia
• Unicellular • Cells lack nuclei and membrane-bound organelles • Distinctive cell walls • Some autotrophs • Some heterotrophs	• Unicellular • Cells lack nuclei and membrane-bound organelles • Distinctive cell walls • Some autotrophs • Some heterotrophs	• Unicellular or multicellular • Cells with nuclei and membrane-bound organelles • Some have cell walls • Some autotrophs • Some heterotrophs	• Multicellular land plants • Cells with nuclei and membrane-bound organelles • Cell walls of cellulose • Autotrophs • Complex organ systems	• Most multicellular thread-like hyphae • Cells with nuclei and membrane-bound organelles • Cell walls of chitin • Heterotrophs (by absorption)	• Multicellular • Cells with nuclei and membrane-bound organelles • No cell walls • Heterotrophs (by ingestion) • Complex organ systems
Escherichia coli	*Acidiphilium sp.*	*Triceratium sp.*	*Banksia ashybi*	*Macrolepiota clelandii*	*Macropus rufus*

Figure 4.5 Three different schemes for the classification of the major groups of cellular life

4.3

Levels of organisation assist classification

■ *explain how levels of organisation in a hierarchal system assist classification*

In the biological system of classification, a group is recognised and given a Latinised name: for example, Plantae (multicellular land plants) and Magnoliophyta (flowering plants). Since the time of Swedish naturalist Carolus Linnaeus (1707–1778) (see Fig. 4.6), the biological system has been conceived as a hierarchy with specific levels or ranks. Thus, the group Plantae has the rank of kingdom and Magnoliophyta is a phylum.

A mnemonic you could use to help remember the correct order of these groups (kingdom, phylum, class, order, family, genus, species) is: **King Ph**il **class**es **ord**inary **families** as **gen**erous and **spec**ial. Of course, you may find it easier to remember one of your own.

Having simple levels of hierarchy for classification makes it an easy and systematic way to name organisms according to their characteristics. Examples of the use of this biological system of classification are shown in Table 4.2.

Figure 4.6
Carolus Linnaeus

The rank order of groups commonly used today is:

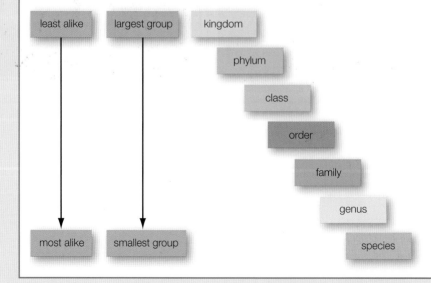

Table 4.2 Biological classifications of Australian organisms

Rank	Platypus	Koala	Dingo	Blue gum	Kangaroo Paw
Kingdom	Animalia	Mammalia	Mammalia	Plantae	Plantae
Phylum	Chordata	Chordata	Chordata	Magnoliophyta	Magnoliophyta
Class	Mammalia	Mammalia	Mammalia	Magnoliopsida (dicots)	Liliopsida (monocots)
Subclass	Prototheria	Theria	Theria	Rosidae	Liliidae
Order	Monotremata	Marsupialia	Carnivora	Myrtales	Liliales
Family	Ornithorhynchidae	Phascolarctidae	Canidae	Myrtaceae	Haemodoraceae
Genus	Ornithorhynchus	*Phascolarctos*	*Canis*	*Eucalyptus*	*Angiozanthos*
Species	*O. anatinus*	*P. cinereus*	*C. lupus* subsp. *dingo*	*E. globulus*	*A. manglesii*

Impact of changes in technology

4.4

- *discuss, using examples, the impact of changes in technology on the development and revision of biological classification systems*

Examples of changes in microscope technology

As you have already discovered in Chapter 2, microscope technologies have changed a lot in the 19th and 20th centuries. The light microscope first revealed that living things were made up of cells. With improvements in light microscopes and the introduction of the electron microscope fine details of the internal structure of cells were seen. In the 1950s, the electron microscope revealed procaryotic cells, leading to a separate kingdom, Monera. In 1967, Fungi were identified as a separate, multicellular eucaryotic kingdom. It was at this time that our classification systems were revised from the two-kingdom system to the five-kingdom system.

Examples of changes in geological technology

Relative dating techniques

Relative dating techniques provide information about the sequence that fossils appear:

- *principle of superposition*—youngest rock layers are found at the top, oldest layers are found at the bottom
- *stratigraphic correlation*—same relative age when similar layers of rock contain similar fossils
  —index fossils (used to date other organisms found in the same rock layers)
  —fluoride-nitrogen relative dating (the older the fossil, the higher the amount of fluoride in comparison to nitrogen).

Absolute and radiometric dating techniques

Absolute and radiometric dating techniques provide information on the age of the fossil:

- *carbon-14 dating*—rate of decay of carbon is used to estimate the age of young fossils
- *potassium-argon dating*—rate of decay of potassium to argon is used to estimate the age of older fossils

- *electron-spin resonance (ESR)*—the amount of accumulation of higher-energy electrons over time is used to date fossils of varying age
- *thermoluminescence*—the amount of accumulated electrons over time, and hence luminescence when heated, is used to date clay products surrounding fossils.

Example of changes in biochemical technology

Changes in biochemical technology have been a more recent occurrence and include:

- *DNA sequencing*—compares variation in the order of bases in gene DNA
- *mitochondrial DNA as a molecular clock*—mutates at predictable rates so can be used to date evolutionary events
- *karyotype analysis*—comparison of chromosomes from photographs
- *DNA-DNA hybridisation*—determines the similarity of DNA from different species to indicate genetic relationship
- Comparison of haemoglobin
  —amino acid sequencing (comparison of sequences reflects degree of difference)
  —electrophoresis (compares electrical charges on protein molecules of closely related species)
  —immune response testing (looks at the reaction between two species' proteins, if they react they are very similar)

Impact on the revision of classification systems

With the development in the last two centuries of microscopy and biochemistry, our knowledge of the kinds of organisms on Earth has increased remarkably. Cell biology has revealed fundamental differences between procaryotes and eucaryotes.

Bacterial classification is changing with the discovery of differences in ribosomal DNA between major groups. Observations of ultrastructure, cell wall biochemistry and photosynthetic pigments of algae show them to be an assemblage of organisms with diverse relationships.

How the major groups are related to one another is still being discovered. This uncertainty of relationships is reflected in continuing changes to classification, including the recent recognition of super kingdoms and the expansion of the number of kingdoms.

Many biology books follow the traditional system that recognises five kingdoms: Monera, Protista, Fungi, Plantae and Animalia. The discovery in 1977 of the two major groups within Monera (Archaea and Eubacteria) by Woese was based largely on differences at a molecular level, not anatomical structure. Two major evolutionary lineages are now recognised among the procaryotes, and they are classified by many taxonomists as the super kingdoms Bacteria and Archaea (see Fig. 4.5).

An example of another recent change to a classification system occurred in 1998 by an international group of botanists. A new classification for the families of flowering plants occurred, now using both the traditional structural appearance of plants as well as the new technique of gene sequencing. Plants were grouped by similarities in their DNA structure. Not only have groups of flowering plants been renamed, but now new evolutionary relationships have been suggested. The Australian family Proteaceae (e.g. banksias) was shown to be related to the family Platanaceae (plane trees).

Binomial system

4.5

- *describe the main features of the binomial system in naming organisms and relate these to the concepts of genus and species*

The ranks or categories of genus and species have a special significance because they form the basis of the binomial system, perfected by Linnaeus. In the binomial system, the nature of each kind of organism is described by two parts: *Homo sapiens* (humans), *Pan troglodytes* (chimpanzee) and *Gorilla gorilla* (gorilla). (It is interesting to note that in 1758, Linnaeus placed chimpanzees in the genus *Homo.*) In the binomial system the genus name is always written first (with a capital letter) and the species name is written second, as in *Eucalyptus viminalis* or *Banksia serrata.* Both the genus and species names are always written in italics.

The taxonomic category, species, is the lowest rank in the Linnaean hierarchy to which all organisms must be classified. Although structurally very similar within the one species, there are still variations between males and females and different ages or maturity levels.

Difficulties in classifying extinct organisms

4.6

- *identify and discuss the difficulties experienced in classifying extinct organisms*

When trying to classify organisms scientists look for similar characteristics to ones already previously classified and dated, from fossil evidence, or with modern present-day living organisms. When trying to classify an extinct organism, there are no modern-day living organisms to compare it to and assist in identifying the characteristics for classifying the organism. Also, an extinct organism may only provide fossil evidence of its existence and the quality of this evidence may vary. Fossil evidence may be incomplete or poorly preserved. This creates problems in determining the organism's characteristics as a whole. The conditions of its past environment may be required to determine the characteristics it may have had to exist in such an environment. Scientists may need to analyse similarities with other organisms living in a similar environment of that time, or even organisms living in a similar environment in the present day.

However, with modern technologies assessing molecular and biochemical evidence now, rather than only the structural focus of classification in the past, we can determine much more information about the organism even when it is partially or poorly preserved. Once this information is collected using all the available scientific technologies, scientists must then put the whole story together around this organism and use and interpret the evidence and test results in order to categorise and classify the organism. Different scientists may make different interpretations from the same information or evidence. Subconscious bias and preconceptions influence scientific interpretation. Arguments still continue over the validity of some evidence.

All classification systems create controversy since they depend upon human decisions about the features that are important characteristics of a group or organism. Classification systems are constantly changing due to the discovery of new evidence and fresh interpretations of current evidence. In addition, significant changes have occurred recently because of the biochemical technologies now available. New genetic links are being found. The more recent systems have changed from the traditional 'premolecular' one due to the recent genetic technology indicating new genetic links.

4.7 Assistance to understanding present and past life

- *explain how classification of organisms can assist in developing an understanding of present and past life on Earth*

There are many alternative views on the evolutionary relationships between different groups of organisms. Although often based on the same evidence, different interpretations have arisen and are constantly changing and conflicting. However, by attempting to arrange and order groups of organisms scientists are able to look at possible evolutionary pathways and relationships. Our present-day organisms are much easier to classify than those that only lived in the past. However, we can look at similarities between modern-day organisms and those from the past to develop an understanding of the differences between present and past life. Some organisms may now be living in similar environments to those in which they lived in the past; however, many organisms have had significant changes in the environments and demonstrated changes in response to this. It is with this in mind that we gain a better understanding of the evolution of organisms over time on Earth and the changes that occurred alongside the environmental changes on Earth.

Answers to revision questions

REVISION QUESTIONS

1. Explain five reasons why scientists need to classify organisms.

2. Construct a dichotomous key (including numbers and letters) to identify the following five organisms (koala, kangaroo, crocodile, kookaburra and echidna).

3. Use Figure 4.4 on page 225, to key out the insect (see Fig. 4.7) below. Include all steps taken to reach your conclusion.

4. Describe two different systems of classification and discuss the advantages and disadvantages of each one.

5. List the seven levels (in rank order from least alike to most alike) used in the biological system of classification.

6. Discuss the impact of two examples of changes in technology on the development and revision of biological classification systems.

7. Describe the main features of the binomial system in naming organisms using a specific example.

8. Identify and discuss the difficulties experienced in classifying extinct organisms.

9. Explain how the classification of organisms can assist in developing an understanding of present and past life on Earth.

Figure 4.7
An insect example

EVOLUTION OF
AUSTRALIAN BIOTA

CHAPTER 1

Australia's past: part of a supercontinent

Evidence for the rearrangement of crustal plates and continental drift indicates that Australia was once part of an ancient supercontinent

Introduction

In the module 'Life on Earth' you studied the origin of living things on Earth, from 4500 million years ago until about 250 million years ago. The Earth's history is subdivided into time spans called eons, eras, periods and ages. There are three main eras: Paleozoic, Mesozoic and Cenozoic. These may seem difficult to remember at first, but they become simpler if we understand their meaning.

Paleo means ancient, *meso* means middle and *ceno* means recent. They end in *zoic,* coming from the word root for *zoology,* relating to animals—the eras are based on the type of animal life present at each time.

You have now studied events that took place prior to and during the Paleozoic era (see Table 2.2 on page 208 in 'Life on Earth'). In this module you will look at events that occurred in the late Paleozoic era, but mainly during the Mesozoic and Cenozoic eras, during which time Australia broke away from a large common land mass and became an island continent, separate from all other land masses. This isolation led to the evolution of Australia's unique flora and fauna of today.

1.1 From Gondwana to Australia—how our continent arose

- *identify and describe evidence that supports the assertion that Australia was once part of a landmass called Gondwana, including:*
  —matching continental margins
  —position of mid-ocean ridges
  —spreading zones between continental plates
  —fossils in common on Gondwanan continents, including Glossopteris *and* Gangamopteris *flora, and marsupials*
  —similarities between present-day organisms on Gondwanan continents

(This dot point addresses outcome P10.)

Wallace, Wegener and drifting continents

Alfred Wallace was a natural scientist in the mid 1800s. He is famous for his work on evolution, but during his travels he also studied how geography affected the distribution of species of plants and animals—something we refer to as **biogeography** today. He studied thousands of plants and animals and kept thorough records of where each was found. He recognised patterns in their distribution and this led to his proposal of two separate zoogeographical areas. The line separating these areas is called Wallace's line—a clear boundary that snakes between the Indonesian islands in the northwest and those in the southeast. He noticed that, despite having similar climate and terrain, the north-western islands (Sumatra, Java and Bali) had fauna, in particular bird species, more similar to those of the Asian mainland, whereas in the southeast, the birds in Lombok and New Guinea were more like those in Australia. Wallace's line divides the Asian and Australian regions. At that point in time, there was no explanation for this phenomenon but Wallace suggested that it reflected that groups of organisms that were isolated evolved to become different. It has only recently been discovered that Wallace's line coincides with a very deep mid-ocean ridge.

In 1915, Alfred Wegener, a German scientist, noticed that identical fossil plants and animals had been discovered on opposite sides of the Atlantic Ocean. Since the ocean was too large for the animals to have crossed it, Wegener proposed the theory of continental drift, suggesting that all the continents had once been connected together, forming a large single mass of land—which he called **Pangaea**—surrounded by one huge ocean. He proposed that the continents then split up into smaller units of land which drifted to their current positions around the Earth. A large amount of evidence to support this theory has been found over the ensuing years.

In this chapter, how the theory of continental drift could account for Australia becoming an island continent will be described, followed by current evidence to support this theory.

How Australia arose from Gondwana

Maps of the world showing the position of continents do not address the fact that the continents may not always have been in those positions. Scientific studies over the past 100 years have led us to believe that continents have indeed moved. It is believed that about 250 million years ago (mya), all the continents were joined together to from one huge land mass called Pangaea. Forces beneath the surface of the Earth caused it to begin to break apart about 225 mya. The sequence of diagrams in Figure 1.1 shows how Pangaea is thought to have eventually given rise to the continents as we know them today. The sequence of events:

- 225 mya: Pangaea, one huge continent made up of all the continents we know today, splits into two large land masses—Laurasia to the north and Gondwana to the south
- 180 mya: Gondwana begins breaking up, as does Laurasia. (Australia is thought to have been part of the southern land mass and so we will follow the breaking up of Gondwana)
- 135–100 mya: Gondwana breaks into three parts:
  —Africa and most of South America (the southern-most tip of South America retained a small land bridge with Antarctica until 60 mya)
  —Antarctica, Australia, New Zealand and New Guinea
  —India (India began drifting northwards, towards the part of Laurasia that would become Asia)
- 80 mya: New Zealand breaks away
- 65 mya: Australia begins to separate from Antarctica (it is still attached by Tasmania)
- 60 mya: final separation of tip of South America from Gondwana
- 45 mya: Australia becomes a separate continent and begins drifting northwards, and therefore becomes hotter and drier.

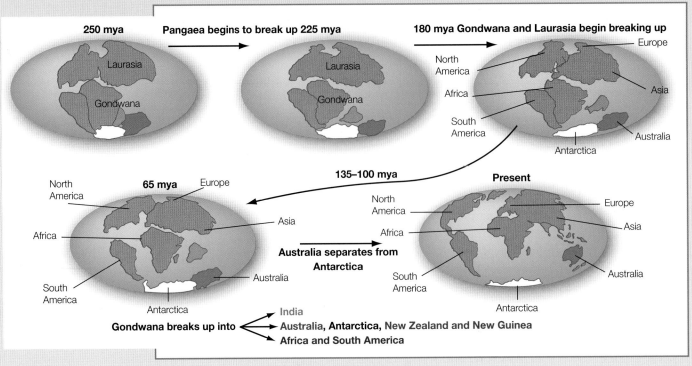

250 mya

Laurasia

Gondwana

Pangaea begins to break up 225 mya

Laurasia

Gondwana

180 mya Gondwana and Laurasia begin breaking up

Europe

North America

Africa

South America

Asia

Australia

Antarctica

65 mya Europe

North America

Africa

Asia

South America

Antarctica

Australia

135–100 mya

Australia separates from Antarctica

Gondwana breaks up into
- India
- **Australia, Antarctica, New Zealand and New Guinea**
- **Africa and South America**

Present

North America

Africa

Europe

Asia

South America

Australia

Antarctica

Figure 1.1 The breaking up of the supercontinent Pangaea to form the present-day continents

STUDENT ACTIVITY

Using the figure provided on the Student Resource CD colour the three parts of Pangaea in the following colours:
- India—red
- Australia—blue
- Antarctica—green.

Use these colours to shade these three regions on the sequence of diagrams showing the separation of Australia over the previous 250 million years, to the present.

SR

Figure to use for this student activity

TR

Solutions to this student activity

Evidence for this theory

- Geological:
  —matching continental margins
  —position of **mid-ocean ridges** and spreading zones between continental plates
- Biological:
  —common fossils
  —similar present-day organisms.

Although the theory of continental drift was proposed over a hundred years ago and has been under discussion since then, finding sufficient evidence to support it has only come about fairly recently.

Background information—the theory of plate tectonics

It was only in the1960s that a theory was proposed that could account for how continents could drift. Harry Hess, a geologist, proposed the **theory of plate tectonics**, which provided a mechanism by means of which continents could move. He explained that continents were carried on large plates beneath the ocean. These plates were positioned on top of the semi-molten interior of the Earth. To understand this idea, picture a cracked eggshell surrounding the contents of a raw egg. The pieces of eggshell would be equivalent to the plates forming the ocean floor. The continents, which are made of lighter rock and are thicker, jut out above the ocean (equivalent to the thicker bits on the cracked eggshell that are sticking up). As the plates move, their boundaries may *collide*; *slide past each other*, so that one slides beneath the other (termed '**subduction**'); or *move apart* (see Fig. 1.2). Picturing the analogy of the egg once again, if two pieces of cracked eggshell move towards each other, they would converge on one side, but a gap would form on the opposite side of each piece, where they moved away from another plate of cracked eggshell.

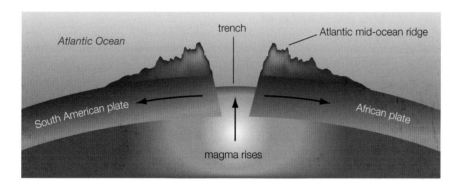

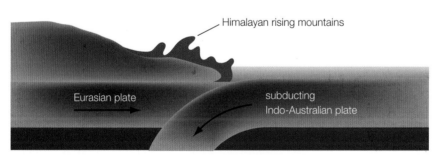

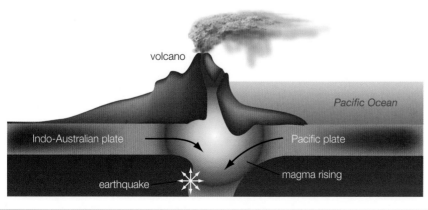

Figure 1.2
Movements of crustal plates: (a) spreading zone, with diverging plates forming a mid-ocean ridge in the Atlantic Ocean (b) converging plates form a subduction zone, with the formation of fold mountains; (c) converging plates forming a collision zone

Studies of the ocean floor have revealed evidence that there are in fact plate boundaries such as those described by Hess. Plates that are moving together are said to be **converging**. Those moving apart are said to be **diverging**. At points where plates collide, the sediments may be folded and distorted to form mountains (e.g. the Himalayan mountains of North India formed when the Indo-Australian plate collided with the Eurasian plate, about 50 million years ago).

🕷 Website which shows an animation of the process of the breaking up of Gondwana and subsequent continental drift: http://kartoweb. itc.nl/gondwana/gondwana_gif.html

Evidence for the existence of Gondwana—geological

Matching continental margins

Looking at the *shapes of the continents* today, it is easy to picture how they may have once fitted together to form one landmass. Scientific studies using computer-generated models of the continents show that they fit extremely well, particularly if the **continental shelf** margins, rather than the shorelines, are used.

Further geological evidence is provided by the similarity of *rock strata* (the layering of rocks) on matching continental margins. Matching layers of rock found on continental margins that fit together suggest that they were once adjoining.

(The student task on the Student Resource CD allows you to check this for yourself by matching current continental margins to build a model of Pangaea.)

Mid-ocean ridges, spreading zones and subduction zones

■ *Mid-ocean ridges* are the sites where two crustal plates meet or move apart. The theory of continental drift suggests that as the continents drift apart magma wells up through the spreading floor and new crust is formed.
  —*Evidence*: There is scientific evidence using **radiometric dating** that the rocks towards the edges of these mid-ocean ridges are younger than those further in from the margins, supporting the idea that rock near the edges of spreading zones are newly laid down.

■ Mid-ocean ridges where plates collide are the sites where most volcanoes and earthquakes occur (to illustrate this complete the problem-solving activity on page 243).

■ At subduction zones, where plates collide and fold, or one plate slides beneath the other, the formation of mountains and the continued movement of present-day continents provides evidence to support the theory of plate tectonics.
  —*Evidence*: The Himalayan mountain range is at a subduction point where India is sliding under the Asian plate. Evidence is provided by the fact that the Himalayan mountains are still slowly rising, supporting the theory of continental drift.
  —*Prediction to validate the theory*: Australia is still moving northwards at a rate of approximately 7 cm per year, which means that it could collide with Southeast Asia in approximately 50 million years if the drifting continues at the current rate.

Evidence for the existence of Gondwana—biological

Fossil evidence

More than 180 mya, when Gondwana existed, the same type of organisms (both plant and animal) would have been distributed across the entire landmass. Many of the species that lived

Extension activity— continental margins

at that time are now extinct and so fossils provide the only record to show that they once existed and how they were distributed. Fossil evidence shows the common occurrence of certain extinct organisms across all of the continents that once formed Gondwana:

- *Glossopteris* and *Gangamopteris* were types of tree ferns that formed the dominant vegetation on Gondwana 280–225 mya, before it split into the continents that we know today (see Fig. 1.3). Fossils show that they had tongue-shaped leaves with a midrib and net venation, and were typically found in swampy habitats. They were the main type of vegetation involved in the formation of coal.

Glossopteris leaf fossils have been found on all the continents that once formed Gondwana (see Fig. 1.4). In Antarctica, *Glossopteris* fossils were found embedded in coal seams and rocks that date back to 250 mya (these were collected by Captain Robert Scott and his team on their final and fatal voyage to Antarctica, found along with their frozen bodies). The fossils that they found were the same as those found in Australia, India, South Africa and South America and those found in South Africa were in rock of the same age.

- Labyrinthodonts were amphibians (now extinct) that resembled salamanders, but differed in that

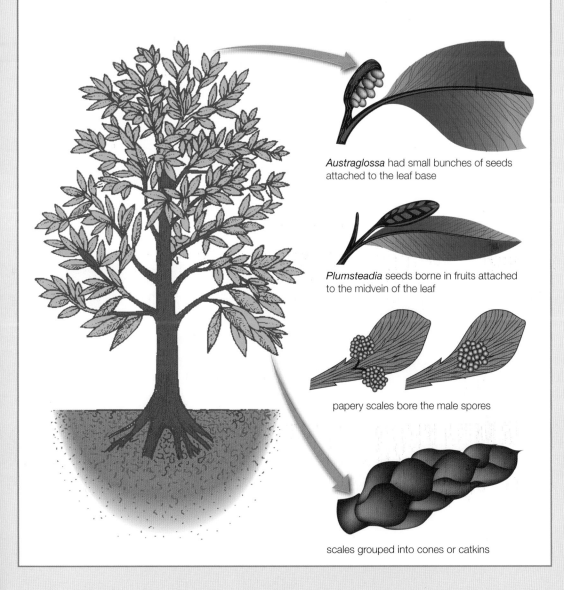

Australglossa had small bunches of seeds attached to the leaf base

Plumsteadia seeds borne in fruits attached to the midvein of the leaf

papery scales bore the male spores

scales grouped into cones or catkins

Figure 1.3
Glossopteris plant with leaves and seed-bearing structures

Figure 1.4 Fossils found in the Hunter Valley, NSW: *Glossopteris*

period are known to have lived in both Australia and South Africa, supporting the theory that these were once part of the supercontinent, Gondwana. Labyrinthodonts were thought to eat insects, the abundance of which at that time is supported by insect fossil remains, as well as evidence provided by the jagged holes present in many of the fossilised *Glossopteris* leaves, where insects probably ate them

■ fossils of *Lystrosaurus*, a sheep-sized reptile that existed 200 mya, were found in South Africa, India and Antarctica (see Fig. 1.5).

It is highly unlikely that creatures like the labyrinthodonts and *Lystrosaurus* would have evolved separately on such isolated continents as Antarctica and Australia and they certainly had no way of crossing the oceans. So the similarities in the fossil records of these continents support the idea that they were once joined.

they had jaws full of teeth and grew to much larger sizes. They lived about 200 million years ago and provided the first evidence that land vertebrates had roamed Antarctica when its climate was warm. Labyrinthodonts of the same

Figure 1.5 Map of the distribution of fossils on Gondwanan continents

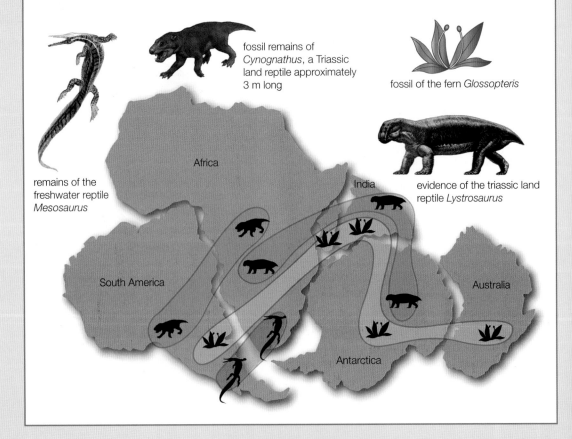

fossil remains of *Cynognathus*, a Triassic land reptile approximately 3 m long

fossil of the fern *Glossopteris*

remains of the freshwater reptile *Mesosaurus*

evidence of the triassic land reptile *Lystrosaurus*

Africa

India

South America

Australia

Antarctica

Evidence provided by the distribution of present-day organisms

Biogeography is the study of the geographical distribution of species, both present-day and extinct. It has long been recognised that Australia and other lands in the southern hemisphere share many similar plants and animals, but it was an English botanist Joseph Hooke (1853), who first pointed out similarities between *southern hemisphere lands* and *rainforest* vegetation *in India*. He also described alpine plant groups in south-eastern mainland Australia and Tasmania as being similar to those found in New Zealand.

Similarities in present-day fauna on Gondwanan continents show that *living marsupials* are found in Australia and South America, suggesting that these continents once were joined.

Flightless birds

The *ratitaes* are an ancient group of flightless birds that are believed to have evolved from a common ancestor on Gondwana. Their present-day southern-hemisphere distribution can be explained as a result of the separation of the southern continents. Each continent except Antarctica is represented by at least one living species: the emu in Australia, the kiwi in New Zealand, the rhea in South America and the ostrich in Africa. The cassowary is found in both Australia and New Guinea, perhaps suggesting that their separation occurred later. Two extinct forms of flightless birds have also been found— the elephant bird in Madagascar (an Indian ocean island off the west coast of Africa) and the moa in New Zealand.

(c)

(d)

(a)

(b)

(e)

Figure 1.6 Flightless birds: (a) emu; (b) kiwi; (c) rhea; (d) ostrich; (e) cassowary

Marsupials

At the time that South America was separating from the Antarctica–Australia landmass 65 mya, marsupials are believed to have been moving from North America into South America. Living marsupials today are found only in South America and Australia. The present-day distribution reflects continental drift if we take the following into account:

- the tip of South America was still very close to Antarctica (60 mya) and so marsupials could cross the land bridge that still existed
- Africa had already split from Antarctica (135 mya) and this could explain the absence of marsupials from that continent
- Antarctica later split from Australia and drifted southwards, becoming too cold for the survival of its marsupials
- Australia became a 'drifting ark' with many marsupials which thrived as it drifted northwards where it was warmer.

(*Note*: Theories such as this are still being investigated by biologists who constantly seek evidence to support their explanation of how the present-day biogeography came about.)

Plants

Southern beech trees, *Nothofagus*, are represented in Australia, Antarctica, South America, New Guinea and New Caledonia in both living and fossil forms, further supporting the idea that these continents once were joined and arrived in their present locations as a result of continental drift.

Figure 1.7
Marsupials:
(a) American opossum;
(b) Australian grey kangaroos

(a)

(b)

STUDENT ACTIVITY

Internet virtual tour of the Australian continental shelf:
Visit this website for a virtual tour of the Australian continental shelf, showing the hidden vista of the ocean floor: www.environment.gov.au/coasts/discovery/flythrough/index.html
Geoscience Australia and the National Oceans Office have produced a virtual tour, showing rugged undersea ridges formed long ago when Australia tore away from Antarctica. A modern technology called a *swath mapper*, installed on a CSIRO marine research ship, was used.

SR — Map to cut and paste

TR — Solutions to this student activity

Problem solving to infer a moving Australian continent

■ *solve problems to identify the positions of mid-ocean ridges and spreading zones that infer a moving Australian continent*

Aim

1. To solve problems.
2. To identify the positions of mid-ocean ridges and spreading zones that infer a moving Australian continent.

Maps showing age of sea floor:
http://zephyr.rice.edu/
plateboundary/TGpart2.ppt
#297,27,slide27
Animation of different types of plate boundaries:
www.classzone.com/books/earth_
science/terc/content//visualizations/
es0804/es0804page01.cfm?chapter_
no=visualization

Background information

The Earth's structure consists of three parts: the core, mantle and crust. The crust is made up of a number of plates sitting on the mantle and any movement is determined by the mantle's convection currents. Both oceanic and continental plates float on the Earth's mantle. The theory of plate tectonics has helped to understand the mechanism for the movement of continents and the formation of mountains and oceans. This theory has led to the discovery of Earth's plate boundaries (see Fig. 1.8) through

methods such as the plotting of earthquakes and volcanoes around the world. Using seismic equipment scientists are able to collect information on the occurrence of earthquakes and volcanoes around the world.

Scientists use the position of earthquakes (see Fig. 1.9) and volcanoes (see Fig. 1.10) to determine plate boundaries and the movement occurring at the edges of different plate boundaries. Hence, they are able to determine effect of the shifting plates on the movement of entire continents.

There are three different types of plate boundaries (see Fig. 1.11):
■ *converging boundary* where two plates are moving towards each other and collide
■ *diverging boundary* where two plates are moving away from each other
■ **transform boundary** where two plates are sliding past each other.

At diverging boundaries, mid-ocean ridges and **sea floor spreading zones** occur due to continents drifting apart and magma welling up from the Earth's mantle to the surface, solidifying to form new crust (see Fig. 1.12). This sea floor spreading creates new oceanic crust. Subduction zones, however, lose part of the oceanic crust at converging plate boundaries. A continental plate collides with an oceanic plate forcing the oceanic plate underneath the continental plate and entering

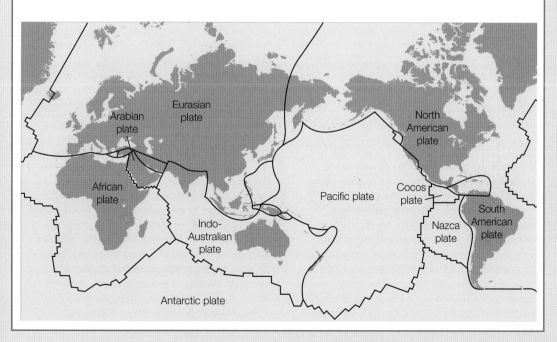

Figure 1.8 World map showing labelled tectonic plates and boundaries

Figure 1.9 World map showing earthquakes that have occurred mainly in the last five years—ranging in magnitude from 2 to 6 on the Richter scale

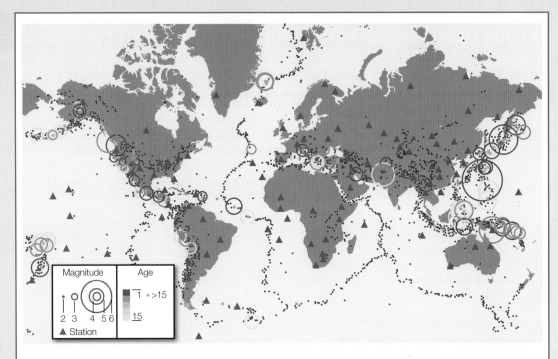

Figure 1.10 World map showing currently active volcanoes

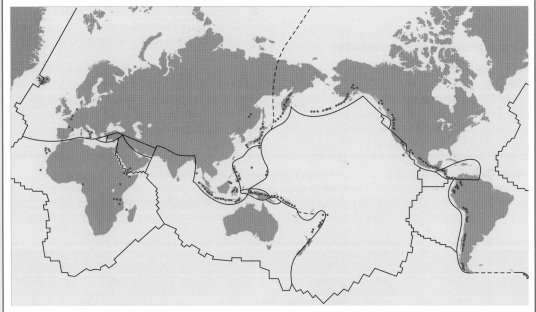

the mantle where it is broken down. This results in forming a **deep ocean trench** (see Fig. 1.12).

Earthquakes and volcanoes commonly occur along plate boundaries due to the pressure and resulting plate movement. Volcanic eruptions occur progressively along the rifts of the mid-ocean ridges.

The occurrence of earthquakes and volcanoes provide evidence for the existence of mid-ocean ridges and spreading zones. The position of the earthquakes and volcanoes pinpoints the area where these are situated deep under the ocean. In Australia, earthquake occurrences have been recorded at different magnitudes (or sizes) ranging from 4 to more

than 6 (see Fig. 1.13). Although the Australian continent does not fall on a plate boundary, small shallow earthquakes occur due to pressure at fault lines. The Australian plate has many north–south concentrations of earthquakes, so this may suggest that it is adjusting to the activity of plate movement in that direction. In eastern Australia, earthquakes occur at a depth of 20 km. Earthquakes less than 5 km are considered as shallow, while those at depths greater than 15 km are deep. Shallow earthquakes cause much damage, but deep earthquakes rarely cause damage. An earthquake exceeding magnitude 7 occurs somewhere in Australia every 100 years.

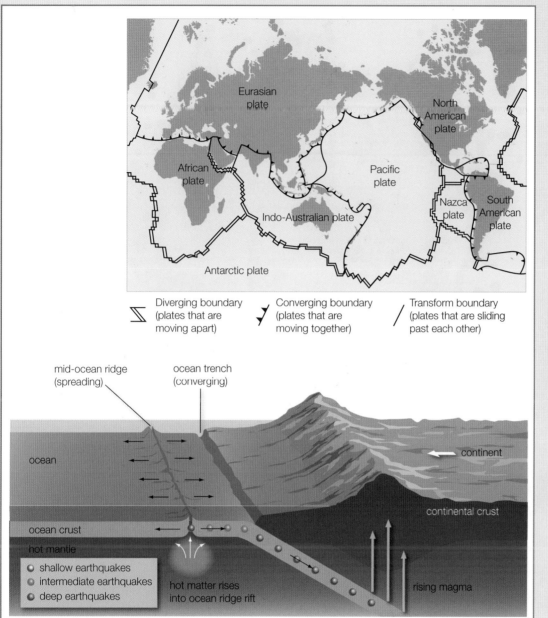

Figure 1.11 World map showing the plate boundaries and their different types of movements—diverging, converging and transform plates

Diverging boundary (plates that are moving apart)

Converging boundary (plates that are moving together)

Transform boundary (plates that are sliding past each other)

Figure 1.12 Formation of a mid-ocean ridge and sea floor spreading

Earthquakes of magnitude 8 and larger normally only occur at plate boundaries. The most disastrous Australian earthquake in the last 200 years was the Newcastle earthquake of 28 December 1989 which had a magnitude of 5.6 and caused $1.2 billion worth of damage.

Australia is far from the edges of the Indo-Australian plate yet volcanoes have been erupting along the eastern part of the continent for the last 33 million years. Australia's volcanoes are not related to the subduction zones that produce volcanoes in New Zealand, Tonga, Samoa and Indonesia. Although there is much evidence for previous existence of active volcanoes in the continent of Australia, at present all volcanoes are dormant or extinct. Active volcanoes are rare in Australia because there are no plate boundaries on the continent.

However, there are two active volcanoes in the Australian territory located 4000 km southwest of Perth on Heard Island and the nearby McDonald Islands. A volcano on the largest of the McDonald Islands erupted some time after 1980 filling some of the island's bays with volcanic material and extending the coastline. Figure 1.14 illustrates the change in size of the island from before the eruption after 1980 to 2001.

When plotting volcanoes of the Australian continent onto a map, we can see that there are several chains with increasingly younger volcanoes to the south (see Fig. 1.15). The age progressions suggest that the hot spot beneath eastern Australia is broad and may take advantage of weak places on the plate to feed magma to the surface. The younger volcanoes

Figure 1.13 Map of Australia showing positions of earthquake occurrences at different magnitudes

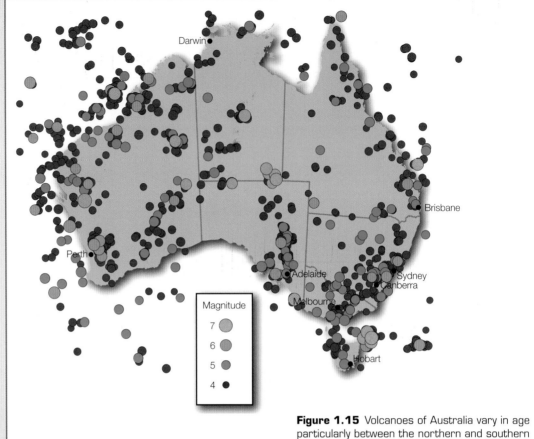

Figure 1.14 Satellite photographs: (a) McDonald islands before eruption in 1980; (b) McDonald islands in 2001, increasing over 1 km in width as the eruption continued

Figure 1.15 Volcanoes of Australia vary in age particularly between the northern and southern volcanoes (ages are shown in millions of years)

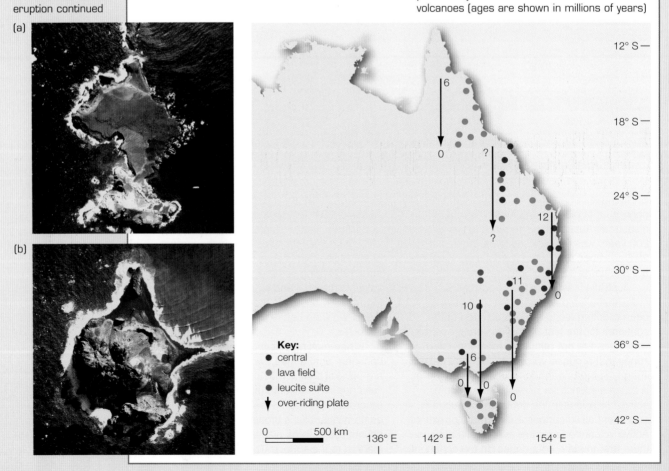

suggest more recent activity regarding the plate movement. In February 2002 an earthquake measuring a magnitude of 4.5 occurred between Victoria and the northwest coast of Tasmania. Scientists believe this region is a hot spot (a place where magma rises to Earth's surface) for the birth of a new active Australian volcano. Although, nothing has formed at present, it may occur sometime in the next 1000 years.

Other than making inferences about continent movement using the position of earthquakes and volcanoes, scientists may also combine other technologies such as geochronology (plotting the ages of the sea floor) (see website listed on page 243) and satellites with the global positioning system (GPS) for determining rates and directions of plate movement.

Method

Read the Background information and observe Figures 1.8 to 1.15 to solve the following problems surrounding the plate tectonics and inferences about a moving Australian continent.

Results—problem solving questions

1. Identify two types of evidence that are used for demonstrating the existence and position of mid-ocean ridges and sea floor spreading.
2. (a) Describe any patterns or trends in the position of earthquakes in Australia in Figure 1.13.
   (b) Explain reasons for the above.
3. Why does Australia not have any active volcanoes on the continent? Explain possible reasons why the two active volcanoes are positioned 4000 km southwest of Australia, towards Antarctica?

4. Identify any pattern or trend in the position of plate boundaries, earthquakes and volcanoes around the world using Figures 1.8, 1.9 and 1.10.
5. Describe the relationship between volcanoes and earthquakes and the position of mid-ocean ridges and sea floor spreading.
6. Discuss what the position of mid-ocean ridges and sea floor spreading may infer about the movement of continental plates, with particular reference to Australia.
7. Using Figure 1.14, determine the possible movement direction that this information suggests about the Australian plate and therefore the Australian continent.
8. Scientists are able to determine the direction and speed of movement of Australia using technology to produce the information in Figure 1.15. Based on the information on the map, which direction is Australia moving?
9. Scientists have suggested that the mid-ocean ridge south of the Australian continent has created 75 km of sea floor over the past one million years. How much should we see created per year from the mid-ocean ridge?

Analysis questions

1. Describe the scientific information that provides evidence towards the movement of crustal plates and therefore movement of continents.
2. Describe how solving problems surrounding the position of mid-ocean ridges and sea floor spreading has assisted in the understanding of continental plate movement.
3. Discuss how determining trends in earthquake and volcano positions can lead to inferences for the direction of movement of the Australian continent.

TR

Extension question

Changing ideas in science—the platypus enigma

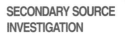

- *identify data sources, gather, process and analyse information from secondary sources and use available evidence to illustrate the changing ideas of scientists in the last 200 years about individual species such as the platypus as new information and technologies became available*

SECONDARY SOURCE INVESTIGATION

PFAs

P3
P5

Introduction

In the late 1700s, expeditions of natural scientists (as biologists were called back then) to distant lands were common. Voyages to

Australia resulted in preserved specimens of the platypus being sent back to England, where scientists of the time were working on refining the classification of living organisms. The first specimen of a platypus was sent

PFA

P3

Scaffolds for breaking
down the question

from NSW to Britain in 1798 by Governor John Hunter, who had an interest in natural history. As more specimens of this 'amphibious mole' began arriving in Britain, they were the cause of much puzzled speculation—an 'enigma' to the scientists of the time. The specimens of this strange, unnamed creature—a mixture of an animal with fur like a mammal, but webbed feet and a bill like a duck—were regarded as a hoax. Scientists at the time thought that perhaps someone had cleverly stitched together the body parts of more than one creature, and sold it to sailors who knew no better!

Thomas Bewick, the first scientist to examine the platypus specimen, wrote that he had resisted any attempt 'to arrange it in any useful mode of classification' and Dr George Shaw, a respected naturalist at the British Museum, suspected that the specimen he had received in 1799 was a 'hoax'. Close and detailed examination showed, of course, that this was indeed a real animal and Shaw went on to name it *Platypus anatinus* in 1799. The original name was derived from Greek (*platypous* = flat-footed) and Latin (*anatinus* = duck-like). The name was later changed to *Ornithorhyncus anatinus,* after the discovery that the name *platypus* had already been given to a beetle (*ornitho* = bird-like, *rhyncus* = snout in Greek). The classification of the platypus continued to confound and interest biologists across the world for nearly 90 years.

Recent discoveries of platypus-like fossils have led to a whole new area of research. In the past, the modern-day platypus was thought of as a 'primitive' species that had survived, but current research and discoveries have led to the suggestion that it is a highly evolved form of an ancestor.

Figure 1.16
The platypus, with its
strange mixture of
characteristics

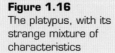

In 1985, a fossil jaw bone and three molar teeth of an early platypus ancestor (*Steropodon galmani*) were found in NSW. The main surprise came when its age was discovered to be 110 million years, making it the oldest mammal fossil found in Australia so far. Another possible ancestor of the platypus, *Obduron,* was discovered in 1991 in South America. It has been dated at about 62 million years old. Three species of *Obduron* have been found in Australia. Another convincing piece of evidence that supports the idea that the southern continents once were joined?

A current area of contention amongst biologists is that of the evolutionary relatedness of the platypus to other mammals. Studies of DNA sequences are being carried out to determine if it is more closely related to marsupial mammals or placental mammals, but there seems to be disagreement about a definite answer. And so the platypus enigma continues!

You will be given two introductory tasks to assist you in researching this complex topic. They will help you answer the third task which directly addresses the topic.

Background information to help get you started

Scientific work and ideas at the time (1798 onwards)

- During the late 18th and early 19th centuries, zoologists, botanists and other natural scientists were tackling the ongoing task of refining the classification systems of the animal and plant kingdoms.
- Evolution—that organisms change over time and came from a common ancestor—was not commonly accepted by scientists in the 1800s; the Darwin–Wallace theory of evolution was put forward and still being argued in the late 1800s.
- Amphibians, Reptiles, Birds and Mammals were all considered to be unrelated groups: their classification was based on evidence obtained from their external structure and their type of reproduction (in particular, whether they laid eggs, or their young were born alive).

Task 1 Structures in the platypus also observed in mammals and other vertebrate groups at that time.

Table 1.2 Features accepted as common to mammals and other vertebrate groups at that time

	Non-mammals	Mammals
Mouth	Bill or beak, no teeth. Birds (bill, duck)	Mouth with lips and teeth
Feet	Webbed feet: Amphibians (which were classified as part of the reptile class back then) and some birds (ducks)	Free digits, not webbed
Reproduction	Internal fertilisation, but egg-laying (oviparous): Reptiles and Birds Internal fertilisation, young develop in soft eggs inside the female's body (ovo-viviparous): some Reptiles	Internal fertilisation, young born alive
Parental care—feeding young:	Young are not suckled—no mammary glands	Mammary glands for milk production; young are suckled
Regulation of body temperature	Ectotherms—body temperature regulated by external sources of heat (e.g. in the environment). Body temperature tends to fluctuate more—Fish, Amphibians and Reptiles Birds are endothermic (see Mammals)	Endotherms—body heat is generated from internal sources. Body temperature is constant and hardly fluctuates with that of the external environment.

Analysing this information

Using the information in Table 1.2 as a guide, research which of the features the platypus shares with mammals or non-mammalian organisms. Using the table on the Student Resource CD list the features evident in the platypus and identify the *class of vertebrates* in which this feature commonly occurs.

Task 2 Platypus hypotheses that have been investigated

Use the details recorded in Task 1 and the statements below to guide your Internet searches, while attempting to do Task 3. As you do your research for Task 3 you should come across information that will help you to determine whether each of the following statements is true or false:
- The platypus gives birth to live young.
- The body temperature falls rapidly when a platypus swims and it has to return to its burrow to warm up. (Radiotelemetry is used to measure platypus temperature and data loggers are used to measure burrow and water temperatures.)
- The platypus is a rare species (studies used tagging, monitoring, radio-tracking and reporting on platypus sightings).
- The platypus has more sets of chromosomes than mammals; its chromosomes may be more similar to those of birds or reptiles (microscopic examination of chromosomes—research at Monash University).

- The platypus does not rely on vision, but on specialised sensory perception to locate food underwater. (Website search hint: type 'electroreception in monotremes' into your search engine.)
- The platypus as we know it is not a 'primitive' animal, it has evolved from an ancestral form.
- Platypus fossil ancestors such as *Obduron* lived before Gondwana split.
- Fossilised platypus ancestors may occur in southern continents other than Australia.
- Studies of evolutionary relatedness show that monotremes (such as the platypus) are more closely related to marsupials than to placental mammals. (Research involves analysis of nuclear DNA and mitochondrial DNA.)

See table: Features common to mammals, non-mammals and the platypus

'Then and now' table

Task 3 Information that you should gather, process and analyse

Use the 'Then and now' table on the Student Resource CD to record the information that you gather, as outlined below.

1. Outline ideas held 200 years ago about the platypus and its relationship to other individual species.
   (a) Use the background information above and the information in Tasks 1 and 2 to answer this question relating to the platypus specifically.
   (b) Also research the knowledge that the Aboriginal people had about the platypus at that time.

2. State the main difficulty involved in trying to classify the platypus.

3. Research the work of scientists in trying to solve the problem of platypus classification and evolution:
   (a) hypotheses posed (see hypotheses listed in Task 2)
   (b) evidence collected and technology used (research each hypothesis and briefly outline the technology used and the scientific findings for each)
   (c) conclusions drawn.

4. Outline the 'new idea' in science—how the platypus is classified today compared with the confusion of 200 years ago.

5. Putting it all together: explain how the ideas of scientists changed and how technology helped scientists to gather evidence to support their current conclusion.

REVISION QUESTIONS

1. List the present-day continents that made up Gondwana in the order in which they separated from the supercontinent.

2. State:
   (a) how long ago Australia is believed to have broken away from Antarctica
   (b) where predictions say Australia will be in 50 million years' time.

3. Explain how the fact that the Himalayan mountains are slowly rising supports the theory of continental drift.

4. Outline two examples of geographical evidence to support the theory that Australia was once connected to the supercontinent called Gondwana.

5. Name two *fossil groups* that are common to Gondwanan continents and identify on which present-day continents each of these fossil groups have been found.

Answers to revision questions

6. Name two *present-day organisms*, one plant and one animal, that are common to Gondwanan continents and identify on which present-day continents each of these fossil groups have been found.

7. Describe ways in which scientists can identify the positions of mid-ocean ridges and spreading zones.

8. Describe how information about the position of mid-ocean ridges and evidence of sea floor spreading can infer that the Australian continent is moving.

(*Note*: The final dot point has been moved from Chapter 1 to a more suitable placement in Chapter 2, page 275.)

CHAPTER 2

The evolution of Australian flora and fauna

The changes in Australian flora and fauna over millions of years have happened through evolution

Variation and evolution

This chapter deals with the changes in living organisms (**biota**) in Australia over millions of years. The theory of evolution cannot be conclusively proven, since it occurred over millions of years, and so scientists gather evidence to support (or backup) the theory. Evidence shows that millions of years ago living things on Earth were simpler and less varied than those living today. As time went on, a greater variety of new types of organisms appeared and often the new organisms were more complex and more advanced than previous ones. In this chapter, we investigate what could have led to the changes and look for evidence in fossil remains of **extinct** organisms, as well as in **extant** organisms (present-day living organisms).

- *discuss examples of variation between members of a species*
- *identify the relationship between variation within a species and the chances of survival of species when environmental change occurs*

Variation

In any population, although offspring resemble their parents, they are not identical to them (and are seldom identical to each other). The term **variation** applies to the *differences* in the characteristics (appearance or genetic make-up) of individuals within a population: not all humans look exactly alike and neither do all dogs or all cats or all elephants. Why do similar members of a population have *differences*? (See Fig 2.1.)

When plants or animals reproduce, the offspring resemble their parents in their basic characteristics. For example, adult elephants give rise to baby elephants and adult humans give rise to human babies, but adult elephants do not give rise to human babies! This rather absurd example gives us the idea that **heredity** is evident within living organisms. Heredity is the transmission of *similar* characteristics from parents to offspring (see Fig. 2.2).

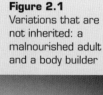

Figure 2.1
Variations that are not inherited: a malnourished adult and a body builder

Environment and variation

The concept of variation itself is not remarkable—differences in nutrition and lifestyle while we are growing up can have noticeable effects on our appearance; underfed children in third world countries have very thin arms and legs and bloated bellies, and those that survive to become adults will probably not reach their full potential height as a result of malnutrition. Body builders who work out at the gym regularly and eat high-protein food, have a resulting increase in body size due to muscle development and increased bone density. These examples of variation are due to interaction with the environment and are *not* passed on from parents to offspring.

Hereditary and variation

There is a second type of variation that *is* remarkable—the type of variation that can be passed on from parents to offspring. Some of the differences between individuals are *hereditary,* passing from parents to offspring. Red hair, blue eyes, close-set eyes and a protruding bottom lip are all examples of variations that seem to run in families. Inherited variations occur in all types of plants and animals (see Fig. 2.3) and it is these inherited characteristics that are important in our studies of evolution. Organisms in a population may vary in

appearance, physiology (functioning) and behaviour. *Heredity* and *variation* are both essential for evolution to occur. Variations which may pass from one generation to another are often produced in a population as the result of *mutation* (a change in the genetic make-up of an individual).

At the time when the scientists Charles Darwin and Alfred Wallace proposed their theory of evolution by natural selection, there was no knowledge of *what* was responsible for the differences in individuals within a population or of *how* such characteristics could be passed on from one generation to the next. Today, with our knowledge from studies of genetics, we know that *genes* on *chromosomes* determine characteristics that are inherited and that *sex cells* or *gametes* carry these characteristics from parents to offspring. When gametes combine, they may bring together two different varieties of the same gene (e.g. the gene for eye colour from each parent). Some genes have more than two varieties within a population; for example, the genes for eye colour and hair colour in humans have many more than two colours. It is these different types of genes in a population that bring about the type of variation that can be inherited in individuals, an essential ingredient in the process of evolution.

Figure 2.2 Heredity— adult animals/humans and their young

(a)

(b)

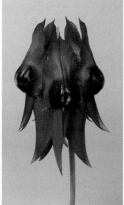

Figure 2.3
Heredity variation:
(a) breeds of dogs—
Jack Russell terrier
(smooth-haired) (left),
Jack Russell terrier
(long-haired) (right);
(b) Sturt's desert
pea—flower colour

Sometimes a combination of the environment influencing hereditary factors may lead to variation—the fur of many arctic animals changes to white when they are exposed to the extremely low temperatures of arctic winters; the ability of the fur to change colour is inherited, but the actual change only comes about if exposed to extreme cold.

So we see that, within a population, variation may have one of three origins:

■ genetic
■ environmental
■ or a combination of both genes and the environment.

Variation and survival of populations

The variation in the **gene pool** of a population (all the *possible varieties* of a gene within a group of *interbreeding organisms*) is important in determining the chances of survival of that population. If there is a sudden *change in the environment*, those individuals in the population that randomly possess a variation that is of *advantage* are more likely to survive the changed conditions. Individuals that do not possess that variation may be unable to compete and survive. Those *that survive are more* likely to reach an age where they can reproduce and pass their favourable characteristic on to their offspring. Individuals with *less favourable variations* will eventually be eliminated from the population as they are out-competed. If individuals within the population become so different that they can no longer interbreed with individuals from the original population to produce fertile offspring, then the population is considered to be a new **species**. Therefore variation in a population is extremely important, because it gives the population a better chance of surviving a sudden environmental change.

Examples:
- if the climate became hotter and drier, those individuals (plant or animal) with a natural variation that allowed them to retain more water would be at an advantage and survive
- if a disease was to sweep through a population, those individuals with a natural resistance to the disease would survive.

These individuals would live to pass on their favourable variation to offspring. We say that these organisms have **adaptations** which enable them to survive—that is, they are adapted to the new environment. (Refer back to 'A Local Ecosystem' where you studied *adaptations* of organisms within your local area.)

The Darwin–Wallace theory of evolution by natural selection

In 1858, based on their independent studies and observation of flora and fauna over many years, both Charles Darwin and Alfred Wallace proposed the same mechanism for evolution—the mechanism of **natural selection**.

Their theory of evolution by natural selection is based on four main points:
1. variation—individuals within a population that reproduce sexually, show *variations* that can be passed from one generation to the next
2. natural selection—**selective pressure** (e.g. change in the environment) puts constraints on organisms (e.g. resources become limited). These constraints are called *selective pressures* and determine which individuals are best suited to the prevailing conditions

Figure 2.4
(a) Charles Darwin;
(b) Alfred Wallace

(a)

(b)

3. **survival of the fittest**—more individuals are produced within a population than can survive; those individuals with favourable variations have a greater chance of survival because they out-compete those with less favourable variations (there is a 'struggle for survival'). Organisms that do survive to reproduce will pass their genetic *variations* on to their offspring

4. **isolation**—if a population is isolated from the original population, interbreeding will be prevented over a period of time. This is necessary for evolution of a new species to occur.

Darwin and Wallace's two main ideas

■ Natural selection and isolation are the *mechanisms* by means of which organisms evolve: the environment selects individuals, *based on variations* that favour their suitability. When resources in the environment become limited, these individuals survive, reproduce and pass on their characteristics.

■ **Speciation**, the formation of a new species, occurs when a population becomes *isolated* from the original group of organisms. Only those individuals that have variations that favour their survival under the changed conditions will reproduce and pass on their characteristics to the next generation. Eventually, the population becomes so different from the original population that they are no longer able to interbreed and produce fertile offspring. A *new species* has been formed.

Variation in living species

■ *Perform a first-hand investigation to gather information of examples of variation in at least two species of living organisms*

FIRST-HAND
INVESTIGATION

BIOLOGY SKILLS

P12.1; P12.2; P12.4
P13

Variation is quite easily observed in different members of the same species. One simply has to look around the classroom to detect variations within the human population.

Introductory task

On a piece of paper, list:
■ at least three inherited characteristics that make us 'human' and distinguish us from all other animals (e.g. speech is one such characteristic). These characteristics represent the heredity aspect of our species
■ as many variations amongst humans as you can (take 1 to 2 minutes to do this, the list is endless). Try to work out which of these variations are hereditary, which are due to environment and which are a combination of genes and the environment
■ in the form of a table, list three genetic variations amongst your class group and count the number of people that possess each type of variation.

Task

You are required to carry out a first-hand investigation to gather information of examples of variation in at least two species of living organisms.

Several examples of variations in Australian species have been listed below, but you may come up with your own and study variations in any population of organisms in which you are interested. Remember that the group of organisms you choose to study must belong to the *same species*. For example you could study variations in race horses (*Equus caballus*), but not variations between the race horse and wild horses, which belong to different species. Information on variations in other pets such as dogs, budgerigars or cats may also be gathered.

Variations within a plant species are also easy to study—look at the number of variations in a selected plant group (e.g. roses, gerberas or carnations) including the colour of flowers, number of petals, distribution of thorns or distribution of leaves.

Aim

Read the information below and then, with your specific choice of plants and animals in mind, write your own aim for this investigation. Record the results of your investigation in a suitable format.

Examples of variation in some Australian species

Magpies vary in body shape and size, beak shape and size and in the colour of plumage (feathers) on their backs, depending on their geographical distribution. Several different suggestions have been made as to the reasons for these variations, but evidence is inconclusive.

Koalas in northern Australia have smaller bodies, shorter hair and a lighter coat colour than koalas in southern Australia. Mammals that are larger tend to survive better in colder climates, because they have less body surface area compared with their volume and so they can conserve heat more efficiently. Wombats, koalas, kangaroos and possums of Tasmania are all larger than those of the same species found further north on the mainland.

Snow gum trees (*Eucalyptus pauciflora*) commonly growing in high altitude areas in and around the Snowy Mountains, tend to be shorter and have a shorter leaf length than trees that grow at sea level. There is a gradual decrease in height and leaf length with an increase in altitude. To prove that these differences are not just variations as a result of environmental differences, seeds from the different gum trees have been grown under standard conditions. Their height showed the same differences as when they were grown at different altitudes, implying that the variation was genetic.

Further references for Australian examples can be found on the Teacher's Resource CD and the Student Resource CD.

Additional information and websites on variation in Australian species

Figure 2.5 Magpies: (a) black-backed variety; (b) white-backed variety

(a)

(b)

2.2 The evolution of Australian flora and fauna

As Australia broke away from Antarctica and began to drift northwards, the continent became hotter and drier. The changes in climate led to changes in the vegetation (flora) on the continent, where only plants that could manage the hot dry conditions survived. Fire became a common feature of the land and this resulted in further change to vegetation. The changed flora led in turn to changes in the animals (fauna) which were dependent on the vegetation for food shelter and protection. All of these changes took place over a very long period of time— spanning millions of years.

Timeline: the formation of the Australian continent

■ *gather, process and analyse information from secondary sources to develop a timeline that identifies key events in the formation of Australia as an island continent from its origins as part of Gondwana*

SECONDARY-SOURCE INVESTIGATION

BIOLOGY SKILLS

P 12.3; P 12.4
P 13.4

Drawing a timeline

Revise instructions for drawing up a timeline (see page 79). The timeline you are required to develop for this part of the course spans millions of years and then the time scale changes to thousands of years. To indicate this, the parts of the timeline should be connected with a dotted line (see the Student Resource CD for sample layout of timeline).

Task

Refer to at least three secondary sources (including the information in Chapter 1, page 79, to develop a timeline which begins when Pangaea split into Laurasia and Gondwana, and ends with the Australian continent in its current position as we know it today). Listed below are some key events to guide you on the type of information that you should include on your timeline:

■ Pangaea splits into Laurasia and Gondwana
■ Gondwana begins to break up
■ Gondwana splits into:
  —Africa and South America
  —Australia, New Zealand, Antarctica and New Guinea
  —India
■ New Zealand separates
■ Australia begins separating from Antarctica
■ Australia becomes a separate continent
■ Australia reaches its current position.
  Later in this module, once you have completed the sections from 'Changing habitats' (page 258) to 'Arrival of humans' (page 267), use the text as well as other secondary sources to add the following interesting events to your timeline:
■ megafauna extinction begins
■ first small groups of people begin to arrive in Australia
■ climate warms and ice caps recede, sea levels rise and Australia is isolated from the rest of the world again

■ moisture-loving plants and animals become restricted to coastal regions
■ flora and fauna evolve in isolation.
  Colour code your timeline to show which events record the change of the Australian continent and which represent the change in biota (living things).

Sample layout of a timeline

 This website allows you to open a web page on each era, with information on Australia's position, climate, vegetation and animals for each. It is written in student-friendly language and is a reliable source: www.lostkingdoms.com/snapshots/geological_time.htm

Websites

Further task: putting it all together

As you continue with this unit of work, you will begin to see connections how the change in Australia's position impacted on the major changes in climate, flora and fauna. A table of comparison has been developed for you on the Student Resource CD which you can complete to create an integrated understanding of how these events fit together.

Student activity—timeline table comparing changes in Australia: formation of the continent, climate change, change in flora and change in fauna

2.3

Changing habitats

- **identify and describe evidence of changing environments in Australia over millions of years**

- 220–110 mya (dinosaur age): Australia was still joined to Antarctica as part of Gondwana. It had a cool, wet climate similar to that of Antarctica today, and could experience semi-darkness for many weeks in winter. A large proportion of inland Australia was covered by shallow sea.
  - —*Evidence*: Fossils of club mosses, which grew in swamps, have been found in Narrabeen shales (200 mya), providing evidence that present-day land in Australia was covered by shallow water.
- 45 mya: Australia became a separate continent with dry land, but great lakes remained in the interior.
  - —*Evidence*: Fossils of aquatic organisms such as crocodiles have been found in areas that are inland and dry in present-day times, suggesting the presence of lakes in the interior in the past.
- 20 mya: Australia was isolated from the other southern continents and as it slowly drifted northwards, its climate began to grow warmer, although it was still wet.
  - —*Evidence*: Tree rings are circular patterns made up of alternating light and dark bands of woody tissue. They reflect rapid growth in summer or warm temperatures and slow to no growth in winter or cool temperatures. Fossilised remains of trees give information in their ring patterns that support the described changes in climatic conditions.

(Additional fossil evidence which supports the more recent changes in climate as described in the following text can be found under 'Evidence

of changes in flora and fauna' in Section 2.5 on page 261.)

- Eight mya: Australia drifted further northwards and became much drier, as the water was captured in large ice sheets when Antarctica froze over.
- 100 thousand years ago (time of megafauna): Australia had drifted very close to its present-day position and the decreased sea levels led to it being temporarily connected by land bridges to Tasmania (in the south) and New Guinea (in the north).
- Up until ten thousand years ago: Australia would go through cycles of higher and lower sea levels as the polar ice caps fluctuated—they would melt and then refreeze. Land bridges would form and then be submerged. Australia's climate also became slightly wetter and drier. (It was thought that humans arrived during this period, approximately 40 thousand years ago, 'island hopping' from the north.)
- Five thousand years ago: Australia's climate continued to dry and most of the interior of the continent gradually became the arid, water-scarce climate that we know today.
  - —*Evidence*: Plants of Australia's present-day open forest, woodland and heath habitats show adaptations to dry, harsh climatic conditions. No fossil remains of these plants (e.g. waratahs) have yet been discovered, implying that they are recent vegetation and were not part of the Australian vegetation millions of years ago. This provides further evidence that Australia's climate as we know it today is very different from that of the past.

(Also refer to Section 2.5 on page 261.)

Variations in temperature and water availability

■ *identify areas within Australia that experience significant variations in temperature and water availability*

Australia's climate has become hotter and drier as the continent has drifted north, resulting in increasing aridity (dryness) as you move inland from the coast. The temperature decreases from the north to the south of the continent. It is interesting to note that there is generally a reduced floral diversity from north to south. Climate can be classified, based on temperature and rainfall. This in turn determines the native vegetation of an area.

Variation in the temperature and/or water availability within the Australian environment has acted as one of the major selective pressures in Australia. Organisms that have random variations that suit them to particular habitats out-compete others.

Figure 2.6 illustrates the climate classification of Australia, showing different climatic regions (temperature) and rainfall zones (water availability).

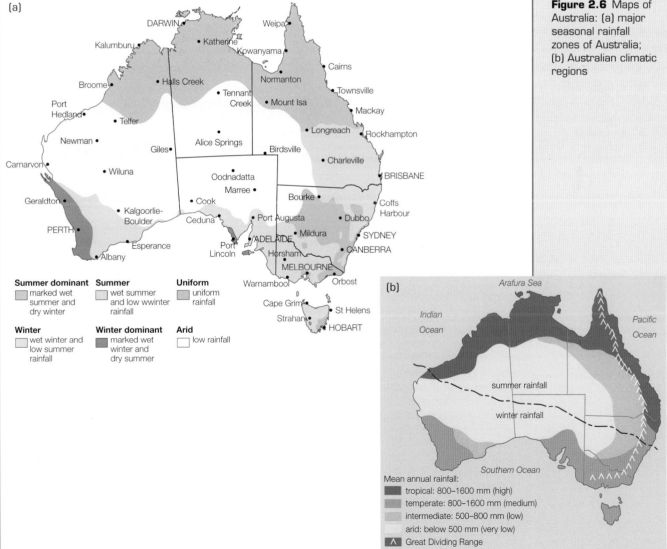

Figure 2.6 Maps of Australia: (a) major seasonal rainfall zones of Australia; (b) Australian climatic regions

STUDENT ACTIVITY

On the map of major Australian ecosystems provided on the Student Resource CD, colour each of the major rainfall areas, using Figure 2.6b and the information in Table 2.1 as a guide.

Map of major
Australian ecosystems

Table 2.1

Ecosystem	Rainfall	Temperature location	Vegetation
1. Rainforest	Annual rainfall is high and occurs throughout the year; tends to be along the coastline	Rainforests may be: ■ *tropical*—hot and humid (e.g. Far North Queensland) ■ *subtropical* (e.g. southern Queensland and New South Wales) ■ *temperate* (e.g. New South Wales to eastern Gippsland, Victoria) ■ *cool temperate* (e.g. Tasmania and Victoria).	Contains many 'primitive' plants that are direct descendents of the earliest Gondwanan species
2. Eucalypt forest ■ wet sclerophyll forests ■ dry sclerophyll forests	Requires more water than woodlands, but less than rainforests	Coastal hills, ridges and slopes of the Great Dividing Range	■ Tall moisture-loving trees, compete for sunlight therefore tall; epiphytes are common ■ Dominated by eucalypts of short to medium height. Has 'hard-leaved' shrubs growing beneath trees
3. Woodlands	Drier than eucalypt forests	Woodlands in drier climates—trees more sparse, western slopes	Many evergreen trees including eucalypts
4. Shrubland: ■ mallee—low to tall shrubland ■ mulga—tall shrubland, dominated by acacias	Semi-arid; hot dry summers, mild winters with unreliable rainfall	Southern parts of Australia, temperate areas (mild summer and cold winter)	Contains saltbush, bluebush, acacias and multi-stemmed eucalypts; low shrublands where soil is saline; open forest with stunted trees
5. Grassland	Semi-arid; typical summer rainfall, heavy storms may occur and frequent fires	Hot and dry; temperature variation is seasonal, but within a narrow range	(Includes cultivated species such as wheat, oats, barley, corn and rice, and pasture for cattle and sheep). Much of the original grasslands are now modified as a result of farming
6. Desert	Very little moisture; rainfall is low (less than 250 mm per year)	Largest temperature fluctuations from hot during the day to freezing at night	Sparse covering of grasses (e.g. spinifex); flowering drought resistant plants

Table 2.1 describes ecosystems from those with regular rainfall (at the top of the table) to those with the greatest variation in rainfall (at the bottom of the table). Temperatures also vary within these biomes. Not shown on the map are the alpine rainforests—these occur between the limit of tree growth (the tree line) and the zone of permanent ice and snow. It also has great variation in (and extremely cold) temperatures. The mean annual temperature in the warmest months is 10°C (south-eastern highlands, Mt Kosciusko, and the 'snow country' of Victoria and Tasmania). Snow gum woodlands grow in these areas at the lower altitudes.

In general the Australian climate can be described as follows:
■ temperature—hot
■ rainfall—low
■ aridity—high
■ topography—flat
■ soils—eroded and infertile
■ fires—frequent.

A hostile environment such as this provides many selective pressures which would lead to natural selection and the survival of uniquely adapted flora and fauna.

Changing flora and fauna

2.5

■ *identify changes in the distribution of Australian species, as rainforests contracted and sclerophyll communities and grasslands spread, as indicated by fossil evidence*

Origins of flora and fauna

If the climate in a habitat changes, the distribution and abundance of living things within that habitat also tends to change. Organisms that are better suited to the new environment survive and new species may evolve.

The distribution and abundance of present-day *plants* in Australia have three main origins:

■ those already on the continent when it split from Gondwana
■ those that dispersed from South East Asia to Australia
■ introduced species (e.g. with the arrival of humans).

The origins of *animals* that led to the present-day fauna (see Fig. 2.9) are:

■ 'original residents'—those that were on the continent when it split from Gondwana (e.g. frogs, reptiles, monotremes, marsupials, emus and lyrebirds)
■ Asian 'immigrants' that arrived when sea levels were low—15 mya and again 40 000–30 000 years ago (e.g. poisonous snakes, back-fanged snakes, rats, mice and bats)
■ those introduced by immigrant traders or late arrival aborigines—4000 years ago (e.g. dingos)
■ those introduced by European immigrants—120 years ago.

This section of the course requires us to start our description of the changes in flora and fauna from the time when rainforests contracted. Some fossil evidence is described in the text below and other fossil evidence is in the secondary source investigation which follows (page 267).

Changing flora and fauna

■ 65 mya: Moist climates supported rainforests which formed the dominant vegetation type (before Australia separated from Gondwana). Rainforests had replaced the towering conifer forests of the previous eras. The climate was wet and warm and there was a large variety of flora and fauna. (Animals included the early relatives of many of the animals we know today, such as koalas, kangaroos, bats, crocodiles and possums.)

■ 45 mya: As Australia separated from Antarctica and began drifting north, the climate dried out and the Australian rainforests contracted, remaining mainly in the coastal regions of Australia. The inland areas which were drying out had more open forests and woodlands. (Animals diversified and varieties that were more similar to the animals we know today developed— kangaroos that hopped rather than walking, large herds of animals that resemble modern-day wombats and carnivorous predators such as the thylacines and marsupial lions.)

—*Evidence*: Fossils at Riversleigh in north-western Queensland demonstrate the change from rainforest to dry habitats (see page 271). They show that inland forests dried up and vegetation changed from forest to woodlands. Some living plants (e.g. the southern beech, *Nothofagus*) are remnants of a time when Australia's climate was much more tropical.

■ As the continent continued to dry out, gum trees and wattles became common in Australia's forests and many wild flowers bloomed. Newly developed fauna included salt water crocodiles and budgerigars. This era formed the link between ancient and modern vegetation. Conifers and cycads were decreasing in importance as flowering plants bloomed.

■ The climate then went through a period of fluctuation from wet to dry and the pattern of forests, grasslands and desserts kept changing. The megafauna abounded, with huge diprotodons (wombat-like animals) and giant goannas *(Megalania)*.

■ Indigenous people arrived and used fire to clear vegetation for movement across the land and to burn off particular areas of bushland.

　—*Evidence*: Fossils show an increase in the incidence of carbon deposits which coincided with the arrival of humans.

■ About 5000 years ago, Australia's climate became consistently drier and has continued to do so. The dry climate meant that there was lightning which gave rise to fires, possibly influencing the change to fire-resistant species which began to flourish.

　—*Evidence*: Carbon deposits dating from that time and the presence of the pollen of fire-resistant species, reveal a 'fire–vegetation' relationship.

■ This further shaped the flora of Australia. Rainforest was replaced by more fire-tolerant open forests, including vegetation such as eucalypts which had woody fruits that burst open as a result of the heat of fire.

Figure 2.7 Fire is an important part of the present and past Australian environment

　—*Evidence*: Unique Australian plants have evolved in isolation, showing adaptations to the dry, harsh climatic conditions (e.g. leathery leaves of eucalypts). Modern Australian plants have a relatively short history with the earliest recorded pollen from a eucalypt dating back to 23 mya.

■ In the interior of Australia, grasslands developed. As sea levels have risen and fallen, Australia has experienced a lot of erosion and so nutrients have leached from the soil, resulting in soils which are nutrient-poor.

(a)

(b)

Figure 2.8 Changing Australian biomes—rainforests contract and grassland and sclerophyll communities spread: (a) rainforest; (b) grassy woodland

continued . . .

(c)

(d)

Figure 2.8 Changing Australian biomes—rainforests contract and grassland and sclerophyll communities spread: (c) grassland; (d) dry sclerophyll

Australia's unique flora

Plants which are able to withstand fire and grow in soils lacking nutrients are now characteristic of Australia. Dry sclerophyll plants (e.g. eucalypts and acacias—also known as wattles) and grasses that typically grow in nutrient-poor soils on rough terrain, such as the slopes of mountains. The term 'sclerophyll' means 'hard leaf' and these plants have evolved in many ways to allow them to use low levels of nutrients. Some developed **mutualistic** relationships involving their roots and soil bacteria and fungi (e.g. waratahs, wattles and banksias). The presence of such diverse species in such barren soils continues in present-day Australian flora. Heathland species are *fire-resistant* and their seedlings only emerge following a fire. Many of their seeds are dispersed by ants.

The arrival of the Europeans with their millions of sheep also influenced Australia's vegetation. A massive loss of edible perennial plants occurred as a result of *over-grazing*, erosion followed and dust storms killed off several native mammals. Europeans began clearing land for agriculture and rainforests diminished further. The modern remaining rainforests in New South Wales are very *diverse* and contain many plants that are direct descendants of the original Gondwanan species.

Modern-day flora is very different from that on other continents, with many species found *exclusively in Australia*. The familiar Australian native species such as wattles, eucalypts, banksias and she-oaks (casuarinas) all possess *adaptations* that allow them to thrive in soils that are nutrient poor, areas that are arid (short of water) and where fire is common.

Australia's unique fauna

Australia has a wide variety of marsupial fauna today, including bandicoots, Tasmanian devils, koalas, wallabies and kangaroos. Most living marsupials today are found in Australia and only a few species exist in South America. There are no remaining living species of marsupials on any northern continents or southern continents other than Australia and South America. This can be explained in terms of the theory of continental drift. When Africa and South America first broke away from Antarctica, the tip of South America was still very close to Antarctica. At that time marsupials are thought to have been evolving in Laurasia. They spread south to Gondwana as the split was occurring and continued southwards, eventually reaching the southern tip of South America, where they could cross the land bridge that still existed, to reach the Antarctica/Australia continent.

—*Evidence*: Marsupial fossils that are 100 million years old have been found in North America and fossil records show that marsupials were common throughout South America.

Once Antarctica drifted southwards and became cooler, the marsupials could not survive. Australia drifted northwards and became warmer and the marsupials flourished in these conditions.

—*Evidence*: This is borne out by living evidence—the fact that Australia has the greatest variety of marsupials that occur in every niche normally occupied by placental mammals on other continents.

The lack of the continuation of marsupial survival on the northern continents is thought to be as a result of them being out-competed by placental mammals. Africa split away from Antarctica before the marsupials crossed from America. It has been suggested that by the time the placental mammals reached the tip of South America, it had separated from Antarctica/Australia and so the placental mammals could not move onto those continents. There is still some speculation over this. The only remaining marsupials in South America are the opossums, which must have been able to survive the invasion of the placental mammals—perhaps they occupied different niches or the opossums were adapted well enough to co-exist with them.

As well as the wide variety and large number of unique marsupials in Australia, there are also large numbers of placental mammals—bats, rodents and seals, dugongs and dolphins. Bats, rats, mice and some poisonous snakes are thought to have been introduced by Aborigines who arrived later than the first group and by traders from Indonesia. Europeans introduced foxes, rabbits and dogs, as well as cattle and sheep.

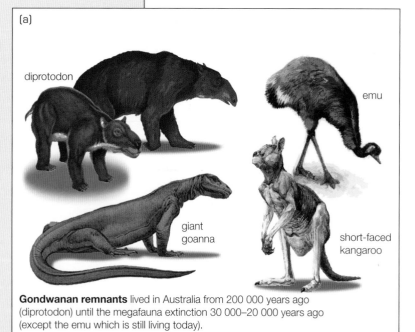

(a)

diprotodon

emu

giant goanna

short-faced kangaroo

Gondwanan remnants lived in Australia from 200 000 years ago (diprotodon) until the megafauna extinction 30 000–20 000 years ago (except the emu which is still living today).

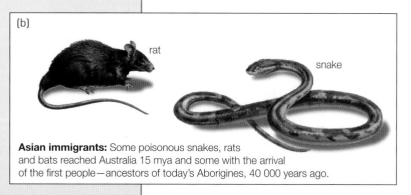

(b)

rat

snake

Asian immigrants: Some poisonous snakes, rats and bats reached Australia 15 mya and some with the arrival of the first people—ancestors of today's Aborigines, 40 000 years ago.

(c)

dingo

Asian: Introduced by Aboriginal immigrants or traders from Indonesia about 4000 years ago

Figure 2.9 Changing Australian fauna: (a) 'original residents'—Gondwanan remnants; (b) and (c) Asian immigrants

continued . . .

(d)

sheep

cattle

dog

rabbit

fox

European: introduced by European immigrants (1778 onwards)

Figure 2.9 Changing Australian fauna: (d) introduced by European immigrants

Figure 2.10 Present-day (a) marsupials; (b) placental mammals

(a)

(b)

2.6 Current theories to account for changes: Climate change versus the arrival of humans

■ *discuss current theories that provide a model to account for these changes*

There is an ongoing debate as to what led to the changes in the Australian flora and fauna, in particular the extinction of the megafauna. Until recently, many supported the theory that it was the result of climate change associated with the last ice age. Opponents of this theory proposed that it was humans alone who have caused almost all extinctions of animals throughout the world. Current researchers tend to think it may be a combination of the two—initiated by the change in climate, with human impact delivering the final blow.

Theory 1: Changes in climate

■ The continent dried out due to the ice age.
■ Rainforests were contracting due to a drying climate. Because rainforests had stored moisture and returned an enormous amount to the atmosphere as a result of transpiration, monsoon rains once penetrated south and kept the rivers and lakes in Australia full. As the rainforests diminished and were eventually replaced by eucalypt forests—these were less efficient at retaining water—less water was returned to the atmosphere.
■ As the climate became hotter and drier, fires broke out, initially due to lightning strikes and drier vegetation caught alight easily. Those plants

and animals that could survive drought and fire, reproduced and flourished, bringing about a change to the flora and fauna.

Arguments supporting climate change

■ Large animals such as megafauna, which were dependent on an ample supply of water, would have died out when water became scarce. They may also have died out because they could not manage the sudden change in temperature, their breeding seasons may have been affected and possibly the plants that they ate became less freely available and/or less palatable.

Arguments against climate change

■ The last ice age was probably similar to previous ice ages. If so, why would the last one have had such an immense effect, when there is no evidence that the previous ice ages had a similar result? Also, the earlier extinctions seem to have occurred before the peak of the last ice age.
■ Climate change today does not seem to select against large, slow-moving species.

Theory 2: The arrival of humans

Aborigines arrived in Australia about 40 000 years ago, probably having 'island-hopped' from the north. They were extremely successful predators.

Figure 2.11
Comparative sizes of extinct megafauna and extant similar species today (extinct species shaded)

Eastern grey kangaroo   Tasmanian devil   Koala

They used fire to burn back the bush—'fire-stick' farming techniques involved burning the bush to regenerate grasses for the animals and for themselves, since increasing an abundance of animals meant that there would be more available for hunting.

Humans hunted the megafauna and, because the larger animals were slower, they were the ones that were killed. The smaller, faster animals that escaped survived to pass their genes on and so the species evolved to become smaller.

It appears that the original indigenous people killed off most of the Australian terrestrial animals that were larger than they were. The introduction of the dingo may have also led to a decrease in the diversity of carnivore predators—it is possible that the dingos drove the thylacine and Tasmanian devil to extinction on the mainland.

Arguments against the arrival of humans
- There is no fossil evidence of kill sites and very little evidence of humans and megafauna coexisting.

- If you consider the size of the animals, there is an overlap in the size of the smallest extinct species and that of the largest present-day species.

Arguments supporting the arrival of humans
- Main evidence for the theory of the arrival of the humans at the same time as the increase in fires is that the increase in carbon deposits in fossils coincides with the time of the arrival of humans (40 000 years ago).

The smaller species of megafauna which became extinct had short limbs which would have made them slow; the largest surviving of our present-day species are also among the fastest (e.g. red and grey kangaroos).

A third theory accounting for the survival of smaller animals is that there were such low levels of nutrients in the soils in Australia, that this may have caused a nutrient depletion throughout the food chain, resulting in smaller animals. The smaller size of mammals in Australia compared with their counterparts on other continents today could provide possible evidence.

Evidence of the evolution of Australian flora and fauna in fossils

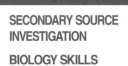

- ***gather information from secondary sources to describe some Australian fossils, where these fossils were found and use available evidence to explain how they contribute to the development of understanding about the evolution of species in Australia***

SECONDARY SOURCE INVESTIGATION

BIOLOGY SKILLS

P14

Aim

1. To gather information from secondary sources.
2. To describe some Australian fossils and where these fossils were found.
3. To use available evidence to explain how fossils contribute to the development of understanding about the evolution of species in Australia.

Background information

Australian fossils found from the past are an invaluable source of information looking at past relationships between organisms, evolutionary changes over time and commonality between extinct and present-day organisms.

This information not only tells us a lot about the changes in organisms over time and their relationships but also the changes in environments over time. Tables 2.2 and 2.3 describe examples of Australian animal and plant fossils found from varying periods of time in the past. One recent discovery in Australia (2007) is the first evidence of a new species of titanosaur, dated at approximately 96–100 million years. The remains, found in Eromanga, Queensland, included a 100 kg thigh bone spanning 1.5 m. This suggests the plant-eating dinosaurs were about 26 m long; by far the largest dinosaur bones ever found in Australia. This is a rare find because it is a complete bone.

For photographs and websites on Australia's latest dinosaur bone

Table 2.2 Fossils of Australian animals

Fossils of Australian fauna

Examples of some fossils of Australian animals are summarised Table 2.2.

Period of time (million years ago)	Where found	Fossil name	Animal group	Description
345–280	Clarencetown, New South Wales	Brachiopod (lamp shells)	Brachiopod	Small organisms, only shell visibly preserved, grooves prominent in shell, semi-circular shell shape
230	Bellambi Colliery, New South Wales	Labyrinthodont	Amphibian	Five narrow toes (unlike dinosaurs that possess three toes) with thin sharp claws, only footprint of organism preserved
225–190	Hornsby, New South Wales	*Cleithrolepsis*	Fish	Small size (relative to a coin), tail fin and upper and lower rear fin are prominent, head structure clear, very clear side-on view of fossil detail
110	Lightning Ridge, New South Wales	*Steropodon galmani* (steropodon)	Mammal	Only fossilised jaw preserved with three molar teeth

Table 2.3 Fossils of Australian plants

Fossils of Australian flora

Examples of some fossils of Australian plants are summarised in Table 2.3.

Period of time (million years ago)	Where found	Fossil name	Animal group	Description
410–345	Mt Pleasant, Victoria	*Baragwanathia longifolia*	Club moss	Swamp plant with small, thin, feathery-type leaves protruding from a long main stem; only leaf and stem structure are preserved
280–225	Hunter Valley, New South Wales	*Glossopteris* (refer to Fig. 1.4 on page 240)	Tree fern	Tongue-shaped leaves with a midrib and net venation, found in swampy habitat; only leaf structure preserved
225–190	Beacon Hill, New South Wales	*Dicroidium*	Tree fern	Forked-frond seed fern; good clear outline of structure of frond preserved
190–120	Gulgong, New South Wales	*Pentoxylon*	Pine	Kauri pine with small, long narrow leaves; only remnants of stems with leaves preserved

Note: For photographs of fossils listed in Tables 2.2 and 2.3 see Mary E. White's *Australia's Prehistoric Plants and their Environment*, Methuen, 1984.

Contribution of fossils towards our understanding of evolution

The development of our understanding about the evolution of species in Australia can be largely contributed to by the discovery of fossils in the following ways.

■ The sequence of fossils in different rock layers indicates the progressive sequence of change in plants and animals over millions of years.

■ The dating of rocks and the disappearance of fossils at a certain date indicates the time period when extinction occurred. This time period could also be linked to events that happened over time which may have caused the extinction. Fossils help scientists to develop theories to explain the reasons for evolutionary changes, such as extinction.

For example, the extinction of *Genyornis*, a bird. Was it the arrival of humans or the melting of ice caps? If it was the melting of the ice caps, we would expect to find fossilised eggs of the *Genyornis* in upper layers of rock, however, they are only found in lower rock layers. This suggests that they may have become extinct after humans arrived, but before the melting of the ice caps. Humans therefore may have been responsible for the extinction of the *Genyornis* (perhaps through hunting).

■ The type of plant and animal fossilised indicates the type of environment in that area in the past. For example, finding fossils of crocodiles in areas that are now flat, arid and have no water suggests that these areas once had water. In turn, in these areas there once existed aquatic organisms.

■ By studying both the evolution of individual species and the terrestrial ecosystems through time, we can better understand our present environment and the conservation status of our fauna. For example, the Tasmanian tiger (*Thylacinus cynocephalus*). Fossil evidence found at Riversleigh provides warnings that the thylacine species decreased as sediments became more recent, showing a decline in diversity and geographic range over time. If we observe such warning signs early, we may be able to prevent extinction from occurring in other species.

Tables 2.2 and 2.3

Websites

www.lostkingdoms.com (*Australia's Lost Kingdoms* online exhibition)
www.amonline.net.au
(Australian Museum online)
www.austmus.gov.au (Australian Museum)
www.abc.net.au/science (ABC Science)
www.museum.vic.gov.au (Museum Victoria—megafauna)

Suggested Australian fossils

Plants

Phyllotheca (horsetail)
Lycopod (club moss)
Gangamopteris (tree fern)

Animals

Kambara implexidens
(Tingamarra swamp crocodile)
Pengana robertbolesi (flexiraptor)
Obdurodon dicksoni (Riversleigh platypus)
Nimiokoala greystanesi
(Riversleigh rainforest koala)
Dromornis stirtoni (Stirton's thunder bird)
Liasis species (Bluff Downs giant python)

Method

Part 1: Gather information from secondary sources

Refer to page 18 ('Searching for information') for suggestions and a reminder on how to go about gathering information from secondary sources.

Part 2: Description of Australian fossils and where they were found

Read the background information and use Tables 2.2 and 2.3 as guides for constructing a table to describe examples of Australian plant and animal fossils. Use the examples provided in Tables 2.2 and 2.3 as a start. Select two more fossil examples of your own and gather information from secondary sources to obtain the information required for your table. Attach a copy of the fossil photograph or drawing to your notes.

Part 3: Contribution of fossils to the development of understanding about the evolution of species in Australia

Read the background information that provides evidence of the contribution that fossils make in the development of our understanding about the evolution of our own Australian species. Search for any further available evidence to explain another possible way that fossils may contribute to our understanding.

Results

■ Construct a table, similar to that provided in Tables 2.2 and 2.3, describing at least three plant and three animal fossil examples.
■ Summarise into points the background information and extra evidence you were able to obtain for explaining the contribution of fossils in the development of understanding about Australian species evolution.

Discussion/conclusion

1. Identify two Australian plant fossils, where they were found and the period of time they existed, and provide a brief description.
2. Do the same as Question 1 for two *named* animal fossil examples.
3. Identify the common observations that were made regarding the quality of the fossil examples.
4. List four ways in which fossils contribute to the development of understanding about the evolution of species in Australia.
5. Select one of the fours ways you discussed in Question 4 and provide an example in support of this point.

Student activity

Comparison of current and extinct Australian life forms

■ *perform a first-hand investigation, gather information of named Australian fossil samples and use available evidence to identify similarities and differences between current and extinct Australian life forms*

Aim

1. To perform a first-hand investigation.
2. To gather information of named Australian fossil samples.
3. To use available evidence to identify similarities and differences between current and extinct life forms.

Recommended class activity

The Australian Museum provides excellent first-hand experience with many Australian fossil samples.

Australia's Lost Kingdoms online exhibition: www.lostkingdoms.com

Method

Part 1: Perform a first-hand investigation

Make careful observations of three Australian fossil samples. Draw each one carefully then copy and complete Table 2.4. If you do not have any Australian fossil samples first-hand, use the secondary information provided in this text, information gathered from other secondary sources and websites or from information gathered during a class visit to the Australian Museum (or equivalent).

Part 2: Gather information of named Australian fossil samples

Use the information provided from the previous investigation (Tables 2.2 and 2.3, and the 'Suggested Australian fossils' list on page 269) and any further secondary source information, to gather details on three named Australian fossils in order to complete Table 2.4 in Part 3. Remember to gather comparative information on the modern day equivalent of each of your three named fossil organisms.

Part 3: Identify similarities and differences between current and extinct Australian life forms

Once you have gathered information on your three named Australian fossils and information on their modern day equivalent organisms, copy and complete Table 2.4 (see Student Resource CD). Two examples have been provided to get you started.

Analysis questions

1. Describe the Australian organism that has the greatest similarities between current and extinct forms.
2. Name the organism with the least similarities. Explain possible reasons why this organism has a greater amount of difference between extinct and current forms.
3. Identify one *named* Australian organism and describe the similarities and differences between its current and extinct forms.
4. Identify the main difference between Australian plants and animals from hundreds of millions of years ago and those found in the present day.
5. Explain why some fossils may have no similar living relative or modern day equivalent.
6. Describe how studying current and extinct Australian organisms can contribute to our understanding of evolution.

Table 2.4

Results

Table 2.4 Comparison between current and extinct Australian organisms

Australian organism (current)	Australian organism (extinct)	When extinct organism lived	Similarities	Differences
Crocodile 	Tingamarra swamp crocodile (*Kambara implexidens*): skull found at Riversleigh in north-western Queensland 	55 million years ago	■ Diet: small vertebrate animals, small mammals, turtles, snakes and fish ■ Environment: swamp area in Queensland ■ Body structure: reptilian scales, long and strong tail, large snout and sharp carnivorous teeth	Tingamarra swamp crocodile was much smaller than the present-day freshwater and saltwater crocodiles being only 1.5 m in length
Platypus	Riversleigh platypus (*Obdurodon dicksoni*): pieces of skull and other skeleton parts found at Murgon in south-eastern Queensland 	23–10 million years ago	■ Diet: insect larvae, yabbies and other crustaceans ■ Environment: freshwater pools surrounded by rainforest ■ Body structure: appears similar ■ Specialist organ: have electric sensors in their bill to find its underwater prey	Riversleigh platypus was larger in size, had a much larger bill, had large teeth (the present-day platypus has no teeth at all)

2.7

Darwin revisited

■ *discuss Darwin's observations of Australian flora and fauna and explain how they related to his theory of evolution*

Darwin travelled the world on a surveying ship, HMS *Beagle*, when he was in his early twenties. The ship circumnavigated the world in five years. During that time, Darwin spent many hours observing the wide range of living organisms in the various countries where they stopped. Many of his famous observations were made in the Galapagos islands and also some important ones in Australia. He read extensively while travelling, including books on geology by Lyell, who proposed that, based on *fossil records*, plant and animal *species changed* and at times died out. He also later read books by Thomas Malthus on organisms' 'struggle for existence' if they outgrew their resources such as their food supply, introducing the idea of *competition*.

His observations in Australia led him to question the conventional ideas of the time and prompted his thoughts on *evolution*. He visited Sydney and the Blue Mountains while the ship had anchored so that the crew could check the clocks at the observatory in Sydney. (Today there is a walking track in the Blue Mountains called 'Charles Darwin's walk'.)

Darwin's observations were meticulously recorded; he collected huge quantities of specimens to send back to England for classification, according to the relatively new Linnaean system. He asked penetrating questions and spent a long period of time 'pondering' what he had observed, leading to his proposal of the theory of evolution by natural selection. Alfred Wallace proposed a similar theory at the same time as Darwin, and both of their work was backed up by research and observations.

Darwin tried to explain the similarities and differences (variations) that he observed. He believed that, rather than similar creatures being created independently, they could arise from a common ancestor, accounting for basically similar organisms changing to become different (**divergent evolution**). He also suggested that organisms that started off distantly related, but were subjected to similar environments could evolve similar features (**convergent evolution**).

Darwin had made numerous observations including these examples of the fact that similar environments in completely different parts of the world seemed to be inhabited by animals having similar adaptations, but obviously belonging to different species. His curiosity drove him to think about possible reasons for the resemblance: why would a creator bother to make two animals, so different in basic design and in the way that they produce and raise their young, rather than just one type of animal, since they live in such similar environments but in different parts of the world.

Observation forced him to discard the idea of fixed, unchanging species and so Darwin set out to try to find an explanation for his observations. He made many observations, both before and after his visit to Australia. Some of his most famous work included the study of finches and other birds, where he found that island populations may differ noticeably from mainland birds as a result of isolation and genetic drift. Darwin would have delayed publishing his ideas even longer if Wallace, who reached the same conclusions as Darwin, had not contacted him. Colleagues arranged for a joint publication of their papers.

For additional information on Darwin

Darwin's observations	How Darwin's observations related to his theory of evolution by natural selection
Magpies and crows are similar to jackdaws in the England, but obviously belonged to different species. The potoroo (rat kangaroo) is similar to the rabbit in England. It is a miniature kangaroo the size of a European rabbit, behaving somewhat like a rabbit, darting about in the undergrowth. The platypus is similar to water rats: 'I . . . had the good fortune to see several of the famous platypus. They were diving and playing in the water; but very little of their bodies were visible, so that they only appeared like so many water rats'.	Darwin's observations of birds, marsupials and monotreme mammals in Australia revealed similarities with mammals in Europe which lived in similar environments. This led him to the idea that organisms could evolve to become similar (convergent evolution). If organisms live in similar habitats, similar variations that they possess would be favoured by natural selection to enable them to survive and breed in those conditions. These favourable variations would then be passed on to the next generation.
Ant lions (same genus but different species?) to that of England: 'I observed a conical pitfall of a Lion-Ant: a fly fell in and immediately disappeared; . . . without doubt this predacious lava [sic] belongs to the same genus, but to a different species from the European one . . . Now what would the Disbeliever say to this? Would any two workmen ever hit on so beautiful, so simple and yet so artificial a contrivance? I cannot think so. The one hand has worked over the whole world'.	Their behaviour is similar to those in England. Same genus but different species?
Eucalypts: Darwin describes, 'the nearly level country is covered with thin scrubby trees, bespeaking the curse of sterility'. He also mentions 'the leaves are not shed periodically'.	In Darwin's observations of plant life in Sydney, he made the link between the harsh environment and the adaptations observed in the vegetation. He also mentions that many of the trees in Australia and other southern continents are evergreen as opposed to those in the northern hemisphere.

Source of quotes: Darwin, 1859, *On the Origin of Species*, http://Darwin.thefreelibrary.com/The-Origin-of-Species

Table 2.5
Darwin's observations in Australia

Figure 2.12
Darwin's observations of Australian flora and fauna

AUSTRALIAN

EUROPEAN

The Huxley–Wilberforce debate

SECONDARY SOURCE
INVESTIGATION

BIOLOGY SKILLS

P12.3; P12.4
P13
P14

■ *present information from secondary sources to discuss the Huxley–Wilberforce debate on Darwin's theory of evolution*

Background information

When Darwin and Wallace proposed their theory for a mechanism by means of which evolution could occur (natural selection), it created an uproar in scientific and religious circles, in fact in society in general. Despite being based on observations of an enormous number of plants and animals over a long period of time, in many different countries (including the Galapagos islands and Australia), the theory of evolution remained an extremely controversial issue. It brought into question the accepted beliefs of the time—that the world was created in six days, each species was created independently of another and that humans were created in the image of God. Darwin and Wallace's proposal that organisms gradually change over time and that one species could change into another, including the human species, was an extremely revolutionary idea.

A famous debate took place between Bishop Samuel Wilberforce and Thomas Henry Huxley at a meeting of the British Association for the Advancement of Science in 1860. The purpose of the meeting was for the reading of a paper with reference to the views of Charles Darwin (and others) at that time, that organisms evolved by the law of natural selection. This paper was to be presented by an academic, Dr John Draper, of New York University. The presentation topic was a highly contentious issue at that time and the reading was well attended, with high-profile members of society

forming part of the audience. Many that attended came to the presentation prepared to put forward and defend their own point of view and that of the sector of society which they represented. Following the reading of the paper, a lively and heated debate took place amongst members of the audience. With a fine display of polished debating skills, witty remarks and now-famous insults, strong opinions were voiced to contest the issue at hand. The reading of the paper faded into insignificance compared with the attention given to the heated debate that followed. Recounts of the famous debate that followed have been found in many forms:

■ as articles reported in journals of the time (*The Guardian, The Athenaeum* and *Jackson's Oxford Journal*)

■ in the form of four contemporary letters: Joseph Hooker to Charles Darwin; Richard Green to Sir William Boyd Dawkins, Balfour Stewart to David Forbes and Huxley himself to Henry Dyster.

However, there is no exact recording of the words used in the verbal exchange that took place and so these sources of information vary in their reporting of the actual wording of the content of the debate, but the general ideas and insults resound. In discussing this debate, it is interesting to look at the different accounts, each supposedly a record of the same happenings at the debate. This provides a good opportunity to evaluate the sources (both at that time and the sources we use) and to try to come up with indisputable facts (if any)

Figure 2.13
(a) Thomas Huxley;
(b) Samuel Wilberforce;
(c) cartoon of Darwin as an ape

(a)

(b)

(c)

about the debate. The debate is also of interest because journal reports of the verbal exchange gave evolution publicity, and knowledge of the theory of evolution spread.

Task

Research at least three current articles which address the Huxley–Wilberforce debate and answer the questions that follow. A table in which to record your answers is provided on the Student Resource and Teacher's Resource CDs.

Research questions

1. Which society hosted the debate? On what date? Where was it held?
2. Who was the invited guest speaker? What was the topic on which he spoke?
3. As a result of what circumstances did Wilberforce and Huxley end up speaking?
4. What position in society did Wilberforce and Huxley each hold?
5. Why was each one invited?
6. Which sector of society was represented by Huxley?
7. Which sector of society was represented by Wilberforce?
8. What was the issue being debated?
9. How well was each prepared for the debate?
10. What arguments were put forward by Huxley and by Wilberforce? (List at least three points made by each in support of their point of view. These should be reported in your own words.)
11. What was the insult that made the debate famous?
12. Was Darwin present at the debate? What was Darwin's reaction to the debate?

There are various accounts of the debate and the famous insult—refer to at least two accounts, preferably three.

To prepare for your discussion of the debate

- Note any inconsistencies in the telling of the story.
- Evaluate the reliability of *three* sources of information about the debate. (These may include current sources that you are using, as well as direct quotations from the letters written at the time of and after the debate.)
  —Note who wrote each source and the qualification of the author or the accomplishments and/or participation of the letter writer in the debate.
  —If you are using an Internet site or book as a secondary source, comment on the scientific standing of the publication.
  —Try to distinguish between fact and opinion in each of the three sources chosen for discussion. State whether the issues that show inconsistency are stated as fact or opinion and, if opinion, provide words used in the source to demonstrate that it is an opinion.
  —State any reasons why the source or authors may have been biased.

Discussion

Discuss the debate: look at recounts of the debate and discuss inconsistencies in how the debate is recorded. Explain how these inconsistencies arose. Acknowledge all your sources in the correct format of a bibliography. Remember to include the access date of any Internet sites.

Student activity and websites

Current research

2.8

- *discuss current research into the evolutionary relationships between extinct species (including megafauna) and extant Australian species*

(This dot point has been moved from Chapter 1.)

Current Australian scientists

As part of the requirements for the preliminary biology course, you are required to identify practising male and female Australian scientists in areas in which they are currently working and gather information about their research. (See skills outcomes P12.3e on page x.)

A large amount of work is being done in the area of evolution using current research techniques. This provides an ideal opportunity for you to meet these requirements. A large amount of work is being done at Adelaide University on extinctions of

Reference details
and websites for
publications

megafauna, impacts of climate change
and evolutionary relationships of extinct
species. You may also like to look into
the work of Tim Flannery—Australian
of the Year, 2007—who has studied
the changing Australian climate and
the impact of the arrival of humans
in Australia. He is renowned for his
research and many publications,
amongst them two books on Australia
past, present and future: *The Future
Eaters* and *The Weather Makers*.

Current research

When researching into the evolutionary
relationships between organisms,
scientists previously had to rely solely
upon fossil evidence and a comparison
of the anatomy of each organism.
Today, current research uses molecular
biology techniques which are far more
complex and accurate in providing
information about genetic relationships
between organisms from different times.
The technique of DNA hybridisation,
DNA sequencing and mitochondrial
DNA sequencing (the details of
these techniques will be covered in
the HSC course) have become even
more sophisticated since the 1980s so
the accuracy of these studies is being
improved all the time.

The Ancient DNA Laboratory at
the Australian Centre for Ancient DNA
(ACAD) is a new research initiative of

the University of Adelaide. Currently,
it is conducting research on the
extinctions of **megafauna** and the
evolutionary relationships between
extinct species. The international
standard facility provides laboratories
isolated from other areas of molecular
biology, so that they are protected from
environmental contamination, allowing
specialist work areas for ancient DNA
studies (see Figure 2.17). Molecular
biology techniques involve areas such
as protein and DNA isolation methods,
DNA sequencing and studies of DNA
damage or modification, in addition
to computer modelling programs.

Research is currently being
conducted on bones of Pleistocene
megafauna and extinct species.
They are also looking at evolutionary
relationships between extinct mammals
and birds, such as the Australian
megafauna (e.g. the thylacine). The
thylacine research project uses ancient
DNA preserved in fossil bones and
museum specimens. The project is
hoping to contrast thylacine information
to that of the Tasmanian devil which
managed to avoid extinction in
Tasmania. Another area of research
is being conducted on conservation
genetics of extant and extinct Australian
lizards and frogs, and also extinct and
endangered Australian marsupials.
An extant species is one defined as in
existence, still existing, and not lost or
destroyed.

The approach of these studies is
to integrate ancient DNA sequence
information with modern data from
archaeology, climate studies and
palaeontology, to analyse a variety
of evolutionary processes.

Figure 2.14
Modern researchers
are studying the
extinctions of
megafauna

**Further information about this
research can be obtained from the
University of Adelaide website:
www.adelaide.edu.au/acad**

Table 2.5 lists examples of extinct and extant (still in existence) Australian species that may demonstrate evolutionary relationships. The difference between the two groups (extinct and extant species) is that the extant group for each species must have possessed adaptations that the extinct group did not, ensuring that it survived to continue the species up to the present day. Although they may share many common characteristics and genetics, there must have been a significant difference to enable one to survive and one not to survive.

Some scientists suggest that the size of the megafauna made it an easy target for hunting due to its slow locomotion, whereas the smaller organisms at the time were unaffected and avoided extinction. Climate change is also thought to be responsible for the extinction of the megafauna. In this case the extant species' adaptations must have been more climate-related. However, it may have been a combination of human impact and climate change that led to the extinction. Perhaps a number of different adaptations saved the extant species from extinction. Fortunately there was enough genetic variation amongst species to allow some organisms to survive to continue the species to present day.

Extinct species	Extant species
Riversleigh platypus (*Obdurodon dicksoni*)	Modern platypus (*Ornithorhynchus anatinus*)
Perikoala palankarinnica *Litokoala kutjamarpensis* *Koobor notabilis* and *Koobor jimbarratti* Giant koala (*Phascolarctos stirtoni*)	Modern koala (*Phascolarctos cinereus*)
Diprotodon optatum	Common wombat (*Vombatus ursinus*)
Australian southern conifer (*Aganthis jurassica*)	Wollemi pine (*Wollemia nobilis*) Kauri pine (*Aganthis* sp.) Norfolk Island pine (*Araucaria heterophylla*)

Table 2.5 Examples of extinct and their respective extant Australian species

REVISION QUESTIONS

1. Describe three examples of Australian fossils and identify where these fossils were found.
2. Identify similarities and differences between current and extinct life forms in two *named* Australian fossil samples.
3. Explain how fossils contribute to the development of understanding about the evolution of species in Australia.
4. Distinguish between *geological* evidence and *biological* evidence.
5. Clarify what is meant by the terms *extinct* and *extant*.
6. Clarify what is meant by 'relict' species and identify an example of relict species form each of the following groups:
   (a) bacteria
   (b) plants
   (c) animals.
7. Define the terms *heredity* and *variation*. Give an example of each.
8. Outline two current theories that could account for the changes in Australian species (including extinction of megafauna). State one argument for and one against each theory.
9. Outline the following points about the Huxley–Wilberforce debate:
   (a) topic of debate
   (b) Huxley's point of view
   (c) Wilberforce's point of view
   (d) the famous insult.

Answers to revision questions

CHAPTER 3

Reproduction and continuity of species

Continuation of species has resulted, in part, from the reproductive adaptations that have evolved in Australian plants and animals

3.1 Cell division and the production of gametes

■ *distinguish between the processes of meiosis and mitosis in terms of the daughter cells produced*

The role of meiosis in maintaining chromosome number

Every species has a characteristic number of chromosomes per cell. For example, humans have 46 chromosomes, camels have 70, tomatoes have 24 and chickens have 78, the housefly has 12 and the fruit fly has 8. Analysis shows that the number of chromosomes does not necessarily reflect the complexity of the organism. The important thing to remember is that the chromosome number is constant for each organism and does not change from one generation to the next.

During **sexual reproduction**, two parents are involved in passing genetic material on to offspring. To prevent the chromosome number from doubling in each successive generation, a mechanism to ensure that each parent contributes only half of his or her chromosomes to the new offspring is necessary. The *importance* of **meiosis,** a type of cell division present in the reproductive organs of both plants and animals, is to ensure that the characteristic chromosome number is maintained during sexual reproduction.

The terms **diploid** and **haploid** refer to the number of *sets* of chromosomes within any cell. In most organisms, the **somatic** (body) cells contain *two sets* of chromosomes, i.e. the diploid number of chromosomes (e.g. in humans this number is 46 or 23 pairs). One set of chromosomes is inherited from the mother (**maternal** chromosomes) and one set is inherited from the father (**paternal** chromosomes). However when a cell involved in sexual reproduction divides by meiosis to produce **gametes** (sex cells), the chromosome number halves—that is, each resulting gamete contains only *one set* of chromosomes and is termed haploid.

■ If $n = 1$ set of chromosomes, then in humans, $n = 23$.
■ Human somatic cells are $2n$ (diploid) and have 46 chromosomes.
■ Human gametes are n (haploid) and have 23 chromosomes.

Meiosis: an introduction

Meiosis is the type of cell division that occurs in the sexual reproductive organs of a plant or animal and it results in the formation of gametes (sex cells). When a cell divides by meiosis, it undergoes two successive divisions—*meoisis I* where the cell divides into two cells and then *meiosis II* where those two cells each divide again, resulting in four daughter cells (called a **tetrad**). Each daughter cell has half the original number of chromosomes that the parent cell had. These resulting daughter cells, or gametes, are:

- **egg cells** and *sperm cells* in animals
- **pollen grains** (in anthers) and *egg cells (inside ovules)* in seed-producing plants.

Gametes are often referred to as 'vehicles of inheritance' because they carry genes from one generation to the next.

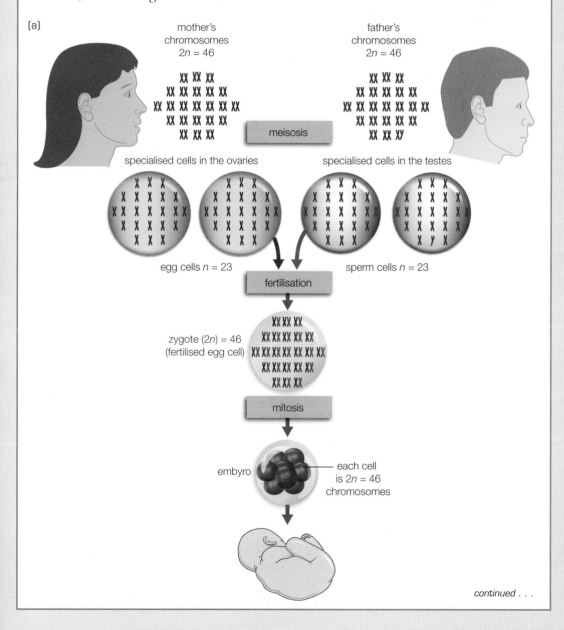

Figure 3.1 Sexual reproduction cycle showing maintenance of chromosome number: (a) in humans

(a)

mother's chromosomes 2*n* = 46

father's chromosomes 2*n* = 46

meisosis

specialised cells in the ovaries

specialised cells in the testes

egg cells *n* = 23

sperm cells *n* = 23

fertilisation

zygote (2*n*) = 46 (fertilised egg cell)

mitosis

embyro

each cell is 2*n* = 46 chromosomes

continued . . .

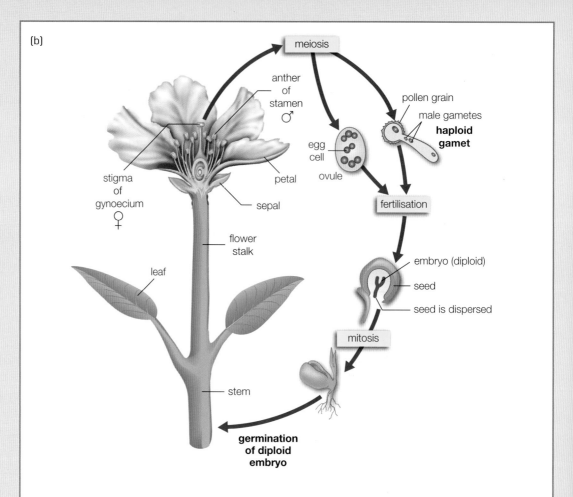

(b)

meiosis

anther
of
stamen
♂

pollen grain
male gametes
**haploid
gamet**

egg
cell
ovule

stigma
of
gynoecium
♀

petal

sepal

fertilisation

flower
stalk

leaf

embryo (diploid)

seed

seed is dispersed

mitosis

stem

**germination
of diploid
embryo**

The process of meiosis

Meiosis can be defined as a reduction
division whereby a diploid cell divides
into four haploid daughter cells (a
tetrad). Meiosis occurs in two stages,
meiosis I (the first meiotic division) and
meiosis II (the second meiotic division).
The reduction in chromosome number
occurs in meiosis I.

Similarities with mitosis

■ The names of the stages—interphase,
prophase, metaphase, anaphase and
telophase—are the same.
■ Interphase occurs first, prior to the
nuclear division. During this stage,
the DNA replicates, so that each
chromosome makes an identical
copy of itself (see Fig. 3.2a).
■ Chromatin material transforms into
chromosomes in the same way
during prophase in the first meiotic
division (see Figs 3.2b and c).

■ The breaking down of the nuclear
material and the formation of the
spindle are the same.
■ Cytokinesis takes place in the same
manner as for mitosis, depending
on whether it is a plant or an animal
cell dividing.

Terminology associated with the process of meiosis

A **homologous pair** of chromosomes
consists of two similar chromosomes:
one of the pair is maternal in origin
and the other is paternal. In humans,
there are 23 homologous pairs of
chromosomes. Each homologous pair
of chromosomes may be termed a
bivalent (see Fig. 3.2).

In the annotated diagrams in
Table 3.1, meiosis is represented in
a hypothetical (model) organism that
has only *two pairs* of chromosomes,
keeping the representation of the
process simple.

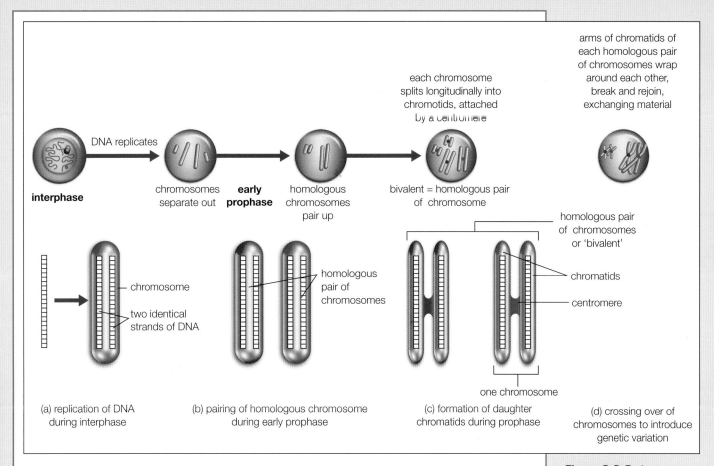

each chromosome splits longitudinally into chromotids, attached by a centromere

arms of chromatids of each homologous pair of chromosomes wrap around each other, break and rejoin, exchanging material

DNA replicates

interphase

chromosomes separate out

early prophase

homologous chromosomes pair up

bivalent = homologous pair of chromosome

homologous pair of chromosomes or 'bivalent'

chromatids

centromere

chromosome

two identical strands of DNA

homologous pair of chromosomes

one chromosome

(a) replication of DNA during interphase

(b) pairing of homologous chromosome during early prophase

(c) formation of daughter chromatids during prophase

(d) crossing over of chromosomes to introduce genetic variation

During meiosis I

1. Chromosomes line up in pairs (one maternal and one paternal chromosome in each pair) during prophase I.
2. **Crossing over** occurs—arms of homologous chromosomes exchange genetic material (this introduces **genetic variation**).
3. Each pair of chromosomes separates (during anaphase I), so that one entire chromosome of each pair moves into a daughter cell. This not only halves the chromosome number in gametes, but it also leads to *genetic variation*, depending on which chromosome (paternal or maternal) of each pair ends up in which daughter cell. This is termed *independent assortment* of chromosomes and produces different combinations of genes in different gametes.

During meisois II

The two daughter cells that result from meiosis I each undergo meiosis II which is similar to mitosis:

■ The centromere divides and chromatids separate from each other (during anaphase), moving to opposite poles (telophase), where a nuclear membrane forms around each set of chromosomes. Cytokinesis follows, resulting in *four* daughter cells (a tetrad), each with *half* the original chromosome number. Genetic variation has also been introduced, since the combination of paternal and maternal chromatin material in each resulting daughter cell is different.

■ Fertilisation—there are many combinations of chromosomes possible in gametes as a result of meiosis, resulting in a variety of gametes forming. Variation is also dependent upon which gametes fuse during fertilisation.

Figure 3.2 Early stages of division in meiosis—interphase and prophase of meiosis I

Table 3.1 Process of meiosis in animal cells

Diagram	Annotation
Early prophase  nuclear membrane breaks down	■ Chromosomes separate out into homologous pairs
Late prophase  crossing over	■ Nuclear membrane breaks down ■ Chromosomes spilt into chromatids ■ Crossing over occurs
Metaphase I  one chromosome pair / other chromosome pair	■ Chromosomes align in pairs in the middle of the cell
Anaphase I 	■ Chromosome pairs separate and each chromosome moves to the opposite end of the cell
Telophase 	■ Two daughter cells form ■ Chromosome number halved
Cytokinesis I 	■ Daughter cells are not identical to each other and have half the original number of chromosomes

continued . . .

Diagram	Annotation
Metaphase II 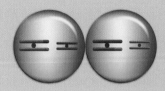	■ Chromosomes align at the equator
Anaphase II 	■ Chromatids ('arms' of chromosomes) move apart, to opposite poles
Cytokinesis II— tetrad  four haploid cells	■ Four resulting daughter cells, not identical to each other and have half the original chromosome number.

Worksheet: outline of annotated diagram of meiosis

Differences between mitosis and meiosis

- *analyse information from secondary sources to tabulate the differences that distinguish the processes of mitosis and meiosis*

Task

SECONDARY SOURCE INVESTIGATION

BIOLOGY SKILLS

P13
P14

1. Analyse information from the following resources to determine the differences between meiosis and mitosis:
   - read the text on pages 280–1 and analyse the annotated diagrams in Figure 3.2 and Table 3.1
   - analyse the comparative diagram sequence on the Student Resource CD
   - refer to at least one other source that shows an animation of the process either on video or on a website, such as: http://biology.about.com/library/blmeiosisanim.htm This website has animations of mitosis and meiosis, showing an animation of mitosis and the first meiotic division occurring side-by side for direct comparison.

2. Draw up a table with the column headings Mitosis and Meiosis and describe the differences that distinguish the processes of mitosis and meiosis. Suggested areas to look at are the:
   - type of cells in which division occurs
   - number of divisions and resulting daughter cells
   - chromosome behaviour in early stages of division—prophase
   - chromosome behaviour in anaphase (first division)
   - end of cytokinesis (mitosis) and cytokinesis II (meiosis)

Comparitive diagram sequence

Completed table comparing mitosis and meiosis

3.2

Reproductive adaptations in animals

■ *Compare and contrast external and internal fertilisation*

In animals, the union of male and female gametes (sperm and ova) can occur outside the body (**external fertilisation**) or inside the body (**internal fertilisation**).

External fertilisation

Many marine organisms carry out external fertilisation as the water environment allows the union of gametes to occur without dehydration. Corals, for example, release large amounts of gametes at the same time in the hope that some will be fertilised and then survive to adulthood. Vertebrate sexual reproduction is thought to have started in the ocean before vertebrates colonised the land.

Bony fish

The females of most species of marine bony fish produce eggs (or ova) in large batches and release them into the water. This is generally followed by the males releasing their sperm into the area of water containing the eggs. This is where the union of gametes (eggs and sperm) occurs. When fertilisation occurs in the ocean gametes tend to disperse quickly so the release of large numbers of eggs and sperm from the females and males must occur almost simultaneously. This is why most marine fish restrict the release of gametes to a few brief and clearly determined periods. Although thousands of eggs are fertilised in a single mating of bony fish, many of the resulting offspring succumb to microbial infections or predation, and few grow to maturity.

Amphibians

The amphibians invaded the land without fully adapting to the terrestrial environment, so their lifecycle is still tied to the water. Gametes from both males and females are released through the cloaca. Among the frogs and toads, the male grasps the female and discharges fluid containing sperm onto the eggs as they are released into the water (see Fig. 3.3).

Internal fertilisation

The invasion of vertebrates onto the land posed a new danger of dehydration. The gametes could not simply be released near each other as they would quickly dry up and perish. This led to the evolving of internal fertilisation and **copulation** where the male gametes are inserted into the female reproductive tract via a penis or similar structure. This allows the union of gametes to occur in a moist environment, even though the animal is on land. The fertilisation environment is not only protected from dehydration and external elements, but it is also protected from predation or dispersal and loss of the gametes. This means that fewer eggs are required to ensure a successful number of offspring survive.

Figure 3.3 Eggs of frogs are fertilised externally

Cartilaginous fish

In contrast to the bony fish, fertilisation in most cartilaginous fish is internal. The male introduces the sperm into the female through a modified pelvic fin.

Reptiles

Most reptiles fertilise their eggs internally and then the eggs are deposited outside the mother's body for development. Male reptiles use a tubular organ, the penis, to inject sperm into the female (see Fig. 3.4). The penis, containing erectile tissue, can become rigid and penetrate far into the female reproductive tract.

Birds

All birds practise internal fertilisation, though most male birds lack a penis.

Figure 3.4 Reptiles such as tortoises carry out internal fertilisation

In some of the larger birds (e.g. swan), however, the male cloaca extends to form a false penis. As the egg passes along the oviduct, glands secrete proteins (egg white) and a hard calcium carbonate shell that distinguishes bird

Table 3.2 Comparison of internal and external fertilisation

Characteristics	Differences External fertilisation	Internal fertilisation	Similarities
Gametes	Large numbers of male and female gametes produced	Large number of male gametes and fewer female gametes produced	Male and female gametes required—sperm and egg (ova)
Union	Occurs in open water environments	Mostly on land, inside the reproductive tract of the female	Sperm fertilise the eggs when they unite
Conception mechanism	Simultaneous release of gametes	Male needs to insert the sperm into the female's reproductive tract via penis or cloaca (copulation)	Sperm will fertilise eggs when in very close proximity to each other, gametes require a watery environment for this to occur
Chance of fertilisation	Low chance of fertilisation because male gametes are released into a large open area where there is less chance of successfully uniting with female gametes	High chance of fertilisation because male gametes are released into a confined space where there is more chance of successfully uniting with female gametes	If male and female gametes are in close proximity to each other, fertilisation will usually occur
Environment for **zygote**	Zygote usually develops externally in a watery environment which is vulnerable to environmental elements such as temperature and predation, infection and rapid dispersal from the area	Zygote usually develops in a very protected environment inside the female's body. Temperature is controlled there is less chance of predation, infection and loss of zygote from the area	Zygote requires a watery environment for development
Number of offspring/zygotes	After many zygotes perish a smaller number of offspring survive; however, the number of offspring produced is usually larger compared to internal fertilisation	Smaller number of zygotes produced because very few perish (higher success rate), therefore, smaller numbers of offspring compared to external fertilisation	All possible gametes will unite to form zygotes where possible during fertilisation
Breeding frequency	Will breed more frequently compared to internal fertilisation due to the lower fertilisation success rate	Will breed seasonally and less frequently due to higher fertilisation success rate	Breeding frequency will depend on the requirements of the species and the favourability of environmental conditions

eggs from reptilian eggs. Most birds incubate their eggs after laying them, to keep them warm.

Mammals

Some mammals are seasonal breeders, reproducing only once a year, while others have shorter reproductive cycles. Cycling in females involves the periodic release of a mature egg (ovum) from the ovary (the process is called ovulation). Mammals require the insertion of sperm into the female's reproductive tract (copulation) for fertilisation to occur.

Table 3.2 summarises the main similarities and differences between internal and external fertilisation.

3.3 Relative success of internal and external fertilisation

- *discuss the relative success of these forms of fertilisation in relation to the colonisation of terrestrial and aquatic environments*

Organisms in aquatic environments are successful in their reproduction and survival as they have adaptations suited to reproducing in this type of environment; however, this also means that they are completely dependent and reliant upon their environment providing the water required for successful external fertilisation. Water protects the gametes from desiccation and possible heat stress. However, in order to survive on land, organisms needed to overcome the dependence on aquatic environments for fertilisation by providing their own enclosed moist environment within the female reproductive tract, protected from the dry terrestrial environment.

Flowering plants have colonised the land by fertilising internally and avoiding gamete desiccation. Reptiles have also colonised the land successfully by producing adaptations to the dry environment by carrying out internal fertilisation and allowing their young to develop inside a waterproof egg to protect from desiccation. Even further, mammals allow internal development of their young after internal fertilisation has occurred. This ensures successful reproduction and survival of the respective species in colonising the land.

External fertilisation

In an aquatic environment

Organisms attempting to carry out external fertilisation in an aquatic environment are usually highly successful. In this environment gametes do not dry out, or dehydrate; however, organisms must produce very large numbers of gametes to compensate for the losses from predation, disease and dispersal to unsuitable environments.

In a terrestrial environment

Organisms attempting to carry out external fertilisation on land are not successful at all due to their complete reliance upon a water environment for fertilisation and the transfer of gametes.

Internal fertilisation

In an aquatic environment

Internal fertilisation is not a necessary adaptation for most aquatic species; however, it is a successful method of fertilisation in this environment. Fewer gametes are required because of the higher chance of the gametes uniting.

In a terrestrial environment

Internal fertilisation has only been possible on land because of overcoming the need for water for fertilisation. This method of fertilisation is very successful as the mechanism for direct transfer of gametes avoids dehydration and loss by dispersal, so fewer female gametes are required. The success of this form of fertilisation is very high as the environment is enclosed in a confined space protecting from predation and disease. Even the driest environments can be colonised successfully by using this method.

Success of internal and external fertilisation in terrestrial and aquatic environments

■ *identify data sources, gather, process and analyse information from secondary sources and use available evidence to discuss the relative success of internal and external fertilisation in relation to the colonisation of terrestrial and aquatic environments*

SECONDARY SOURCE INVESTIGATION

BIOLOGY SKILLS

P13
P14

Aim

1. To identify data sources, gather, process and analyse information from secondary sources.
2. To use available evidence to discuss the relative success of internal and external fertilisation in relation to the colonisation of terrestrial and aquatic environments.

Method

Part 1: Identify data sources, gather, process and analyse information from secondary sources

Refer to page 18 ('Searching for information') for suggestions and a reminder of how to approach the identifying of data sources, and gathering, processing and analysing information from secondary sources.

Part 2: Discuss the relative success of internal and external fertilisation

Read and summarise Sections 3.2 and 3.3, then copy (see Student Resource CD) and complete Table 3.3, describing the advantages and disadvantages of internal and external fertilisation in terrestrial and aquatic environments.

Discussion/conclusion

1. Describe the characteristics of fertilisation in terrestrial organisms that provide an advantage for successful colonisation.
2. Briefly describe the characteristics of fertilisation in aquatic organisms that provide an advantage for successful colonisation.
3. Discuss the relative success of internal and external fertilisation in relation to the colonisation of terrestrial and aquatic environments.

Table 3.3

Table 3.3 Advantages and disadvantages of internal and external fertilisation in colonising terrestrial and aquatic environments

Results

Environment	Type of fertilisation	Internal fertilisation	External fertilisation
Terrestrial environment	Advantages		
	Disadvantages		
Aquatic environment	Advantages		
	Disadvantages		

3.4

Mechanisms of fertilisation and development

- **describe some mechanisms found in Australian fauna to ensure**
  - **—fertilisation**
  - **—survival of the embryo and of the young after birth**

Marine animals

Staghorn coral

Many marine animals, such as the staghorn coral, achieve fertilisation by simply shedding millions of gametes into the sea (see Figure 3.5). Environmental cues, such as water temperature, tides and day length, help synchronise the reproductive cycle. When one coral starts to spawn, pheromones released along with gametes will stimulate nearby individuals to spawn, resulting in co-ordinated spawning over a wide area. During the mass spawnings of coral on Australia's Great Barrier Reef, the number of gametes shed is so great that, for a time, the sea turns milky. Within one day, fertilised eggs develop, forming into swimming larvae. After a few days at the surface, the larvae descend to find a suitable site to form a new colony. Although millions of staghorn coral larvae are produced, almost all are eaten by predators. Of the few remaining, only a tiny fraction develop into adulthood.

Amphibians

Southern gastric brooding frog

Southern gastric brooding frog eggs are fertilised externally in a watery environment. The female releases eggs and, after they are fertilised by the sperm, the female swallows them. Rather than leaving the eggs to develop alone and unprotected, the young develop internally in the female's stomach. In the stomach, digestive secretions cease and the eggs settle into the stomach wall, where they are protected and absorb nutrients from the mother. This gastric brooding appears to last about 6–7 weeks, during which time the female does not eat. When the young frogs are ready, they are regurgitated through the mouth (see Fig. 3.6). Therefore, these animals have external fertilisation but internal development. This is an extreme example of parental care which was discovered in forests north

Figure 3.5
Staghorn coral
(*Acropora yongei* sp.)
releases bundles
containing sperm
and egg

Figure 3.6 Young froglets emerging from the mouth of a female southern gastric brooding frog (*Rheobatrachus silus*) after developing in their mother's stomach

of Brisbane in 1974. This mechanism would provide some protection for the underdeveloped young from predation, infection or dispersal which can significantly reduce the success of offspring survival.

Birds

In birds, fertilisation is internal but the fertilised egg undergoes the majority of its development externally. Parental care is often needed continuously. Most species brood their eggs, with parents taking turns so that each can go and feed.

Brush turkey

The megapodid birds, such as the brush turkey, build a mound from twigs, soil and leaf litter, into which they place their eggs (see Fig. 3.7). Heat from decomposition of the leaf litter keeps the eggs warm, but the male parent frequently tends the nest, adding or removing material to control the temperature of incubation.

Figure 3.7 A brush turkey maintains the temperature of its nest by adding or removing leaf litter

Reptiles

Crocodile

Fertilisation occurs internally, then the female crocodiles (*Crocodylus porosus*) lay clutches of eggs in sandbanks beside the sea or river (see Fig. 3.8).

When an egg hatches, the offspring resemble a miniature adult, able to crawl up from the buried nest to the surface to find its way to the water and to feed independently. Female crocodiles produce small numbers of large yolky eggs containing sufficient food reserves for more elaborate development.

Figure 3.8 Development in a yolky egg gives rise to tiny crocodiles (*Crocodylus porosus*)

Monotremes

Duck-billed platypus and the echidna

Both of these monotremes are oviparous, meaning their eggs after being fertilised internally are deposited outside the mother's body to complete their development. They incubate their eggs in a nest (see Fig. 3.9) or specialised pouch, and the young hatchlings obtain milk from their mother's mammary glands by licking her skin (as monotremes lack nipples).

Figure 3.9 The duck-billed platypus with its young in a nest

The echidna does not lay eggs into a nest, but places them into an abdominal pouch where they stay for about seven weeks. The young suck milk from the skin of the mother's stomach.

Marsupials

Marsupials from the kangaroo and wallaby family have an extraordinary ability to control embryonic development. Mothers are pregnant for a short period of time and can become pregnant again just after giving birth. Since milk production lasts much longer than the pregnancy, it is necessary to delay development of the new embryo until the pouch becomes vacant. This delay in development, embryonic diapause, is controlled by the suckling of the young in the pouch.

Red kangaroo

The red kangaroo carries out both internal fertilisation and internal development of the young after birth. In good conditions, it can have three offspring at different stages of development. The female may have an older young out of the pouch but still being suckled, a newborn young in the pouch, and an embryo in

diapause in the uterus. At this stage, the mother will be producing two different types of milk simultaneously. The newborn young will get low fat/high carbohydrate milk from nipples of the mammary glands (see Fig. 3.10), while the young outside the pouch will feed from a different nipple (or teat) to get large volumes of high fat/low carbohydrate milk. In times of drought the mother may be unable to produce sufficient milk to sustain a growing young in the pouch. If the young dies, a new young will enter the pouch a month later. Since the newborn young is so small, it only needs a small quantity of milk for the first few weeks in the pouch. This ensures there is always a young ready when the drought ends. This 'production line' approach allows very rapid population growth when conditions are good. However, under prolonged drought conditions, breeding stops and only begins again when rain triggers a hormonal response in the female. This very effective mechanism controls the rate of reproduction depending on the favourability of the environmental conditions at the time.

Figure 3.10
Kangaroos give birth to small foetuses which complete their development in a pouch

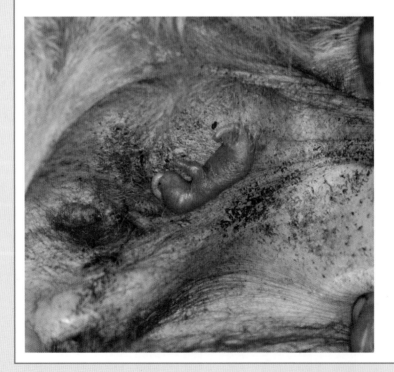

Reproductive adaptations in plants

3.5

- *describe some mechanisms found in Australian flora for:*
  —pollination
  —seed dispersal
  —asexual reproduction
  with reference to local examples

Pollination

Pollination is the process required by plants for sexual reproduction. Flowering plants (**angiosperms**) and conifers (**gymnosperms**) sexually reproduce by fertilising internally like many animals do; however, the sperm is contained in pollen grains which prevent it drying out. Gymnosperms have only one mechanism for pollination, wind, whereas angiosperms, which evolved much later, have a high proportion of species (about 65 per cent) that use animals such as insects, birds and mammals as agents for pollination. Some angiosperms still continue as wind-pollinators.

To understand the process of pollination in flowering plants we first must understand the basic structure of the flower as the reproductive organ of sexually reproducing plants. The flower contains reproductive parts that are female (**carpel**) and male (**stamen**), as well as other non-sexual parts, illustrated in Figure 3.11. In order for fertilisation to occur in the flower, the male gametes (**pollen**) from the **anther** must firstly be deposited on the female part of the flower called the **stigma**. This process is called **pollination**. Once pollen has been deposited on the stigma, it germinates and is transferred down the **style**, within a pollen tube, to the **ovules** contained in the **ovary** (see Fig. 3.12). In flowering plants, fertilisation occurs in the ovary.

Figure 3.11 Top view and longitudinal section of a typical flower—male parts labelled in blue and female parts in red

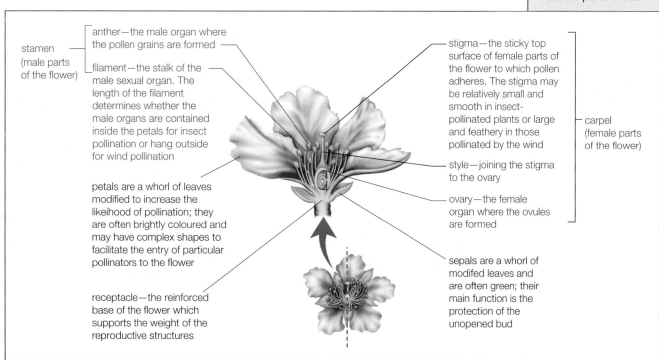

stamen (male parts of the flower)

anther—the male organ where the pollen grains are formed

filament—the stalk of the male sexual organ. The length of the filament determines whether the male organs are contained inside the petals for insect pollination or hang outside for wind pollination

petals are a whorl of leaves modified to increase the likeihood of pollination; they are often brightly coloured and may have complex shapes to facilitate the entry of particular pollinators to the flower

receptacle—the reinforced base of the flower which supports the weight of the reproductive structures

stigma—the sticky top surface of female parts of the flower to which pollen adheres. The stigma may be relatively small and smooth in insect-pollinated plants or large and feathery in those pollinated by the wind

carpel (female parts of the flower)

style—joining the stigma to the ovary

ovary—the female organ where the ovules are formed

sepals are a whorl of modifed leaves and are often green; their main function is the protection of the unopened bud

Figure 3.12 The pathway pollen tubes take as they travel from the stigma to the ovary

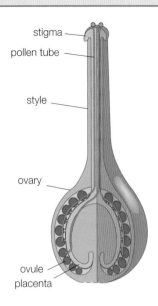

stigma
pollen tube
style
ovary
ovule
placenta

Pollination is the process of the pollen being transferred to reach the stigma. Pollen may be carried to the flower by wind or by animals, or it may originate from the same individual flower. When pollen from a flower's anther pollinates the same flower's stigma, the process is called **self-pollination**. When pollen from a flower's anther pollinates a flower's stigma from a different plant, the process is called **cross-pollination**. In most species the pollen is produced at a different time from when the stigma can receive it, so that plants are not usually pollinated by their own pollen. In most instances, flowers are pollinated with pollen from other plants of the same species. This ensures greater variation in the offspring.

Pollination by wind

Early seed plants were pollinated passively, by the action of the wind. As in present-day conifers, great quantities of pollen were shed and blown about, occasionally reaching the vicinity of the ovules of the same species. Individual plants of any given species must grow relatively close to one another for such a system to operate efficiently. Otherwise, the chance that any pollen will arrive at the appropriate destination is very small. The vast majority of wind-blown pollen travels less than 100 m. This short distance is significant compared with the long distances pollen is routinely carried by certain insects, birds and mammals.

Many angiosperms are wind-pollinated. The flowers of these plants are small, greenish, odourless, and with reduced or absent petals. Such flowers often are grouped together in fairly large numbers and may hang down in tassels that wave about in the wind and shed pollen freely (see Fig. 3.13). Wind pollination is very inefficient so large quantities of pollen are produced. Different pollen grain structures ensure compatibility with the same species (see Fig. 3.14). Wind-pollinated species do not depend on the presence of a pollinator for species survival.

Figure 3.13 Grass flowers such as *Lolium perenne* are small and pollinated by wind

Wind is responsible for pollinating many Australian plant species, especially the grasses (Fig. 3.13). In these species the anthers are very long and produce large amounts of light pollen, which is easily picked up by the wind passing over the flowers. Usually the stigmas are also very large and spread out to receive pollen carried by wind.

Table 3.4 Summary of features of wind-pollinated flowers

Feature of flower	Wind-pollinated flowers
Petals	Small and inconspicuous, usually green or dull in colour
Scent	Usually absent
Nectar	None
Anthers	Anthers protrude outside the flower so pollen is easily blown off by the wind; abundant amount of pollen is produced
Stigma	Stigma protrudes from the flower, it is often long, feathery and sticky to increase surface area for trapping the wind-borne pollen
Pollen	Very small grains, light and powdery; large amounts produced

Pollination by animals

Flowers that attract animals are more effective in ensuring the transfer of pollen. This is of a considerable advantage since a one-to-one relationship between a plant and animal species reduces wastage of pollen by ensuring that it is deposited on the correct flower. Animals that act as pollinators search in flowers for a meal of nectar (a sugary liquid secreted by nectaries in the flower) or pollen. Flower scent, colour, markings, shape and nectaries are important in attracting animals, all differing between each flower species depending upon the type of animal they are attracting.

Pollination by birds

Bird-pollinated plants must produce large amounts of nectar because if the birds do not find enough food to maintain themselves, they will not continue to visit flowers of that plant. Flowers producing large amounts of nectar have no advantage in being visited by insects because an insect could obtain its energy requirements at a single flower and would not cross-pollinate the flower.

Figure 3.14 These flowering plants produce pollen that is carried by wind, each with different pollen grain structures (inset): (a) grass; (b) ash; (c) daisy; (d) wattle. Grass, ash and daisy pollen are all major causes of hayfever. Wattle pollen occurs in large clusters

(a) Grass   40 µm

(b) Ash   40 µm

(c) Daisy   30 µm

(d) Wattle   50 µm

Bird-pollinated flowers produce much less pollen than wind-pollinated plants. Such plants are rarely fragrant because birds have little sense of smell. Red does not stand out as a very distinct colour to most insects, but it is a very conspicuous colour to birds. The red colour of the flower signals to birds the presence of abundant nectar and makes that nectar as inconspicuous as possible to insects.

Birds may not be attracted by scents but are often attracted to bright flowers like the red flowers of the New South Wales waratahs. Waratah flowers are long, tubular and slightly curved; their rate of nectar production is relatively high and they are commonly visited by nectar-feeding honeyeaters (Fig. 3.15). Many Australian flowers are pollinated by birds, especially the honeyeaters such as wattle birds and noisy miners. Birds play a larger role in Australia compared to Europe where almost all flowers are pollinated by bees.

An example of a flower with a complex mechanism for securing pollination from birds is the 'bird of paradise' (*Strelitzia reginae*). The flowers indicate that there is something attractive on offer. Part of the flower acts as a perch, and the action of the bird inserting its beak to collect nectar forces the pollen to become exposed to the bird's neck. In this way the pollen is carried from plant to plant.

When one flower is pollinated it folds back exposing the next flower. Each one opens out in succession like a fan.

Table 3.5 Features of bird-pollinated flowers

Feature of flower	Bird-pollinated flowers
Petals	Usually large and colourful, red or orange, often form a tubular shape, sometimes no petals at all
Scent	Rarely fragrant because birds have little sense of smell
Nectar	Large amounts of nectar produced in nectaries at base of flower
Anthers	Commonly lower than stigma, sometimes not enclosed by any petals and often colourful
Stigma	Higher than anthers, sometimes not enclosed by petals and often colourful
Pollen	Sticky or powdery pollen, small amount produced

Pollination by insects

Among insect-pollinated angiosperms, the most numerous groups are those pollinated by bees. Like most insects, bees initially locate food by odour, and then orient themselves on the flower or group of flowers by shape, colour, and texture. Flowers that bees characteristically visit are often blue or yellow. Many have stripes or lines of dots that indicate the location of the nectaries, which often occur deep within the specialised flowers. Some bees collect nectar, which is used as a source of food for adult bees and occasionally for larvae. One effective example of bee-pollinated flowers is the grass trigger-plant (*Stylidium graminifolium*) (see Fig. 3.16). When a bee crawls inside the flower to collect nectar the *Stylidium* is triggered to stamp pollen onto the bee in the exact spot where the stigma from another flower will pick it up.

Flower shape can restrict access to pollen and nectar to only those insects that have the appropriate tools

Figure 3.15 The red colour of the New South Wales waratah (*Telopea speciosissima*) attracts birds as pollinators

or abilities. For example, the nectar at the base of a long tubular flower may only be accessed by insects that have long mouthparts, such as butterflies, moths, flies and bees with long, lapping 'tongues'.

The flower shape can be so restricting that a certain type of behaviour may be required to access the pollen. For example, 'buzz pollination' is needed to pollinate many *Hibbertia* species. It is practised by the blue banded bee (Fig. 3.17) and a number of native Australian carpenter bees, and involves the bee holding onto the plant and vibrating to get the pollen out.

Some other examples of Australian native plants that are pollinated by bees are:
■ bottlebrush (*Callistemon*)
■ eucalyptus

■ grevillea
■ *Hibbertia scandens*
■ lemon scented tea tree (*Leptospermum polygalifolium*).

Native daisies use bright colours (often yellow) and sweet nectar to attract butterflies such as the Australian painted lady to assist pollination (Fig. 3.18).

Cycads found in central Australia rely on thrips for pollination (Fig. 3.19). The thrips are attracted by scent, which is then used to direct the insects to the male and female cones. Thrips are very small insects that cannot carry many pollen grains, so the plant needs to attract large numbers of them. A male cycad, cone laden with pollen, will emit a strong and pungent scent that will attract as many as 50 000 thrips. Female cones also emit a scent once they are ready to receive pollen, which then attracts the pollen-laden thrips.

Figure 3.16 Grass trigger-plants (*Stylidium graminifolium*) are triggered to stamp pollen onto bees

Figure 3.17 Blue banded bee showing off her beautiful iridescent furry stripes

Figure 3.19 Cycads in central Australia rely on thrips for pollination

Figure 3.18 Native daisies are pollinated by butterflies such as this Australian painted lady

Table 3.6 Summary of features of insect-pollinated flowers

Feature of flower	Insect-pollinated flowers
Petals	Usually large and colourful (yellow or blue), may be shaped to encourage specific pollinators
Scent	Often present because insects are highly attracted to scents
Nectar	Sometimes produced at base of petals so insect must enter the flower to reach the nectar
Anthers	Enclosed within flower, commonly lower than stigma
Stigma	Enclosed within flower, sticky, and commonly higher above the anthers
Pollen	Relatively large grains and often sticky; small amount produced

Pollination by deceit

There are some orchids whose flowers mimic the shape and colouring of female insects. The mimics are so realistic that male insects will attempt to mate with the flowers, thereby pollinating them. For example, the hammer orchids of Western Australia have a flower that resembles a wingless insect with shiny eyes, hairy thorax and a fat body (see Fig. 3.20). The flower is held outwards by a hinged arm. When triggered by an insect, the hinged arm moves towards the flower, effecting pollination.

Figure 3.20 A hammer orchid (*Drakaea glyptodon*) mimics the female wasp whereby the male is deceived and mates with the orchid flower, effecting pollination

Pollination by other animals

Other animals including bats, possums and small rodents may aid in pollination. The signals here are also species specific. These animals also assist in dispersing the seeds and fruits that result from pollination. Small Australian mammals like pygmy possums, some of the marsupial mice (e.g. *Antechinus stuartii*) and the honey possum (from Western Australia) pollinate common plant species found in the bush and gardens throughout Australia. These plants may include many of the *Callistemon* (bottlebrush), *Banksia* and *Grevillea* species (Fig. 3.21).

Figure 3.21 Australian examples of species pollinated by birds: (a) *Banksia ericifolia*; (b) *Grevillea banksii*

(a)

(b)

The Australian honey possum is one of the few mammals that specialises in eating flower nectar (see Fig. 3.22).

Figure 3.22
Australian honey possum (*Tarsipes rostratus*) feeding on and pollinating *Dryandra quercifolia*

Comparison of wind, bird and insect-pollinated flowers

Table 3.7 Comparison of features of wind, bird and insect-pollinated flowers

Feature of flower	Wind-pollinated flowers	Bird-pollinated flowers	Insect-pollinated flowers
Petals	Small and inconspicuous, usually green or dull in colour	Usually large and colourful, red or orange, often form a tubular shape, sometimes no petals at all	Usually large and colourful (yellow or blue), may be shaped to encourage specific pollinators
Scent	Usually absent	Rarely fragrant because birds have little sense of smell	Often present because insects are highly attracted to scents
Nectar	None	Large amounts of nectar produced in nectaries at base of flower	Sometimes produced at base of petals so insect must enter the flower to reach the nectar
Anthers	Anthers protrude outside the flower so pollen is easily blown off by the wind; abundant amount of pollen is produced	Commonly lower than stigma, sometimes not enclosed by any petals and often colourful	Enclosed within flower, commonly lower than stigma
Stigma	Stigma protrudes from the flower, is often long, feathery and sticky to increase surface area for trapping the wind-borne pollen	Higher than anthers, sometimes not enclosed by petals and often colourful	Enclosed within flower, sticky, and commonly higher above the anthers
Pollen	Very small grains, light and powdery; large amounts produced	Sticky or powdery pollen; small amount produced	Relatively large grains and often sticky; small amount produced

Seed dispersal

After successful pollination and fertilisation of the flower, the seed develops. It is an advantage for a plant to spread or disperse its seeds over a wide distance. This prevents overcrowding from occurring within the same plant species and increases the chances of survival in situations of environmental change, such as in cases of fire or disease. Seeds are dispersed by wind or by animals such as insects, birds and mammals and are designed to disperse in many ways (Fig. 3.23). Australian native plants have evolved a variety of adaptations to aid in the effective and successful dispersal of their seeds.

Figure 3.23 Some of the many adaptations of seeds to facilitate dispersal—seeds have evolved a number of different means of moving distances from their maternal plant

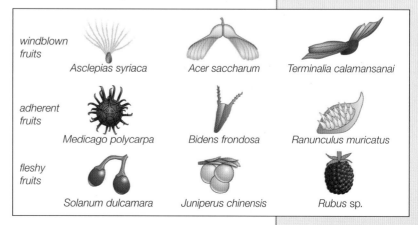

297

Wind

Some seeds are aerodynamically designed to be blown long distances by the wind; for example, *Flindersia*, leptospermums, melaleucas (with very fine, light seeds), native daisies (with feathery pappus) and casuarinas.

Animal

Other seeds have structures, such as hooks or barbs, which cling or stick to the fur or feathers of animals. This is so they can be carried long distances before they fall to the ground (e.g. *Pisonia* (birdlime tree) and *Pittosporum*).

Others seeds are enclosed in bright coloured fleshy fruits. Red is a very conspicuous colour to birds, so any fruit or berries containing seeds are highly likely to be dispersed by birds (e.g. lilypillys). Smaller birds are also interested in purple berries such as the tree violet (*Hymenanthera dentate*).

Seeds can pass through the digestive system of mammals or birds which unknowingly transport the seeds to new locations so that they germinate at the spot upon which they were defecated. Some seeds must pass through the gut of an animal to be able to germinate; for example, the nitre bush (*Nitraria billardieri*) depends on emus. Another example is the mistletoe; it has sticky seeds which are deposited on trees by mistletoe birds.

To bribe ants, wattle (*Acacia)* seeds have some lipids attached to their outsides. The seeds (like *Lomandra* and *Grevillea*) are carried by the ants to their nests where they consume the lipids but leave the hard seeds underground, safe from fire. Wattles then can flourish and grow after the hottest of bushfires.

Fire

Eucalypts, banksias and many other Australian plants store their seeds until the fire destroys the branch or the entire plant. This allows the capsules to open, releasing the seeds for dispersal, usually by the wind. This provides a significant advantage to the seeds as the fire clears land areas and invites recolonisation by new plants. The fastest dispersing and germinating plants can colonise more areas of land. Not all banksias and eucalypts store their seeds waiting for fire. *Banksia integrifolia* and *Eucalyptus melanoxylon* release their seed once it is ripe. This is seen as a primitive feature compared with other species that are actually more suited to their environment and have more effective colonising mechanisms.

Of the dry simple fruits, some open and shed their seeds at maturity (Fig. 3.24). A follicle, which opens on the lower side, is formed from a single carpel (e.g. *Grevillea*). Pods or legumes open on two sides (e.g. *Acacia*). The dry fruits of many Australian native plants have thick, woody walls that protect the seeds from the heat of bushfires. In eucalypts, the top of the ovary forms three to four valves, the fruit splits open and a capsule matures and dries out. After a fire, the heat causes the valves of fruits held in the tree canopies to open quickly and seeds are released. The woody follicles of banksias are very fire resistant and open after fire. Each follicle has two seeds attached to a wing-like structure, the separator. When the separator is sufficiently wet by rain following a fire, it expands, pulling the seeds out of the fruit (Fig. 3.24d). Seeds will otherwise remain protected in the fruit until rain falls, providing suitable conditions for germination.

Water

Some seeds rely on water dispersal, such as the water gum (*Syzygium francisii*) and the mangrove (*Avicennia marina*). Seeds may float small or large distances from the parent plant along rivers and estuaries or across seas.

Explosion

Finally, some seeds are violently propelled from the base of the fruit in an explosive discharge. Seeds are ejected from the pod at high speeds caused by the drying and contraction of the pod. Some seeds such as the *Acacia cultriformis* can be thrown up to 2 m by this method; other plants such as the *Viola betonicifolia* also use such a mechanism to send their seeds distances.

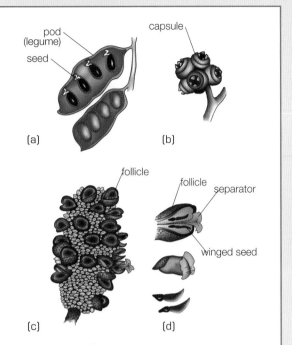

Figure 3.24 Dry fruits include: (a) pods or legumes, *Acacia*; (b) capsules, *Eucalyptus*; (c) follicles, *Banksia*; (d) follicle separator in *Banksia*, which when wet pulls the two seeds out of the fruit

Table 3.8 Summary of different seed dispersal mechanisms

Method of dispersal	Description of dispersal mechanism(s)	Examples
Wind	■ Seeds may have different mechanisms or adaptations to assist the dispersal over long distances by the wind	■ Fine, light aerodynamic seeds (e.g. melaleuca and casuarina) ■ Winged seeds (e.g. hakea) ■ Feathery pappus (e.g. native daisies)
Animal	■ Seeds may possess hooks or barbs to catch seeds on the outside of animals ■ Fruits may be eaten and seeds then carried in the gut and deposited in faeces in a new area	■ Colourful fruit (e.g. lilypilly and tree violet) ■ Sticky fruit (e.g. mistletoe) ■ Burrs or hooks (e.g. birdlime tree and pittosporum) ■ Through animal gut (e.g. nitre bush)
Fire	■ Some seeds are stored until fire causes pods to open	■ Eucalypts, banksias, acacias and grevilleas
Water	■ Seeds may float on water and be dispersed distances over different water bodies (rivers and oceans)	■ Water gums and mangroves
Explosion	■ Some seed may be ejected from pods at high speeds. The pods explode when ripe and shoot seeds away from the parent plant.	■ Acacias and viola

Asexual reproduction

Asexual reproduction is the making of a new individual without the use of sex cells or gametes. Only one parent is required for the mitotic cell divisions to occur. Some types of asexual reproduction are:

- binary fission
- budding—e.g. *Hydra* (Fig. 3.25a) and coral
- spore formation—e.g. moss, fungi and ferns
- vegetative propagation—e.g. plant cuttings like roses
- regeneration—e.g. starfish (Fig. 3.25b) and earthworms
- parthenogenesis—e.g. Binoe's gecko (*Heteronotia binoei*).

Plants that reproduce asexually clone new individuals from portions of the root, stem, leaves or ovules of adult individuals. The asexually produced individuals are genetically identical to the parent. There are different types of asexual reproduction in plants as follows.

Vegetative reproduction

New plant individuals are simply cloned from parts of adults, such as runners, rhizomes and suckers.

Runners

Some plants reproduce by means of runners which are long, thin stems that grow along the surface of the soil. In the cultivated strawberry, for example, leaves, flowers and roots are produced at every other node on the runner. Just beyond each second node, the tip of the node turns up and thickens, producing new roots and a new shoot that continues the runner. Another example is spinifex grass (Fig. 3.26) which has long stems that grow horizontally along the surface of the soil. At each node, leaves and roots are produced that can be subdivided into new plants.

Figure 3.25
Asexual reproduction:
(a) budding: in *Hydra*, a new individual is produced from a bud that branches from the side of the parent body wall;
(b) regeneration: a single arm of the sea star (*Linkia mulifora*) regenerating four new arms to form an entire new individual

(a)

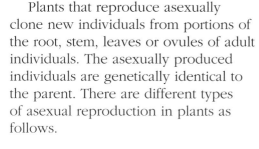

Figure 3.26 Spinifex grass (*Spinifex hirsutus*) has long stems that grow along the surface of the soil, producing new leaves and roots (plantlets) at each node

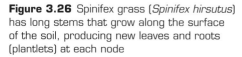

(b)

Rhizomes

Underground horizontal stems, or rhizomes, invade areas near the parent plant with each node being able to give rise to a new flowering shoot. Corms (swollen stems, e.g. gladioli), bulbs (swollen leaf bases, e.g. daffodils) and root tubers (e.g. begonias and potatoes) (Fig. 3.27) are stems specialised for storage and reproduction. The eyes or 'seed pieces' of the potato give rise to the new plant. Rhizomes are characteristic of ginger, ferns such as bracken fern and many grasses.

Figure 3.28 The common reed (*Phragmites australis*) spreads rapidly by suckering through aquatic habitats

Figure 3.27 The potato develops new tubers from swollen regions of the stem

Suckers

The roots of some plants produce 'suckers' or sprouts, which give rise to new plants. Trees and shrubs that sucker, such as reeds (Fig. 3.28), wattles and blackberries can spread quickly into a vacant patch of habitat after disturbance. When the root of a dandelion is broken when someone tries to pull weeds out from the ground, each root fragment may give rise to a new plant.

Budding

Budding is one of the more unusual forms of asexual reproduction seen in plants. Budding involves the development of a new individual as an outgrowth of the parent plant. For example, *Kalanchoe* produces buds along leaf margins, which can break off and form new plants, eventually growing to adult size (Fig. 3.29).

Apomixis

In certain plants, such as kangaroo grass (*Themeda triandra*), lemon and orange trees (*Citrus*) and dandelions, the embryos in the seeds may be produced asexually from the parent plant. The seeds produced in this way give rise to individuals that are genetically identical to their parents.

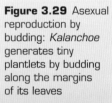

Figure 3.29 Asexual reproduction by budding: *Kalanchoe* generates tiny plantlets by budding along the margins of its leaves

By reproducing asexually this way these plants also gain the advantage of seed dispersal, an adaptation usually associated only with sexual reproduction, as well as the rapid multiplication of plants.

Totipotency

An entire plant can be generated from a single plant cell. This property is known as totipotency and was first demonstrated in 1958 by F. C. Steward. He took a section of carrot root and, under sterile conditions, cut out tiny pieces of tissues. These were then cultured in a shaking flask in a liquid containing nutrients and minerals, plus green coconut milk to stimulate growth. Free cells separated off from the pieces of tissue and divided, forming embryo-like structures. These structures could be transplanted on to a solid medium and grown in the light into new carrot plants that were clones of the original plant.

Other examples of Australian plants reproducing asexually

- *Polystichum prolifeum*
- *Asplenium bulbiferum*
- *Viola hederacae*
- *Dichondra repens*
- Ground orchids

Features of pollination in native flowering plants

FIRST-HAND INVESTIGATION

BIOLOGY SKILLS

P11
P13
P14
P15

■ *plan, choose equipment or resources and perform a first-hand investigation to gather and present information about flowers of native species of angiosperms to identify features that may be adaptations for wind and insect/bird/mammal pollination*

Aim

1. To plan, choose equipment or resources.
2. To perform a first-hand investigation.
3. To gather, and present information about flowers of native species of angiosperms.
4. To identify features that may be adaptations for wind and insect/bird/mammal pollination.

Equipment

Suggestions for equipment are:
- variety of native angiosperm (flowering plant) specimens
- hand lens
- scalpel
- forceps
- Petri dish
- binocular microscope
- labelled diagrams of longitudinal sections of flowers (bisected/cut in half) showing different arrangements of reproductive structures demonstrating the different types of pollination, such as those in Figure 3.30
- key to Australian flowering plants.

Method

Part 1: Planning the first-hand investigation

This investigation will involve each group collecting four different native flower specimens, bisecting each flower longitudinally and scientifically drawing and labelling each flower's structure. You will then be expected to make deductions about each flower's method of pollination using the characteristics you observed in each flower.

Write up and describe the method you would undertake to carry out this investigation. Once you have completed this, list the equipment your group will need to carry out the method (see equipment suggestions). This includes selecting the four native flower specimens that your group will be observing. A suggestion would be to choose some from your own garden or ask your teacher about native flowers available in the school grounds.

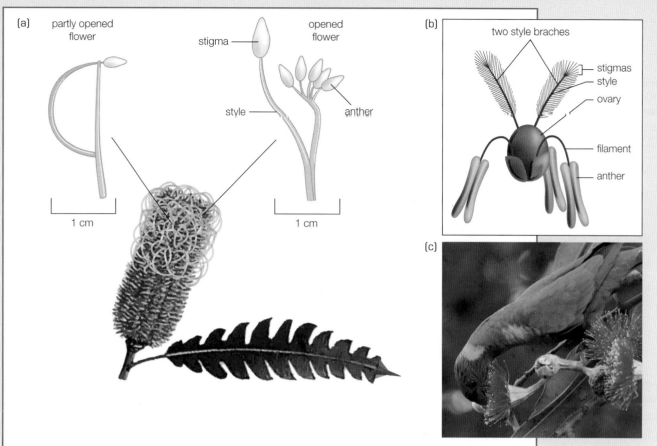

(a) partly opened flower — opened flower — stigma — style — anther — 1 cm — 1 cm

(b) two style braches — stigmas — style — ovary — filament — anther

(c)

Part 2: Gather information about flowers of native species of angiosperms

Read Section 3.5 on pollination in plants (including Table 3.7), observe the flower diagrams in Figure 3.11, and research information from a variety of other secondary sources, in order to obtain information on the four native angiosperm flowers that you have selected for this investigation.

Part 3: Present information about native flowers and identify features that may be adaptations for wind and insect/bird/mammal pollination

After writing up your investigation method and equipment list, carry out your investigation and present your information in two ways:

- Draw large, clear, fully-labelled diagrams of your four longitudinally bisected flowers (see Fig. 3.31).
- Copy (see Student Resource CD) and complete Table 3.9 with one example already provided.

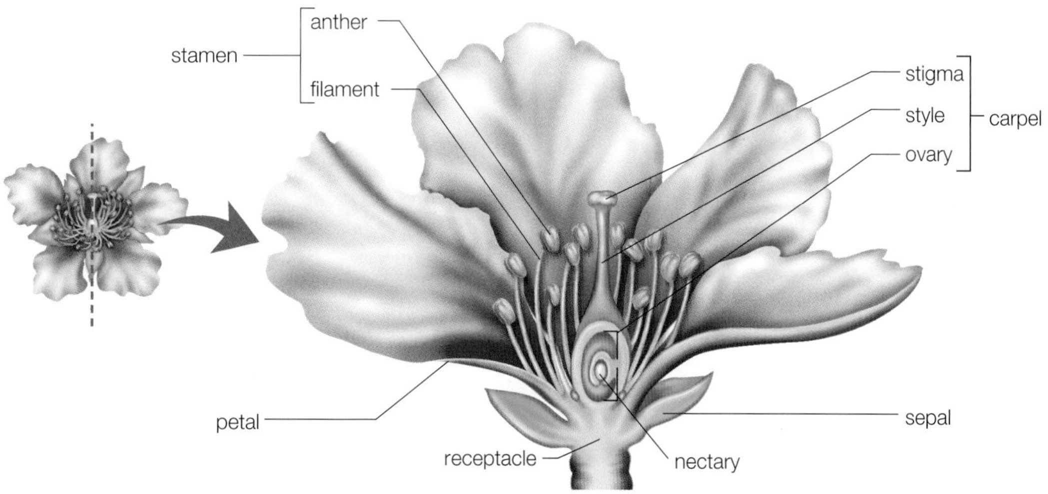

Figure 3.30
Longitudinal sections of flowers: (a) *Banksia ericifolia*—insect pollination; (b) Grass flower with large stigma surface—wind pollination; (c) Eucalypt flower

stamen — anther — filament — stigma — style — ovary — carpel — petal — receptacle — nectary — sepal

Figure 3.31 Scientific, labelled drawing of the structure of a flower

Results

Table 3.9 Features of native flowers that may be adaptations for pollination

Feature of flower	Scientific and common name of selected plant specimen	
	Banksia ericifolia (banksia) partly opened flower — stigma — opened flower style — anther 1 cm — 1 cm	
Flowers	No single flowers but large number grouped together (inflorescence)	
Sepals	None	
Petals	Yellow	
Scent	Sweet	
Nectar	Present	
Anthers	Brightly coloured	
Stigma	Long and above anthers	
Type of pollination and justification	Insect pollination because yellow coloured petals suggest bee pollination, brightly coloured anthers suggest insect attractor, nectar reward for insects, and high number of flowers allow many insects to pollinate at the one time	

Table 3.9

Discussion/conclusion

1. **Identify** the structures of native flowers that may be adaptations to wind pollination?
2. **Identify** the structures of native flowers that may be adaptations to insect/bird/mammal pollination.

3. (a) Was any particular type of pollination more common than others in your specimens? If so, which one?
   (b) Can you **explain** possible reasons for this?
4. Briefly **describe** how and why the features of wind-pollinated flowers differ from insect- and bird-pollinated flowers.

Evolutionary advantages of sexual and asexual reproduction

3.6

■ *explain how the evolution of these reproductive adaptations has increased the chances of continuity of the species in the Australian environment*

As organisms have evolved from aquatic environments moving onto the land, the evolution of reproductive adaptations has ensured survival of species. Over time, organisms have continued to develop and become more specialised in their reproductive adaptations surviving in harsh arid Australian conditions with extremes of drought and fire.

Asexual reproduction

Reproduction is not necessary for individual success, but for the continuation of the species. There are a number of strategies of reproduction depending on the environment of the organism. Organisms that reproduce asexually do not have to rely on another individual organism to provide gametes and are at an advantage when sudden or unexpectedly favourable conditions arise because they can quickly reproduce themselves (with offspring identical to the parent). This can become a competitive edge if the organism lives in an environment that is often disturbed, and they are particularly well suited to a certain environment or habitat. Asexual reproduction among plants is far more common in harsh environments where there is little margin for variation. The main disadvantage to asexual reproduction is if extremely harsh conditions arise, the whole group of species is particularly vulnerable to these conditions, or to disease, parasitism and predation.

Sexual reproduction

Sexual reproduction produces offspring that are genetically different and possibly better adapted to new and changing environmental conditions than their parents. This gives the species a better chance at surviving in ever-changing environments. However, sexual reproduction is often a more energetically expensive process, compared to asexual reproduction, and may be the first thing an organism abandons in times of hardship.

External fertilisation

The chances of successful external fertilisation are increased by the synchronisation of the release of gametes, reproductive cycles and the mating behaviours of each species. External fertilisation and development means that parents spend less time looking after the young, but more gametes have to be produced to ensure that some eggs get fertilised. The advantage of this method is the high dispersal of young. The gametes are thrown into the sea and fertilised eggs are carried away to settle in an area different to their parents. This reduces competition for food and living space for the parent generation, and allows quick recovery of populations away from damaged areas.

Internal fertilisation

Organisms that use internal fertilisation tend to be more adapted to terrestrial environments and reproducing successfully on land. Fewer gametes are produced because there is a much higher rate of fertilisation and survival.

The move to internal fertilisation and development has demonstrated new adaptations for reproduction on land, which may have started with the ovules of flowers becoming enclosed in the ovary to provide adequate protection from desiccation.

Parental care

Parental care varies between aquatic and terrestrial organisms. Many aquatic species simply abandon the fertilised eggs and leave them to risk development in the open sea. This means that less energy is put into caring for the young and the survival rate of the young is much lower; therefore, more eggs have to be produced to compensate. Mammals are generally **viviparous** (give birth to live young); fish, birds, some reptiles and many invertebrates are **oviparous** (egg-laying). Oviparous animals will devote varying amounts of energy to caring for their eggs. Some oviparous animals brood their eggs until they hatch to increase the survival rate of their eggs, while others will stand guard over a nest of eggs until they hatch.

Plants

In plants, self-pollination expends less energy in the production of pollinator attractants and can grow in areas where the kinds of insects or other animals that might visit them are absent or very few. These plant species contain high proportions of individuals well-adapted to their particular habitats.

In cross-pollinators, animal agents such as insects, birds and mammals have become a more effective way of transferring pollen to the stigma. As flowers become increasingly specialised, so do their relationships with particular groups of insects and other animals.

Many features of flowering plants seem to correlate with successful growth under arid and semi-arid conditions. The transfer of pollen between flowers of separate plants, sometimes over long distances, ensures cross-pollination and may have been important in the early success of angiosperms. The various means of effective fruit dispersal that evolved in the group were also significant in the success of angiosperms. As early angiosperms evolved, all of these advantageous features became further elaborated and developed, and the pace of their diversification accelerated. In addition to insects, birds and mammals now assist in pollination and seed dispersal.

Reproduction—Australian species

Individual Australian species have variations in their reproductive structures and mechanisms. If the environment changes, the individual species most suited to the change will survive and reproduce, passing on their characteristics to their offspring. As the Australian environment becomes more arid (drier and often hotter), organisms possessing reproductive adaptations that enable their young to survive should be able to increase in numbers.

Adaptations for colonisation and survival

Reproductive adaptations are needed for successful colonisation and survival in the Australian environment. Australia has many areas of harsh arid conditions (and sometimes drought or fire), making it difficult for effective fertilisation and development unless the organisms possess adaptations suitable for successful reproduction in harsh environments. Reproducing offspring in times favourable to the organism— suitable climate and resources, available water and food supply— increases the chances of continuity of the species.

Possessing adaptations for survival and the ability to flourish after extreme harsh conditions pass (eg. drought or fire) also increases chances of continuity of the species. Species need to survive the harsh times and maintain their population numbers (without becoming extinct) until conditions improve, then utilise adaptations to rapidly increase species numbers afterwards (e.g. the kangaroo with its embryo on standby). Many Australian plants possess adaptations to harsh conditions like

fire, for example hakeas have woody seed pods able to survive the high temperatures of fire. The pods do not usually open unless stimulated by the heat of fire, landing on soil enriched by ash from the fire. This means that the seed is not released and dispersed until environmental conditions are favourable for rapid increase and therefore continuity of the species. Hakeas also regenerate and ensure continuity of the species after fire by growing lignotubers from the fire-damaged plant.

Conditions under which asexual reproduction is advantageous

3.7

- ■ *describe the conditions under which asexual reproduction is advantageous, with reference to specific Australian examples*

Advantages of asexual reproduction

- ■ Only one parent is required so energy is not wasted on producing large numbers of gametes or on finding a mate. This is advantageous:
  —in arid conditions or where environmental conditions are not as favourable; for example, spinifex grass survives and reproduces successfully by sending out runners in harsh sand dune conditions such as high temperatures, high salinity and wind erosion (Fig. 3.26)
  —when food supply may be short and there is a need to use less energy to reproduce
  —when there is a small mating population or time constraints on finding a mate.
- ■ It is a relatively quick process and large numbers of offspring can be produced rapidly. This is an advantage when rapid recovery is needed after a decline in numbers (e.g. after a bushfire or drought). The colony wattle (*Acacia murrayana*)

(Fig. 3.32) can send up shoots from the outer roots which grow into separate plants if the parent shrub dies. This allows for regrowth to occur quickly. It often forms colonies from root suckers. Another example is the scaevola misty blue (*Scaevola striata*) (Fig. 3.33) which can also recolonise damaged areas such as sand dunes by reproducing asexually. It is a hardy groundcover that is adaptable to most soil types and full sun.

Figure 3.32
Colony wattle

Figure 3.33
Scaevola

- If there is no variation in the environment then the identical offspring will always be adapted to their surroundings and survive to reproduce successfully. Corals, such as the grooved brain coral (Fig. 3.34), reproduce by budding when conditions are favourable, however if the environment does change (eg. a new disease or pest enters) the entire species may rapidly decline and die out.
- Asexual reproduction is advantageous when environmental conditions are stable. In this situation the offspring of the parent plants are identical, having features that make them suited to the environment and likely to survive to reproduce themselves. This type of reproduction allows rapid colonisation after harsh conditions such as fire or drought which may decrease the species population numbers. Many Australian plants have adaptations for survival in this situation where reproduction is stimulated.

Figure 3.34 Grooved brain coral

Table 3.10 Summary of examples of the advantages of asexual reproduction in Australia

Australian example	Asexual reproduction mechanism	Why it is an advantage
Spinifex grass	Reproduces by runners which put out individual stems and roots at nodes along the ground	Reproduces successfully in harsh conditions, requires less energy to reproduce by runner, very rapid method of reproduction
Colony wattle	Sends up shoots from outer roots which grow into separate plants	Rapid method where large numbers can be reproduced quickly, an advantage when rapid recovery is needed after a decline in numbers (e.g. fire or drought)
Grooved brain coral	Reproduces by budding, where a new bud can form a new individual separate from the parent plant	An advantage when there is no variation in environmental conditions, it will always have adaptations to its surroundings and survive to reproduce successfully, all offspring are genetically similar to parents.

REVISION QUESTIONS

1. Identify the type of cells which divide by meiosis.

2. Explain the importance of meiosis.

3. Describe what is meant by the following terms:
   (a) homologous chromosomes
   (b) diploid and haploid
   (c) chromatid.

4. Draw a cell with three pairs of chromosomes:
   (a) towards the beginning of meiosis, while it is undergoing crossing-over in one of its pairs of chromosomes
   (b) after the first meiotic division
   (c) after the second meiotic division.

5. Compare the number of chromosomes in a cell that has completed meiosis with that of the parent cell.

6. Explain why the daughter cells that result from meiosis are not genetically identical to each other.

7. State one similarity and two differences between the processes of mitosis and meiosis.

8. Describe the similarities and differences between the characteristics of internal and external fertilisation, using two examples of each.

9. Compare the relative success of internal and external fertilisation in relation to the colonisation of terrestrial and aquatic environments.

10. Describe two mechanisms found in Australian fauna to ensure the survival of the embryo and young after birth. Give one example of each.

11. (a) Describe the pollination mechanism found in two *named* Australian plants.
    (b) Identify two ways that the reproductive structures differ between wind-pollinated and insect-pollinated plants.

12. Identify two different methods used for seed dispersal in *named* Australian plants.

13. Describe three types of asexual reproduction in plants. Give one Australian example for each.

14. Explain how the evolution of reproductive adaptations has increased the chances of continuity of the species in the Australian environment.

15. Describe the conditions under which asexual reproduction is advantageous using a *named* Australian example.

Answers to
revision questions

CHAPTER 4

Understanding past environments helps us to understand the future

A study of palaeontology and past environments increases our understanding of the possible future range of plants and animals

4.1 Understanding past environments

■ *explain the importance of the study of past environments in predicting the impact of human activity in present environments*

Introduction

As we have already discussed in the module, the history of the Australian continent has dictated, to a large extent, the composition of the Australian **biota**. Isolated as an island continent after its separation from Gondwana, Australia evolved unique (**endemic**) animals and plants. As the climate became more arid during the last 30–40 million years, the typically Australian element of the biota evolved under the new environmental conditions. With the proximity to islands and other land masses, it was also possible for people to enter northern Australia, at least 40 000 years ago. The impact on the Australian environment of the first humans, especially through the use of fire, continues to be widely debated. Europeans arrived more recently. The impact of Europeans on the Australian environment was rapid and extensive.

Extinction rates and declines in abundance and range of native flora and fauna have been highest in regions where settlement first occurred, attributable to inappropriate land and water use, habitat loss and fragmentation, over-exploitation of both terrestrial and marine resources, and the spread of introduced herbivores, predators, weeds and disease. Human impact on the Australian environment has greatly accelerated over the last 200 years due to extensive forest and woodland clearing, changes to water regimes and introduced animals and plants.

Studying past environments and the extinction of species before, during and after the arrival of humans assists in predicting the impact of changes to the present environment. The impact of humans can only be measured when the state of the past environment is known. This is why the study of past environments is highly important in predicting human impact in present and future environments.

A great deal can be learned about current rates of extinction by studying the past, and in particular the impact of human-caused extinctions. These extinctions are thought to have been caused by hunting, and indirectly by burning and clearing forests, although some scientists attribute these extinctions to climate change.

The majority of historic extinctions have occurred on islands. Of the 90 species of mammals that have gone extinct in the last 500 years, 73 per cent lived on islands (and another 19 per cent lived in Australia). Island populations are often relatively small, and thus particularly vulnerable to extinction.

In recent years, however, the extinction crisis has moved from islands to continents. Most species now threatened with extinction occur on continents, and we can now predict that these areas be affected the most from the extinction crisis in the next 100 years (see Fig. 4.1 and Table 4.1).

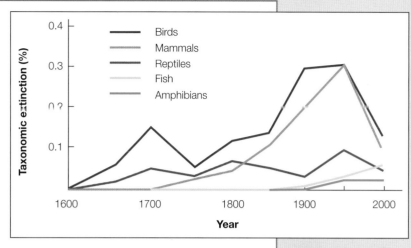

Figure 4.1 Trends in species loss: the graph presents data on recorded animal extinctions since 1600

Table 4.1 Recorded extinctions since 1600

Taxon	Recorded extinctions				Approximate number of species	Percent of taxon extinct
	Mainland	Island	Ocean	Total		
Mammals	30	51	4	85	4000	2.1
Birds	21	92	0	113	9000	1.3
Reptiles	1	20	0	21	6300	0.3
Amphibians[a]	2	0	0	2	4200	0.05
Fish	22	1	0	23	19100	0.1
Invertebrates	49	48	1	98	1000000+	0.01
Flowering plants	245	139	0	384	250000	0.2

[a]Many species may be on the verge of extinction.
Source: Reid and Miller (1989)

Some people have argued that we should not be concerned because extinctions are a natural event and **mass extinctions** have occurred in the past. Indeed, mass extinctions have occurred several times over the past half billion years. However, the current mass extinction event is notable in several respects. Firstly, it is the only such event triggered by a single species—humans. Although species diversity usually recovers after a few million years, this is a long time to deny our descendents the benefits of biodiversity. In addition, it is not clear that biodiversity will rebound this time. After previous mass extinction events, new species have evolved to utilise resources available due to species extinctions. Today, however, such resources are less available due to human activities destroying habitats and removing the resources for their own use.

Biologists can estimate rates of extinction both by studying recorded extinction events and by analysing trends in habitat loss and disruption. Since prehistoric times, humans have had a devastating effect on biodiversity almost everywhere in the world.

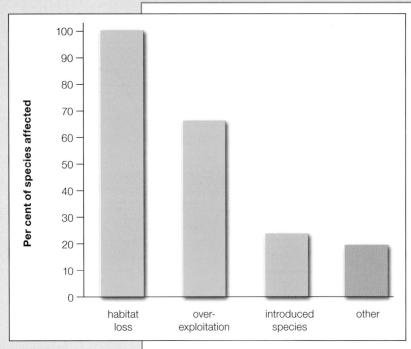

Figure 4.2 Factors responsible for animal extinction

Factors responsible for extinction

A species being rare does not necessarily mean that it is in danger of extinction. The habitat it utilises may simply be in short supply, preventing population numbers from growing. In a similar way, shortage of some other resource may be limiting the size of populations. Studying a wide array of recorded extinctions and many species currently threatened with extinction, conservation biologists have identified a few human factors that seem to play an important role in many extinctions: over-exploitation, introduced species, disruption of ecological relationships, loss of genetic variability, and habitat loss and fragmentation (see Fig. 4.2 and Table 4.2).

Table 4.2 Factors responsible for extinction of different animal groups

| Group | Percentage of species influenced by the given factor[a] | | | | | |
	Habitat loss	Over-exploitation	Species introduction	Predators	Other	Unknown
Extinctions						
Mammals	19	23	20	1	1	36
Birds	20	11	22	0	2	37
Reptiles	5	32	42	0	0	21
Fish	35	4	30	0	4	48
Threatened extinctions						
Mammals	68	54	6	8	12	–
Birds	58	30	28	1	1	–
Reptiles	53	63	17	3	6	–
Amphibians	77	29	14	–	3	–
Fish	78	12	28	–	2	–

[a]Some species may be influenced by more than one factor, thus, some rows may exceed 100%
Source: Reid and Miller (1989)

The extent of environmental change is best illustrated by the number of large animals that became extinct during the period in which people have been in Australia. Most extinctions occurred between 35 000 and 15 000 years ago, at a time when conditions were driest during the last glacial period. Although climate has been considered to be the main cause of these extinctions, there is no evidence to suggest that climatic conditions at this time were more extreme than during the previous glacial phases. However, increased environmental stability resulting from human burning and associated vegetation changes may have sufficiently altered stream flow and lake levels to produce a more drought-prone environment. A generalised summary of the development of the present vegetation and environment in south-eastern Australia is shown in Figure 4.3.

Predicting which species are vulnerable to extinction

To assess whether a particular species is vulnerable to extinction, conservation biologists look for changes in population size and habitat availability. Species whose populations are shrinking rapidly, whose habitats are being destroyed, or which are endemic to small areas can be considered endangered.

Historical information is critical in explaining the present state of the environment and therefore also in predicting the future. In environmental management this usually involves two aspects:

1. baseline information for some point in the past which has relevance to the ecosystem being managed. In Australia, this has most often been taken to be the moment of first European settlement, as representing the state of the environment prior to the impacts and changes brought by that settlement

2. measurements of change since that point in time. Recognition of different rates and types of change in the past is a crucial foundation to understanding change in the present and to managing human activity into the future.

An example of this approach to environmental management is the Parramatta River. It has been commonly assumed that the mangroves along the banks of the Parramatta River (the central river flowing into Sydney

Figure 4.3 Summary of vegetation and environment changes in south-eastern Australia over the last 23 million years

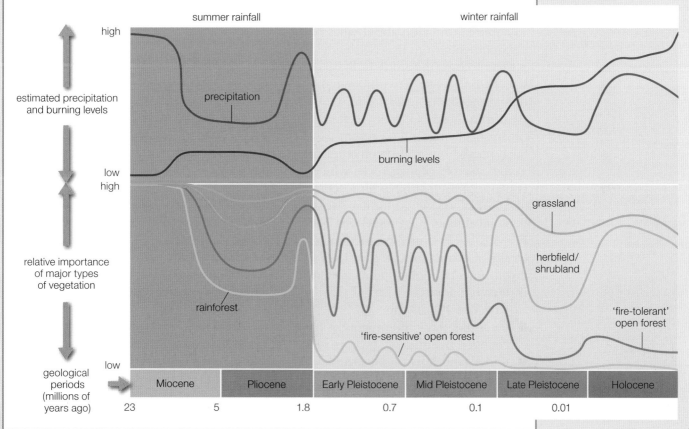

Harbour) are remnants of former more extensive growth, destroyed in the course of the city's European history. This assumption, based on the present distribution of mangroves, has been overturned by research conducted by L. C. McLoughlin in 1987 and 2000 exploring historical sources back to initial settlement, finding major changes in the vegetation from the early 1800s up to 2000. The earliest data showed mangroves confined to patches only in lower areas of the river and other prominent vegetation such as saltmarsh. The latest data showed increased mangrove growth to eventually line all available sections of the riverbanks and invade and replace saltmarshes. Such assumptions have had a significant influence on foreshore planning, management and restoration activity. Fortunately, this study has now provided more accurate information about the human impact on vegetation such as mangroves along the Parramatta River.

4.2 Distribution of flora and fauna in present and future environments

■ *identify the ways in which palaeontology assists understanding of the factors that may determine distribution of flora and fauna in present and future environments*

Palaeontology

Fossils are the remains of once-living organisms that were adapted to their environments. They can provide valuable information about what past environments were like. Through palaeontology, the study of fossils, we can predict the environmental requirements of organisms in the past from those of closely-related organisms living in the present day. Such predictions will be most reliable in the case of younger rocks which contain fossils that have representatives alive today. As we go further back in geological time, the predictions become less reliable because we encounter fossils of extinct groups about whose environmental requirements nothing is directly known.

Palaeontologists can:
■ determine how the organisms have changed over time
■ understand how organisms may be related
■ understand why some organisms have become extinct
■ see the effect of species extinction on other organisms
■ recognise changes in past distribution of organisms in order to provide information about how the distribution may be currently changing.

The environmental information obtained from fossils may be as simple as whether the rocks in which they occur were deposited in the sea, in a brackish estuary, in fresh water, or on the land. For example, rocks containing fossils of corals, brachiopods or echinoderms must have been deposited in the sea because living representatives of those groups are found only in the sea today. Similarly, fossils of land-dwelling animals such as kangaroos indicate deposition on land or in an adjacent body of fresh water.

Fossils of reef-building corals indicate that the rocks in which they occur were deposited in warm, shallow seas. At the present-day, reef-forming corals are found in tropical seas and only at depths of less than 200 m where sunlight can penetrate the water to reach the photosynthesising algae within their cells.

Fossil evidence may provide clues about the interactions of organisms with each other, biotic and abiotic factors of past ecosystems, and evidence of climate change in past environments. This fossil evidence therefore provides us with the factors that may have determined the distribution of fauna and flora in the past and hence distribution in present and future environments.

Factors determining distribution of Australian marsupials

By looking at the evidence for changes between past and present climate, and observing the changes in the distribution of organisms that formed fossils over time, we can understand more about the factors that may determine distribution of flora and fauna in present-day environments and predict the movement and distribution of organisms in the future. South American marsupials have been eventually dominated by placental mammals over time. This could indicate the possibility of a similar occurrence with our Australian marsupials.

Palaeontologists can compare past life to modern groups of organisms to discover genetic relationships and the age of different groups. The fossil record for kangaroo-like marsupials in Australia extends back 45 million years ago (see Fig. 4.4) to a time when rainforest was widespread. As the Australian plate drifted north, aridity increased and grasslands and open forests became more common. The number of living species of grazing macropod kangaroos that adapted to a diet of grasses reflects success in the drying environments, while the once common browsing (leaf-cutting) sthenurine kangaroos have declined, possibly due to reduced availability to low-leaf foliage. The living kangaroo most similar to the ancestor of all kangaroos is the musky rat kangaroo, *Hypsiprymnodon moschatus* (see Fig. 4.5), which lives in rainforests and eats a variety of foods. It has simple, rounded molars for crushing soft food items. Species of *Macropus*, such as the red kangaroo (see Fig. 4.6) have high-crested molar teeth that efficiently shear and grind food into

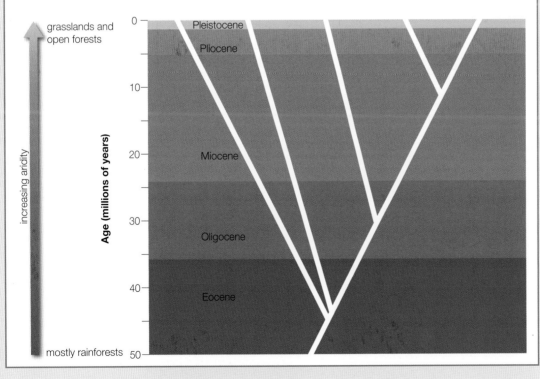

Figure 4.4
A phylogenetic tree representing a simplified view of the relationships of kangaroos

Figure 4.5 The musky rat kangaroo, *Hypsiprymnodon moschatus*

Figure 4.6 The red kangaroo, *Macropus rufus*

a paste. This allows a high proportion of nutrients to be extracted from relatively poor-quality grasses.

Hypsiprymnodon, which retains many ancestral kangaroo features, does not hop bipedally and has a less specialised foot structure, differing from all other kangaroos in retaining the first toe. Kangaroos such as *Macropus* species have a hopping form of locomotion and can achieve speeds greater than 50 km/h as a result of the reduction of the number of toes. This is an advantage in the grasslands for avoiding predators.

Fossil sites in Australia

There are a number of fossil sites around Australia (see Fig. 4.7) that are rich in fossil samples providing information about the species and environment in that area in the past. Table 4.3 summarises the fossil findings in each area, which in turn can provide evidence and assist in indicating the possible factors that may determine distribution of plants and animals currently and in the future, by telling us factors that influenced them in the past.

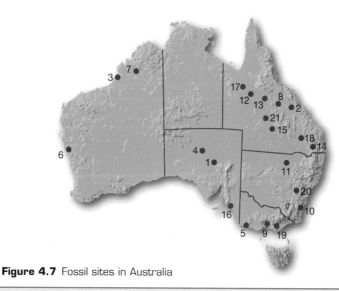

1 Andamooka
2 Bluff Downs
3 Broome
4 Coober Pedy
5 Dinosaur Cove
6 Geraldton
7 Gogo
8 Hughenden
9 Inverloch-San Remo
10 Koonawarra
11 Lightning Ridge
12 Maxwelton
13 Minmi Crossing
14 Murgon
15 Muttaburra
16 Naracoorte
17 Riversleigh
18 Roma
19 Strzelecki Ranges
20 Wellington
21 Winton

Figure 4.7 Fossil sites in Australia

Table 4.3 Australian fossil sites and their findings

Fossil site	Age of fossils (years ago)	Examples of fossil specimens	Past flora and fauna proposed from evidence
Lightning Ridge	110 million	■ *Steropodon galmani* (small monotreme) ■ *Tachyglossus aculeatus* (short-beaked echidna) ■ *Ornithorhynchus anatinus* (platypus) ■ *Muttaburrasaurus langdoni* (herbivorous dinosaur) ■ *Rapator ornitholestoides* (carnivorous dinosaur)	Forests of ferns and pines and the beginnings of flowering plants, herbivorous and carnivorous dinosaurs, and monotremes.
Murgon	55 million	■ Placental mammals (condylarth) ■ Crocodiles ■ Snakes ■ Frogs ■ Salamanders ■ Marsupial mammals ■ Rainforest plants	Rainforests, no dinosaurs but reptiles, songbirds, large numbers of marsupial mammals, few placental mammals, and amphibians lived near the streams.
Koonwarra	115–118 million	■ *Australurus plexus* (mayfly nymph) ■ *Tarwinina australis* (flea) ■ *Ginkgoites australis* (ginkgo leaf)	Large freshwater lake with fish, plants, insects, crustaceans, spiders, birds and crabs.
Inverloch	115 million	■ *Ausktribosphenos nyktos* (placental mammal) ■ *Koolasuchus cleelandi* (large amphibian)	Dinosaurs and mammals.
Riversleigh	25 million—40000	■ Parrots ■ *Emuarius* (ancestral to emus and cassowaries) ■ Marsupials ■ Crocodiles ■ Snakes ■ Lizards ■ Turtles ■ *Yalkaparidon*, dasyurids, thingodontans (marsupial mammals)	Rainforests with high diversity of animals such as birds, reptiles, large numbers of marsupials, kangaroos, possums, wombats and monotreme mammals.
Bluff Downs	5 million	■ Bluff Downs giant python (*Liasis* sp.), ■ Koala-like mammal (*Koobor jimbarretti*) ■ Ancestral dasyurids	Lakes and streams, wetland area with rich diversity of animals such as birds, reptiles and mammals, many very large animals (e.g. giant python 8 m long).
Wellington caves	5 million—30000	■ Diprotodon (giant marsupial kangaroo) ■ *Thylacoleo* (marsupial lions)	Leaf-browsing decreasing and grass-grazing marsupials increasing (kangaroos, wallabies and tree-kangaroos) over time, and grasslands expanding.
Naracoorte	300000	■ *Macropus giganteus* (eastern grey kangaroo) ■ *Wallabia bicolour* (swamp wallaby) ■ *Sthenurus brownei* ■ *Protemnodon brehus*	Large range of vertebrates: megafauna—large-sized animals, marsupials, birds, reptiles and frogs.

Extension information: visit www.amonline.net.au/fossil_sites/index.htm for Australian fossil sites in detail (Australian Museum online).

317

Reasons for evolution, survival and extinction in Australian species

■ *gather, process and analyse information from secondary sources and use available evidence to propose reasons for the evolution, survival and extinction of species, with reference to specific Australian examples*

Introduction

The history of life on Earth over the past 3500 million years has been characterised by a dramatic increase in biological diversity. This increase in diversity did not occur in a gradual way over time but it is the collective result of a number of rapid evolutionary spurts followed by occurrences of mass extinction. This rise and fall of extinctions over time has resulted in an overall increase in biological diversity.

Mass extinctions are often followed by occurrences of diversification, during which biological diversity is restored and eventually increased. This occurs by the rapid evolution of the survivors of the extinction, which adapt to repopulate the environment space left by the extinct organisms.

The reasons for the mass extinction of the dinosaurs, including Australia's megafauna, continue to be debated amongst scientists around the world. There are three main theories as to why this extinction occurred.
1. *Human impact*—hunting large and slow, easy prey.
2. The last *ice age*—with significant amounts of water trapped in ice, sea levels drop and climates become drier and colder. Habitats would be destroyed and food would be scarce. With such extreme conditions larger animals would find it the most difficult to survive.
3. A combination of both *human impact and the ice age*—habitats destroyed and scarce food supply leads to large declines in numbers and human hunting directly leads to the final extinction.

Other approaches that species take that are more successful for survival and evolution are retreating into a suitable environment without being affected by environmental changes, or evolving through adaptations to the change in environment. Variation in species creates better opportunities for survival in changing environments.

Aims

1. Gather, process and analyse information from secondary sources and use available evidence.
2. Propose reasons for the evolution, survival and extinction of species, with reference to specific Australian examples.

 A list of Australia's lost kingdoms: www.lostkingdoms.com/facts/ index.cfm

Method

Part 1: Gather, process and analyse information from secondary sources and use available evidence

For a reminder on how to undergo this process go to page 18 ('Searching for information'). Read Section 4.1 and then select two Australian examples for each of the following categories:
■ species that have evolved and changed (e.g. made adaptations) over time
■ species that have survived unchanged (e.g. by retreating) over time
■ species that have become extinct over time.

Start your search using the website listed above, selecting examples from the most recent periods such as Holocene (10 000 to present) and Pleistocene (1.6 million—10 000 years ago) for those organisms that may have survived and evolved over time, and the less recent periods for organisms that became extinct. Other suggested websites can be found in the secondary source task in Section 2.6 of this module.

Once you have gathered information from secondary sources and processed and analysed the relevant information for your six selected examples, use the evidence available to attempt Part 2.

Part 2: Propose reasons for the evolution, survival and extinction of species, with reference to specific Australian examples

Copy (see Student Resource CD) and complete Table 4.4 by proposing reasons why each of your six selected Australian species examples evolved, survived or became extinct over time. Four examples have been provided: two extinct examples, the thylacine (Tasmanian tiger), which became extinct in 1936, and the diprotodon that became extinct 50 000 years ago coinciding with the arrival of humans into Australia; one survival example, the spotted cuscus (*Phalanger maculates*); and one evolution example, the kangaroo (*Macropus* species).

Results

Table 4.4 Reasons for the evolution, survival and extinction of some Australian species

Table 4.4

Australian example	Species evolved, survived or became extinct	Proposed reasons
Thylacine (Tasmania tiger)	Extinct	■ Introduction of the dingo 3500 years ago ■ Hunted by farmers in Tasmania (seen as a predator to sheep and chickens) ■ Tasmanian government bounty for threat to farming ■ Disease ■ Possibly all of the above reasons in succession
Diprotodon optatum (marsupial mammal—megafauna)	Extinct	■ Human hunting ■ Climate change ■ Or a combination of both
	Extinct	
	Extinct	
Phalanger maculates (spotted cuscus)	Survived	■ Retreated with the rainforest as the climate changed
	Survived	
	Survived	
Macropus species (kangaroo)	Evolved	■ Developed adaptations to the changes that occurred in the environment (such as change from forests to grasslands with increasing aridity) ■ Reduced number of toes and can hop bipedally at high speeds away from predators in open grassland, unlike earlier ancestors ■ Also change from those with leaf-browsing diet to those that were grass-grazers
	Evolved	
	Evolved	

4.3 Understanding and managing the present environment

■ *explain the need to maintain biodiversity*

Biodiversity refers to the variety of all forms of life, the diversity of the genetic characteristics they contain and the ecosystems of which they are components. Genetic diversity within a species is what allows populations to adapt to changes in the environment. To maintain biodiversity we must maintain genetic diversity within the species as well as maintaining the living species within their habitats.

The value and benefits of maintaining biodiversity fall into four main categories:

■ direct economic value of products we obtain from species of plants and animals, and other groups:
  —bioresources such as food, fibre, timber and medicines
  —the potential of any undiscovered bioresources is significant (e.g. ants possess specialised glands for producing antibiotics to reduce disease in their colonies). These discoveries hold great potential for therapeutic use
■ indirect economic value of benefits produced by species without consuming them
  —ecosystems underpin many of our natural resources and provide services such as healthy soil, clean water and crop pollination

■ ethical—ethically all species have a right to exist just as humans do
■ aesthetic value—humans enjoy the beauty of the natural environment, and countries like to conserve their heritage to be passed down.

At an international level, biologists are working together to discover and record all the types of organisms on Earth, the world's biodiversity, so that it can be synthesised into a classification system that will reflect our knowledge of all of life. Today, despite the fact that we realise that biological diversity is one of our most precious resources, species are being lost at a rate 100 to 10 000 times faster than before human intervention. We do not know the exact rate of extinction of species, but in Australia since European settlement in 1788, 75 per cent of rainforests, 99 per cent of south-eastern temperate lowland grasslands and 60 per cent of south-eastern coastal vegetation have been cleared, resulting in significant loss of species. The dramatic loss of species caused by human activity worldwide is known as the **biodiversity crisis**. Each species represents an immense amount of genetic information. At present, we do not know what will be the ecological impact of a continuing loss of species.

Current effort to monitor biodiversity

■ *process information to discuss a current effort to monitor biodiversity*

Aim

1. To process information.
2. To discuss a current effort to monitor biodiversity.

Background information

The Australian government's *Environment Protection and Biodiversity Conservation Act 1999* was created to meet Australia's obligations as a signatory to the 1992 Convention on Biological Diversity. This Act protects all native fauna and provides for the identification and protection of threatened species. In each Australian state and territory, there is a statutory listing of threatened species. At present, 380 animal species are classified as either endangered or threatened under the *EPBC Act*. In fact, a complete cataloguing of all the species within Australia has been undertaken; it is a significant step in the conservation of Australian fauna and biodiversity. In 1973, the federal government established the Australian Biological Resources Study (ABRS), which coordinates research in the taxonomy, identification, classification and distribution of Australian flora and fauna.

Monitoring biodiversity

Monitoring is essential to any action plan to conserve biodiversity. It provides information on how the management plan for conserving biodiversity is performing. If species continue to decline in population numbers, then the current management plan is not effective and requires change. Changing management involves assessing human-related activities such as mining, forestry and recreation, as well as conservation-related activities such as those existing in reserves and national parks.

Ideally all species in existence would benefit from having their populations monitored; however, this would be a very time-consuming and costly process, and funding is limited. So, at present, monitoring programs tend to focus on threatened species or ecosystems, or ecosystems that provide economic value through industry (e.g. mining, logging and tourism).

You carried out abundance (or population) measurements of selected species in your biology field trip earlier this year. If you were to repeat your data collection at the same time next year, it would be possible to see changes over that time and allow you to comment on the population status or health of species biodiversity in that area.

Many government funded groups are monitoring biodiversity, using more complex versions of the methods you may have used during your field trip. These programs are collecting abundance measurement and information regarding human impact on different organisms such as native plants, birds, reptiles, mammals and amphibians. Amphibians have been identified as good indicators of environmental health because they are very sensitive to changes in their environment (their skin is permeable to both liquids and gases). Because of this sensitivity their decline within an ecosystem may also indicate that other components of the ecosystem are in a state of decline or bad health. Information can also inform us about the nature of the population decline and the potential threats causing decline.

The following case study attempts to provide a balanced argument and highlights the complexities and difficulties in managing and conserving biodiversity.

Method

Part 1: Process information

Refer to page 18 for suggestions and reminders on how to go about processing information. Read Section 2.10 (page 50) in the Local Ecosystem module discussing the nationally endangered Baw Baw frog (*Philoria frosti*) and the case study below, and process your information to begin completing Part 2.

Part 2: Current effort to monitor biodiversity

Copy (see Student Resource CD) and complete Table 4.5 and use this to attempt to discuss the points for and/or against this current effort to monitor biodiversity. You will be using this task to address PFA P5 in describing the scientific principles employed in a particular area of biological research. In doing so, we are looking at one of the current critical issues surrounding Baw Baw frog research.

SECONDARY SOURCE INVESTIGATION

PFAs
P5
BIOLOGY SKILLS
P12
P14

For websites on how different groups and government departments are monitoring biodiversity

Case study—Baw Baw frog

The Baw Baw frog is one of Australia's most critically endangered amphibians and it is Victoria's only endemic amphibian, being restricted to the Baw Baw Plateau in the Central Highlands of Victoria. The dramatic decline of the species in recent years coincides with a global decline in amphibian populations. This decline in amphibian populations also corresponds with the overall global biodiversity crisis. Since 1996, the Australian Government has provided funding to support research and population monitoring on the species through its endangered species program. Population monitoring is a key action in the draft National Recovery Plan (2002–2006) and State Action Statement for the species. These plans identify research and management actions required to conserve the species and are supported by legislation at both state (*Flora and Fauna Guarantee Act 1988*) and federal (*Environment Protection and Biodiversity Conservation Act 1999*) levels. The effort to protect and conserve the Baw Baw frog and its habitat is also likely to benefit the conservation of a number of other threatened species and communities located in the same area as well as contributing to the identification of the likely causes of decline.

An additional advantage of the Baw Baw frog monitoring program is that it is designed to distinguish between 'real' population trends and natural population fluctuations that can occur in response to variable factors such as weather and food resources. The longevity of the monitoring program (24 years as of 2007) and the selection of an appropriate number of survey sites in which to census, provides the program with sufficient power to detect 'real' trends in population size over time. However, monitoring programs that are designed to measure 'real' trends in population often require many surveys or repeated measurements of abundance over a season or year and as a result can be very expensive.

This high cost can inadvertently prevent monitoring from occurring when funding is in short supply or not made available.

Presently, the causal mechanisms responsible for the decline of the Baw Baw frog are still yet to be formally established. Currently, the most plausible reasons for the decline of the species are:
- an introduced pathogen (eg. chytrid fungus)
- climate change (this species is very sensitive to small changes in temperature and relative humidity).

Other potential threats to the frog include:
- habitat clearance and fragmentation (e.g. alpine resort development and timber harvesting)
- introduced flora and fauna (e.g. cattle, deer, rabbit, cat, fox, and dog)
- forestry activities (e.g. roading and timber harvesting)
- fire (inappropriate fire regimes)
- increased UV-B radiation
- atmospheric pollution (e.g. industrial and agricultural emissions)
- multiple and interacting factors (combinations of the above factors).

Approximately 70 per cent of the current total population of adult Baw Baw frogs occurs in state forest where the primary land use has been identified as timber harvesting. Since the discovery of the Baw Baw frog in state forest in 1996, forestry activities have been identified as a serious threat to the long-term survival of the species. The sensitivity of the Baw Baw frog, due it its narrow ecological requirements, suggests that forestry activities may impact through:
- direct destruction of frogs and habitat
- changes to climatic and water conditions from activities in and adjacent to frog habitat (e.g. stream-flow amount is reduced up to 25 years following timber harvesting)
- sedimentation of breeding habitat following activities in and adjacent to frog habitat
- fragmentation of populations, and/or destruction or modification of dispersal habitats.

The case of conserving the Baw Baw frog presents a very interesting issue because the frog is situated within an area of very high timber-resource value. The Victorian Government has been faced with the challenge of implementing components of biodiversity and resource-use legislation

For further information from websites on the Baw Baw frog

Figure 4.8
Baw Baw frog

that are in conflict with each other. The biodiversity legislation states that the Baw Baw frog will be protected and its long-term survival ensured, whilst legislation relating to the timber harvesting industry states that the government will guarantee resource security for specified volumes of timber from the area where the Baw Baw frog is located. Due to the substantial economic value of native forest timber to local and state economies in Victoria ($3000 million in commercial turnover each year and provision of many jobs), scientists are often faced with the predicament of having to demonstrate negative impacts of resource use on the prospects of survival of a species before conservation initiatives are implemented. Demonstrating such impacts, or lack there of, can often take considerable periods of time when researching impacts on biodiversity. These lengthy periods of time can sometimes discourage governments from wanting to invest in such projects.

In the case of the Baw Baw frog, the Victorian State Government allocated funding to undertake research involving experimental timber harvesting within the habitat of the frog to investigate the potential negative impacts of forestry activities on the species. This research was identified as an action in both the draft National Recovery Plan (2002–2006) and State Action Statement for the species. However, due to concerns expressed by the Australian Government and local environmental groups over the risk of conducting the research on an endangered species with a very small distribution, the research was discontinued. This case provides an example of when concerns expressed by external stakeholders and the public can influence decisions made by governments.

Due to the extended drought across south-eastern Australia, the management of water resources within Melbourne's water catchments has more recently become a significant issue in Victoria. Because the Baw Baw frog lives in Melbourne's primary water catchment (the Thomson Catchment and Reservoir) and that re-growth forest following timber harvesting activities can significantly reduce run-off of water up to 25 years after harvesting, a future decision by the Victorian Government to protect the habitat of the frog by excluding timber harvesting may be influenced more so by the increasing economic value of water within the Thomson Water Catchment compared with the harvesting of timber.

Source: Dr Greg Hollis, Senior Biodiversity Officer, Department of Sustainability and Environment

Results

PFA P5: Describe the scientific principles employed in a particular area of biological research

Table 4.5 Current issues, research and development in biology—monitoring the Baw Baw frog

Issue: Should forestry activities continue in the habitat of the Baw Baw frog?	
Points FOR issue (i.e. in favour, such as benefits, positive outcomes and positive consequences)	**Points AGAINST issue** (i.e. negative factors such as expense, limited uses and negative consequences)

Table 4.5

Discussion/conclusion

Summarise your overall points for and against the issue, summing up the discussion with a conclusive statement.

Analysis questions

1. Describe two ways in which the Australian government is involved in monitoring biodiversity.

2. Discuss why it is important for governments to support biodiversity monitoring initiatives.
3. Explain the importance of amphibians like the Baw Baw frog when aiming to monitor biodiversity.
4. List the possible threats towards the rapidly declining Baw Baw frog population.
5. Discuss the main points for and against the effort to monitor biodiversity in Baw Baw frog habitats.

REVISION QUESTIONS

Answers to revision questions

1. Explain the importance of the study of past environments in predicting the impact of human activity in present environments.
2. Identify the ways in which palaeontology assists our understanding of the factors that may determine the distribution of flora and fauna in present and future environments.
3. Explain the need to maintain biodiversity.
4. Using a *named* Australian example, identify a reason for the extinction of the species.
5. Discuss an example of a current effort in Australia to monitor diversity.

Glossary

abiotic	pertaining to the physical and non-living components
absorption	the uptake of substances across a boundary; in cells it is the uptake of water or nutrients from the surroundings
abundance	the number of individuals (size) of a population
acidophiles	types of bacteria (Archaea) that survive in highly acidic environments (acid-loving)
active transport	movement of any molecules through a membrane against the concentration gradient; energy is required for this process
adaptation	an alteration in structure, function or behaviour, that is hereditary, by which a species or an individual improves its condition in relation to its environment
aerobic respiration	the process of respiration carried out in the presence of oxygen
allelochemicals	the chemicals released by allelopathic organisms
allelopathic	having characteristics of allelopathy
allelopathy	the inhibition of growth in one species of plants by chemicals produced by another species
alveoli	air sacs with extremely thin walls, within the lungs
ammonia	a toxic form of nitrogenous waste, excreted in a dilute form together with large volumes of water
anaerobic	not requiring oxygen
anaerobic respiration	the process of respiration carried out without the presence of oxygen
angiosperms	flowering plants
anoxic	no free oxygen
anther	the part of the flower that houses developing male reproductive cells (pollen)
anthrax	contagious disease in cattle, caused by a bacterium
apical meristems	growing point of a plant containing cells that divide by mitosis, located at the tip of a root or the tip of a stem
appendix	a small projection from the digestive tract; remnant of the caecum
aquaporins	membrane proteins that act as pores, allowing water to move through by osmosis
aquatic environment	a water environment (e.g. freshwater, marine or estuarine)
Archaea	one of the super kingdoms of procaryotes (bacteria)
arid	areas lacking sufficient water or rainfall, and commonly high temperatures
arteries	blood vessels with thick walls that carry blood under pressure, away from the heart towards other organs of the body
assimilation	the conversion of absorbed simple substances into more complex molecules, which then become part of the structure of an organism
ATP (adenosine triphosphate)	serves as a major energy source within a cell to drive a number of biological processes
autotrophs	organisms able to synthesise their own food, by photosynthesis
Bacteria	one of the super kingdoms of procaryotes (Eubacteria)
beneficial interactions	when one or more organisms benefit from a relationship
bilayer	two layers—an outer and inner layer—of phospholipids forming the cell membrane
biodiversity	the number, relative abundance and genetic diversity of organisms on Earth
biodiversity crisis	the dramatic loss of species caused by human activity worldwide
biogeography	the study of the geographical distribution of species, both present-day and extinct
biomass pyramid	diagrammatic representation of the amount (weight) of organisms in a particular area at a particular time

biomarkers	chemicals that are produced by only one group of organisms providing evidence of their existence in the past
biome	large regional system characterised by major vegetation type (region with similar ecosystems grouped together)
biosphere	the part of the Earth and atmosphere in which living things are found
biota	the flora and fauna of a given habitat or region; the sum total of all living things on Earth
biotic	pertaining to living features (e.g. organism abundance, distribution, or interactions)
bivalent	a pair of homologous chromosomes which become apparent in the first meiotic division during crossing over
blood vessels	arteries, veins or capillaries that carry transport fluid (blood)
botanist	person who studies plants
breathing	a mechanical (physical) process involving muscles and the skeleton in animals, which enables an organism to inhale and exhale
caecum	enlarged organ at the end of the small intestine, where microbial fermentation occurs in herbivores to assist with the digestion of cellulose
cambium	meristematic tissue found in the stems of plants that divide by mitosis to allow secondary growth (increase in width)
canines	teeth that are sharp and pointed to help hold and kill prey and for tearing meat from the bones
capillaries	the smallest of blood vessels with very thin walls, which carry blood between arteries and veins
carbohydrates	a class of organic compounds made up of the elements carbon, hydrogen and oxygen, with a 2:1 ratio of hydrogen to oxygen. This class includes sugars and starches
carbon fixation	a chemical process whereby carbon dioxide is combined with hydrogen to form carbohydrates
carnassial	teeth that have sharpened cutting edges to effectively slice and shear meat, characteristic of carnivores
carnivores	organisms that eat or consume other animals (meat-eaters)
carnivorous	organisms that have the ability to consume animals only
carpel	the female reproductive organ of a flowering plant; it encloses ovules; it ripens to become a fruit
carrier protein	a small organic molecule that facilitates movement of substances with low lipid solubility, or substances moving against the concentration gradient, across a cell membrane
cell	the basic unit of all living things, made up of protoplasm (cytoplasm and a nucleus) surrounded by a cell membrane and, in plants and some organisms, a cell wall
cell cycle	the repeating sequence of growth and division through which cells pass
cell division	the process by which cells divide into two, by either keeping the chromosome number the same (mitosis) or halving their chromosome number (meiosis)
cell membrane	the boundary surrounding the protoplasm of any cell. Also termed *cytoplasmic membrane*, *plasma membrane* or *plasmalemma*
cell sap	the solution of water and dissolved contents inside the vacuole of plant cells
cell theory, the	a generally accepted scientific theory that cells are the basis of all living things and can only arise from other cells
cellulose	insoluble organic, complex polysaccharide (carbohydrate) that is the main component of cell walls in plant cells
centromere	an organelle present in animal cells, responsible for forming fibres called the spindle during cell division
channel protein	protein that spans the lipid bilayer of a cell membrane, to allow the passage of ions, water and chemicals of low lipid solubility
chemical digestion	the breaking down of food into its basic monomer compounds by the chemical action of digestive enzymes
chemical reactions	a sequence of steps by means of which substances interact and are transformed into other substances, involving an energy change
chemical respiration	see respiration

chlorophyll	green pigment found in all green plant cells, responsible for light capture in photosynthesis
chloroplasts	organelles found in green plant cells; responsible for the process of photosynthesis
chromatin material	nuclear material, made of DNA protein, which stores the hereditary information as linear sequences of genes in cells. Chromatin shortens and thickens into chromosomes at the start of cell division
chromosomes	thread-like structures made of DNA, visible in dividing cells as a result of the shortening and thickening of chromatin material at the start of cell division; a coloured (stainable) body, observed in cells that are dividing; it is made up of chromatin material and contains a linear sequence of genes
closed circulatory system	a system of blood flow where the transport fluid is pumped around the body through a series of blood vessels which divide into capillaries in the tissues; blood is eventually returned to the heart without having left the closed system of vessels at any point
colonial organisms	a colony of single-celled organisms
commensal	organisms can be involved in this type of relationship where one organism benefits and the other is unaffected
commensalism	a symbiotic interaction between two species where one benefits and the other is unaffected
community	groups of different populations in an area or habitat
companion cells	a type of phloem cell in plants, which controls the functioning of the sieve tubes
compound microscope	a microscope which passes an image through two lenses to increase magnification, forming an inverted image, but having the advantage of revealing detail that is too small to be observed clearly with the naked eye or with a simple microscope
concentration gradient	difference in the concentration of a substance in two regions (that may be separated by a membrane)
consumers	heterotrophic organisms that ingest other organisms in a food chain
continental shelf	the gently sloping undersea area surrounding continents, at depths up to 200 m, after which the continental slope drops steeply to the ocean floor
convergent evolution	the process whereby organisms that do not have recent common ancestors develop similar features or adaptations because they live in similar habitats
converging	moving towards each other
copulation	mating between sexes; associated with internal fertilisation
cosmos	the universe ordered as a whole
cristae	folds of the inner membrane of a mitochondrion, to increase its surface area for the location of groups of respiratory enzymes
cross-pollination	the pollination of a carpel by pollen from a different individual
crossing over	the mutual exchange of similar segments of chromatids which occurs between homologous chromosomes during meiosis I
cyanobacteria	a photosynthetic eubacterium
cytokinesis	the division of the cytoplasm that follows nuclear division during mitosis or meiosis
cytoplasm	all the cell contents excluding the nucleus, including the cytosol (molecules in a gel-like solution) and the organelles
daughter chromatids	two strands of chromatin, which are held together by a centromere, to form one chromosome; forming after DNA replication at the start of cell division
daughter chromosomes	two strands of chromatin that move apart during cell division, as a result of the centromere dividing
decomposer	an organism, such as fungi and bacteria, that consumes and breaks down organic matter for energy, releasing inorganic nutrients
deep ocean trench	when a continental plate collides with an oceanic plate, the oceanic plate is forced underneath, forming a deep trench in the ocean floor
degraders	organisms that feed on dead organisms and organic wastes
dependent variable	a factor which changes during an experiment (variable), as a result of the experiment. It is the observed or measured outcome that depends on other factors that have been changed in the experiment
detrimental interactions	when one or more organisms are harmed or disadvantaged from a relationship

detritivores	animals that eat organic litter or detritus (a type of degrader)
detritus	organic debris produced during the decomposition of animals and plants
dicotyledons	a class of flowering plants which generally have net-veined leaves, a tap root system and two cotyledons or seed leaves in the developing embryo
digestion	the breakdown of complex, usually insoluble food into simpler, smaller soluble molecules that can be easily absorbed
differentiate	develop by a process of specialising in structure; refers to the maturation of a cell so that it can perform a particular function
diffusion	passive movement of any molecules along a concentration gradient, until equilibrium is reached
diploid	having two sets of chromosomes
disaccharides	sugar molecules (e.g. sucrose, maltose and lactose) made up of two similar monosaccharide units
distribution	where a species occurs
diverging	moving away from each other (moving apart)
divergent evolution	evolving (changing in structure) to become different from another organism or a common ancestor
DNA (deoxyribose nucleic acid)	a nucleic acid that is the hereditary material of an organism
dorsiventral	tissue arrangement in a leaf, where the upper (dorsal) surface has a different arrangement to that of the lower (ventral) surface
double membrane	two unit membranes (two lipid bilayers) that surround some organelles (e.g. nucleus and chloroplasts)
ecology	the study of the relationships that living organisms have with each other and their environment
ecosystem	community together with its environment; any environment containing organisms interacting with each other and the non-living parts of the environment (e.g. rainforest ecosystem)
ectoparasites	parasites that live on the surface of their host
egestion	the elimination of undigested food from an organism
egg cells	female gametes or sex cells that have half the original chromosome number of that organism
electron micrographs	photographs of images seen under an electron microscope
electron microscope	a microscope that uses the wave properties of electrons to magnify an image more than 200 times larger than that of a light microscope, allowing the viewing of the ultrastructure of things
embryonic cells	immature, undifferentiated cells that have the ability to divide and become any other cell type
endemic	a species that is unique to a specific geographic region; it is assumed to have evolved there
endoparasites	parasites that live internally in their host
endoplasmic reticulum (ER)	cell organelle made up of a system of flattened membranes, functioning in transport within a cell
environment	both living and non-living surroundings of an organism
epidermis	outermost layer of cells, usually protective in function
epiphytes	plants that grow on another plant for support (not parasitic)
equilibrium	a balanced or stable state or equal distribution of particles
estuarine environment	a water environment (usually at the mouth of a river to the sea) that fluctuates between freshwater from the river and saltwater from the sea
estuary	where the mouth of a river meets the sea
eucaryotic cells	cells with a membrane-bound ('true') nucleus and other membrane-bound organelles
evolution	the change in a population over a period of time. It implies that organisms were not created independently of each other, but may have arisen from a common form which changed over time
excretion	the elimination of wastes produced during metabolism (e.g. nitrogenous wastes and carbon dioxide)
extant	still presently living
external fertilisation	fertilisation, or the union of gametes, occurring outside the organism's body
extinct	no remaining members of the species (species has died out)

extinction	when a species or group of organisms has died out, or been wiped out of existence
extremophiles	organisms adapted to living in extreme conditions (e.g. extreme temperature, pressure or chemical concentration, such as high acidity or salinity)
facilitated diffusion	movement of molecules across a cell membrane along a concentration gradient, assisted by carrier proteins in the membrane
fermentation	a change brought about by micro-organisms such as yeast, which convert grape sugar into ethyl alcohol
flaccid	the limp state of a plant cell when its contents have shrunk as a result of water loss (plasmolysis)
fluid mosaic model	current generalised model for the structure of all cell membranes
food chains	sequences of organisms from producers to consumers along which energy flows in an ecosystem; usually with three or four trophic levels
food web	a number of interacting food chains in an ecosystem
fore-gut fermentation	microbial breakdown of food which occurs in the stomach of some herbivores, resulting in cellulose digestion before the food reaches the intestines
fossils	the preserved remains of organisms or traces of organisms (e.g. footprints)
gametes	haploid sex cells such as egg cells and sperm cells which fuse during fertilisation
gaseous exchange	the exchange oxygen and carbon dioxide with the external environment in plants and animals
gene pool	the range of genes (and their variations) present in a population
genetic variation	differences in various traits or features that are genetically determined amongst members of a population
geological timescale	information from fossil evidence to provide a timescale illustrating the different periods of time that different organisms existed
geology	the scientific study of the origin, history and structure of the Earth as recorded in rocks
glycogen	the main storage form of polysaccharide carbohydrates in animal cells
granum	a group or stack of photosynthetic membranes (lamellae), containing chlorophyll, in chloroplasts of plant cells
grassland	habitat where the dominant vegetation is grass, very few shrubs and trees, typically a low or sporadic rainfall area
growth	increase in the size and/or complexity of an organism as a result of cell division and/or cell enlargement
guard cells	bean shaped epidermal cells in leaves that surround a stomate or pore and control the opening and closing of that pore
gymnosperms	conifers
habitat	a place where an organism lives
Hadean eon	period of time approximately 4.5 to 3.8 billion years ago
haemolymph	the transport fluid in an open circulatory system such as that of insects (equivalent to blood)
halophiles	bacteria (Archaea) that survive in high saline environments (salt-loving)
haploid	the condition in a cell of having only one set of chromosomes which are unpaired; half the usual number of chromosomes (e.g. gametes are haploid)
heart	muscular, rhythmically contracting pump that forms part of the circulatory system in animals; responsible for circulating blood
herbivores	organisms that eat or consume plants only
herbivorous	organisms that have the ability to consume plants
heredity	similarity between parents and offspring, as a result of the inheritance of genes, carried on DNA molecules, by offspring from their parents
heterotrophs	an organism that cannot make its own food and so must consume other living organisms to obtain organic nutrients
hindgut fermentation	microbial breakdown of food which occurs in the caecum of some herbivores, resulting in cellulose digestion in this enlarged organ at the end of the small intestine

homologous pair	two similar chromosomes in a cell, one paternal and one maternal in origin, that carry alleles of the same genes in the same sequence and that pair up during meiosis
hydrothermal vents	cracks in the Earth's surface that release water at high temperatures, caused by magma under the crust
hypothesis	a possible solution to a scientific problem, based on accumulated scientific information, suggesting a general principal that can be tested experimentally
incisors	teeth used for biting or gnawing; well developed in herbivores
independent variable	a factor in an experiment that is changed by the experimenter and affects the final outcome of the experiment
ingestion	the intake of food into a digestive tract (multicellular organisms) or into a cell (unicellular organisms)
inorganic	molecules or compounds that do not contain carbon
inorganic compounds	chemical compounds that are part of the inanimate, non-living world, and are not produced by living organisms and do not contain hydrocarbon chains (the combined elements of carbon with hydrogen)
interference competition	where organisms harm each other while obtaining a resource, even if that resource is not in limited supply
internal fertilisation	fertilisation, or the union of gametes, occurring inside the organism's body
internal gills	organs of gaseous exchange inside the body of aquatic animals, such as fish
interphase	the stage preceding mitosis or meiosis, during which the replication of DNA occurs
interspecific competition	individuals of different species striving for the same resource that is in limited supply
interstitial fluid	(also known as tissue fluid) a fluid that lies in the spaces between cells, bathing them
intraspecific competition	individuals of one species striving for the same resource that is in limited supply
isobilateral	arrangement in a leaf, where the upper (dorsal) surface has a similar arrangement to the lower (ventral) surface
isolation	the effects of separation that prevent individuals from interbreeding
isotonic	describes solutions that have the same concentration of dissolved substances and therefore the same osmotic pressure
kidneys	main organ of excretion of nitrogenous wastes and maintenance of body fluid composition
large intestine	(colon) the last part of the digestive tract where absorption of water and minerals typically occurs
lenses	for a microscope—a transparent, biconcave structure that produces a magnified image of an object
lenticels	small raised pores present in the outer cork layers of woody stems
lignin	chemical substance associated with some plant cell walls (e.g. xylem tissue), making them stronger and impermeable to water
light	radiation in the visible spectrum
light-independent phase	second stage of photosynthesis where the products of the first stage are combined with carbon dioxide to make sugars
light phase (photolysis)	first stage of photosynthesis where light energy is used to split water
lipids	a group of long chain chemical compounds that are high in energy, insoluble in water and form the main component of the bilayer of cell membranes
macromolecules	large complex molecules made up of smaller respeating subunits
macroscopic	large enough to be seen with the naked eye
magnification	the ability of a lens or microscope to enlarge an image
malaria	a widespread disease caused by a parasite in the blood cells, transmitted by the female *Anopheles* mosquito
marine environments	saltwater environments (e.g. ocean or sea)
mass extinctions	extinctions that have occurred on a large scale (e.g. dinosaurs)
maternal	derived from the mother (female parent)

matrix	the internal, fluid-filled space in mitochondria, containing enzymes for the final chemical reactions of chemical respiration
megafauna	extremely large animals, most of which are extinct today
meiosis	a process of cell division that is considered to be a reduction division because it halves the number of chromosomes in the resulting gametes (egg and sperm cells) that it produces
meristem	localised region of cells that are actively dividing (undergoing mitosis) in plants
mesophyll	tissue found in the middle layer of a leaf, made up of palisade cells and spongy cells
metamorphosis	a rapid and distinct change in form during the life cycle of an organism, where the larva changes into an adult
meteorite	a meteor (stony or metallic mass) that survives the intense heat of atmospheric friction and reaches the Earth's surface
methanogens	type of bacteria that uses hydrogen gas and carbon dioxide to generate energy and make sugars
microbial fermentation	the breakdown of food by the action of bacteria
microfossils	fossils of single-celled anaerobic procaryotes
microscopic	too small to be seen with the unaided eye; visible with a microscope
microvilli	microscopic finger-like projections on the surface of a cell, to increase its surface area for the uptake of nutrients, particularly epithelial cells in animals that are involved in absorption
mid-ocean ridges	ridges along plate margins in the ocean crust that slowly release magma
midrib	the main vein of a leaf
mimicry	where individuals of one species have characteristics (e.g. visual or behavioural) that resemble those of another species
mitochondria	(singular: mitochondrion) an organelle in all eucaryotic cells, responsible for cellular respiration and therefore energy production in a cell
mitosis	the process of cell division whereby somatic (body) cells undergo a single nuclear division, giving rise to two genetically identical daughter cells
molars	teeth used for chewing, well developed in herbivores
molecular technologies	techniques used in the branch of genetics that deals with hereditary transmission and variation at the molecular level (e.g DNA sequencing)
monocotyledons	class of flowering plants which generally have parallel veined leaves, a fibrous root system and one cotyledon or seed leaf in the developing embryo
monomers	small unit molecules forming the basis of larger, more complex polymer molecules
monosaccharides	simple sugars, with molecules either containing five carbon atoms (ribose and deoxyribose) or six carbon atoms (glucose, fructose and galactose)
multicellular	made of many cells
mutualism	the symbiotic interaction between two species where both benefit from the association (e.g. lichen)
mutualistic	to be in a relationship where both species benefit
nanobacteria	see nanobes
nanobes	filament-type structures found in rocks; they are able to withstand radiation, cold and acidic conditions
natural selection	the process by which certain members of a population that are more suited to prevailing environmental conditions survive and reproduce (their chances of survival are influenced by how successfully their genetic make-up enables them to withstand changes in the environment)
niche	place of a species within a community, involving relationships with other species
nuclear envelope (nuclear membrane)	the boundary that separates the nucleus from the cytoplasm, consisting of two phospholipid bilayers, forming a double membrane
nuclear sap	the semi-liquid, slightly viscous background material of the nucleus in which chromatin material is found. Also known as nucleoplasm
nucleolus	a dark-staining round or oval body inside the nucleus of a cell, responsible for the formation of ribosome sub-units

nucleoplasm	see nuclear sap
nucleotides	a monomer or subunit of nucleic acids that has a distinct structure made up of sugar, a phosphate and a nitrogenous base
nucleus	an organelle that contains the genetic information of the cell (chromosomes) and controls most of the cell's functioning
oesophagus	long tubular structure in the digestive tract, carrying food from the mouth to the stomach
oil-immersion lenses	high power objective lenses on light microscopes that are designed to give clearer resolution when a drop of oil is placed on top of the coverslip of the specimen being viewed
omnivores	organisms that consume both plants and animals
open circulatory system	a transport system in small invertebrates where the transport fluid is pumped out of blood vessels into the surrounding tissues, where it bathes the cells directly, before flowing back into vessels and returning to the tubular heart
organ	a body structure composed of a variety of different tissues that work together to perform a function as part of a system
organic	carbon-containing molecules or compounds
organic compounds	chemical substances that are synthesised in living organisms and contain atoms of the elements of carbon and hydrogen
organism	a thing that is or once was alive and can carry out most of the functions that characterise being alive. Plants, animals, microbes and fungi are all organisms
osmosis	the movement of water molecules from a region of high water concentration to a region of low water concentration through a selectively permeable membrane
osmotic pressure	a measure of the solute concentration in a solution, that in turn results in water moving into a solution by the process of osmosis, increasing its pressure; as the concentration of the solute rises, the osmotic pressure rises
ostia	pores in the tubular heart of an organism with an open circulatory system
outgassing	emission of gases
ovary	the female reproductive organ where eggs are produced and which in flowers, contains ovules
oviparous	releases eggs that are fertilised externally
ovules	contained in the ovary of flowers and develop into fruit after fertilisation
oxic	having oxygen
'oxidising' atmosphere	an atmosphere that doesn't contain free hydrogen
palaeontology	the scientific study of fossils and all aspects of extinct life
palisade cells	elongate plant cells that contain chlorophyll; main photosynthetic cells in plants
Pangaea	the single land mass that existed more than 250 million years ago, made up of all the continents joined together, surrounded by one huge ocean
parasite	an organism that lives and feeds on or in another organism, the host, which is usually larger than the parasite
parasitic	characteristic of a parasite
parasitism	the symbiotic relationship between two species where one benefits (parasite) and the other is unharmed (host)
parfocal	refers to the microscope objective lenses that are designed to keep an image in focus when changing from low to high power: if an image is in focus with one objective lens, when the eyepiece is rotated it will remain in focus eliminating the need to adjust the coarse focus knob or lower the slide during changeover
passive movement	movement of molecules along a concentration gradient (from high to low concentration), requiring no energy input
pasteurisation	method of partial sterilisation (of wine or milk) by heating it to a temperature just below its boiling point, discovered by Louis Pasteur
paternal	derived from the father (male parent)

penicillin	a chemical compound produced by the mould penicillium; an antibiotic used to reduce bacterial infections
peptide bonds	a chemical bond (force of attraction) that occurs between amino acids in a polypeptide chain
pericycle	a special layer of meristematic tissue in plant roots which is responsible for the development of branch roots
permeability	see permeable
permeable	a term used to describe a membrane or other barrier that allows molecular substances to pass through it
phagocytosis	cell eating; a type of endocytosis whereby solid particles are engulfed by a cell by invagination of the cell membrane, forming a vacuole
phase contrast microscope	a microscope that takes advantage of the fact that light changes phase when it passes through structures of different densities, enhancing the contrast necessary to view a specimen
phloem	the vascular tissue in plants that transports organic nutrients (food) from where they are manufactured, up and down the plant
phospholipids	lipid molecules that have a polar (charged) phosphate end; this is the main type of lipid molecule forming the bilayer of membranes in cells
photosynthesis	food-making chemical process in plants that uses carbon dioxide, water and the energy of light, in the presence of chlorophyll, to manufacture organic molecules (mainly sugars) with oxygen as a by-product
physiological	to do with the functioning of an organism
pinocytosis	cell drinking; a type of endocytosis whereby liquid particles are engulfed by a cell by invagination of the cell membrane, forming a vacuole
plan sketch	a scientific diagram using single, solid lines to show the distribution of something (e.g. cells in an organ or plants and animals in their habitat)
plasmolysis	a condition in plant cells where the cell contents shrink as a result of water loss, causing the cell membrane to pull away from the cell wall and the cell to become flaccid
pollen	the collective term for pollen grains
pollen grains	small, granular male reproductive structures produced by anthers in seed-bearing plants
pollination	the process in which pollen of flowering plants is transferred to the stigma for fertilisation
polymers	very large molecules made up of a chain of similar smaller molecular subunits (monomers) joined together
polypeptide	a molecule consisting of a single chain of many amino acids joined together by peptide bonds. Polymers are the chains of which proteins are made
polysaccharides	a complex carbohydrate consisting of many monosaccharide (single sugars) units joined together
population	groups of organisms of the same species living in the same area at a particular time
pores	openings or breathing holes on the surface of a plant or animal body, through which gases or liquids can pass
potometer	apparatus used to measure the rate of transpiration in plants
predator	an organism that catches and kills another organism for food
predator–prey relationship	the relationship between a predator and its prey
premolars	cheek teeth for chewing
prey	something that is hunted or caught for food
primary consumers	organisms first in the food chain to consume other organisms (herbivores)
procaryotic autotrophic organisms	unicellular organisms with cells lacking membrane-bound nuclei, that carry out photosynthesis (e.g. cyanobacteria)
procaryotic cells	cells that do not have their DNA enclosed by a membrane or form a proper nucleus; they have no membrane-bound organelles within the cell; procaryotic organisms are usually unicellular (e.g. bacteria)

procaryotic heterotrophic organisms	unicellular organisms with cells lacking membrane-bound nuclei that obtain their energy from organic molecules in their environment (e.g. bacteria)
procaryotic organisms	bacteria; small cells that lack membrane-bound organelles such as a nucleus, mitochondria or chloroplasts
producers	plants that make their own food through the process of photosynthesis (autotroph); constitutes the first trophic level in a food chain
profile sketch	drawing illustrating a side-on view of an area, showing the distribution of organisms along a line
protein	a complex macromolecule consisting of polypeptide chains of amino acids, containing the element nitrogen as well as other elements commonly found in organic molecules
protoplasm	the entire contents of a cell including the cytoplasm and nucleus
pyramid of numbers	diagrammatic representation of the numbers of organisms at each level of the food chain
pyramids of energy	the diagrammatic representation of the energy flow through a food chain
quadrats	square frames (usually 1 m × 1 m) used in estimating abundance in plants or slow-moving animals
qualitative (results)	results that are made by observation and recorded as a description
quantitative (results)	results that are measured and recorded as numbers (quantities)
rabies	contagious viral infection that enters the body through an animal bite
radioactive	unstable, emitting particles (known as radioactive decay)
radioisotopes	unstable forms of a molecule which emit radioactive particles
radiometric dating	a method of estimating the age of objects or material using the decay rates of radioactive components
rainforests	a type of ecosystem characterised by a dense canopy of trees, ferns and other plants in enormous variety; found in a high rainfall area
'reducing' atmosphere	an atmosphere that contains free hydrogen
reliability	increased by using a variety of secondary sources when gathering information; occurs when the same experimental method yields the same or similar results when repeated by other people
resolution	the ability of a lens (or microscope) to distinguish between two very closely positioned structures as distinct and separate images
resource competition	where organisms utilise a resource that is in short supply
respiration	chemical reaction in the mitochondria of cells, whereby energy is released from organic compounds (especially carbohydrates)
respiratory surface	a body surface that is in contact with the external environment and has become specialised for the exchange of oxygen and carbon dioxide
saline	containing salt
salinity	the amount or concentration of dissolved salt
sampling technique	an ecological technique used to estimate species populations by the collection and/or counting samples of the population
scavengers	animals feeding on dead organisms
scientific method	the procedure for carrying out a valid scientific experiment, sometimes referred to as 'fair test'
scientific theory	a scientist's explanation of observed behaviour in terms of a model that has familiar properties. It cannot be proved or disproved experimentally; it can only be supported or refuted by evidence
sea floor spreading zones	zones where continents drift apart, releasing magma up to the surface, solidifying and forming a new crust, hence spreading the sea floor
secondary or tertiary consumers	organisms second or third in the food chain to consume other organisms (carnivores or omnivores)
selective pressure	a change, usually in the environment, that causes some organisms with a particular variation to survive and reproduce and those without it to decrease in number
selectively permeable	describes a membrane or other barrier that allows only certain substances to pass through
self-pollination	when pollen from a flower's anther pollinates the same flower's stigma (or the stigmas of flowers of the same individual plant)

semi-permeable	see selectively permeable
sexual reproduction	a method of producing offspring that involves the fusion of male and female gametes (sex cells) to form a zygote, containing a combination of genetic material from both parents
sieve plates	perforated end wall of a sieve tube element of phloem tissue, with pores that allow strands of cytoplasm to pass through
sieve tube elements	elongate, main cellular components of phloem tissue which are responsible for transporting food
simple microscope	a microscope that magnifies an image through only one lens, not a series of lenses, where the resulting image is not inverted (e.g. a stereo dissecting microscope)
sinuses	cavities
small intestine	organ of the digestive tract where most chemical digestion and absorption of digested food occurs
smallpox	an infectious disease caused by a virus
solvent	liquid basis of a solution which allows another substance (solid, liquid or gas) to dissolve in it
somatic	a body cell or any diploid cell that is not involved in sexual reproduction and cannot form gametes
speciation	how new species arise; the formation of new species
species	a group of organisms of similar appearance within a population; the members of which can interbreed to produce fertile offspring
spiracles	external openings on insect bodies, often containing valves, to regulate intake and outlet of air in tracheal tubes
spongy cells	irregularly shaped cells in the mesophyll of leaves for gaseous exchange and photosynthesis
spontaneous generation	a view that life can arise from non-living things, independent of any parent being present
stamen	the male reproductive organ of a flower comprising the anther and filament
starch	a complex, insoluble polysaccharide that is not sweet to the taste and is one of the commonest forms of energy storage in plant cells
stem cells	undifferentiated cells, either embryonic or adult, that can divide and give rise to other cells
stigma	the female part of the flower that receives pollen grains, leading to fertilisation
stomach	organ of the digestive tract where food is stored, physically digested by churning and some chemical digestion of protein occurs
stomata or **stomates**	(singular: stoma) an opening or pore located in the epidermis of plant parts (usually leaves and green stems) through which gases such as water vapour, oxygen and carbon dioxide can enter and leave
stroma	colourless fluid cavity of the chloroplast in which grana are embedded and starch may be stored
stromatolites	a concentrically layered rock, the layers being formed by the successive growth of thin mats of cyanobacteria
style	the pathway for pollen tubes between the stigma and ovary in flowering plants when pollen grains are received on the stigma
subduction	the sliding of one crustal plate beneath another when crustal plates converge and meet
sugars	sweet-tasting carbohydrate molecules that may be monosaccharides or disaccharides
survival of the fittest	the relative ability of an individual organism to live long enough to produce fertile offspring and pass their genes on to the next generation
sweat	watery secretion of mammalian skin in which salts and some nitrogenous wastes are lost, dissolved in water, to cool the body
symbiosis	interactions in which two organisms live together in a close relationship that is beneficial to at least one of them
symbiotic interactions	see symbiosis
system	integrated group of organs that work together to perform a common function
taxonomy	the classification of organisms in an ordered system that indicates natural relationships
terrestrial environment	an environment existing on land
tetrad	a group of four cells that are formed as a result of the meiotic division of one cell

theory of plate tectonics the theory that continents are carried on large crustal plates positioned on top of the semi-molten interior of the Earth, but beneath the ocean

thermoacidophiles organisms that grow best in high temperatures and highly acidic environments

thermophiles organisms that grow best in hot conditions between 30°C and 50°C (high temperature-loving)

thylakoids chlorophyll—containing flattened membranes in chloroplasts

timeline a linear representation of important events in the sequence in which they occurred, whereby each event is drawn on the line to a scale, reflecting the time that has elapsed between each event

tissues groups of cells that have a similar structure and perform a common function in multicellular organisms

tracheae windpipe that transports air between the throat and lungs in vertebrates

tracheal tubes air-conducting tubes in the respiratory system of insects

tracheoles the smallest branches of air-conducting tubes in the respiratory system of insects, carrying air directly to and from the cells of the insect's body

transects a narrow strip that crosses an entire area when studying the distribution of a species

transform boundary where two plates are sliding past each other

translocation mechanism of transport of food (organic nutrients) in the phloem of plants

transpiration evaporation of water vapour from a plant through the stomates/stomata of leaves

transpiration stream mechanism of transport of water and dissolved nutrients in the xylem of plants

transport medium the fluid in which substances are carried within a living organism

transport system a system of vessels arranged to carry substances from one part of a body to another

trophic involving the feeding habits of different organisms in a food chain

trophic (feeding) levels the position of an organism in a food chain (e.g. primary producer or secondary consumer)

tropical areas of hot and humid climate

turgid state of a plant cell in which the contents are swollen with an increased volume of fluid in the vacuole, causing the cell wall to stretch and become rigid, giving mechanical support in plant tissues

turgor firm state of a plant cell where the cell wall is stretched by an increased volume of water in the vacuole and protoplasm

unicellular made of one cell only

urea water soluble nitrogenous waste product most commonly excreted in terrestrial animals as part of urine

uric acid almost insoluble nitrogenous waste product excreted as a white sludge by animals which are adapted to habitats where water is scarce

urine fluid containing excretory wastes such as urea; expelled from vertebrates

vaccine serum or plasma that is administered to people or other animals to produce and immune reaction to disease-causing organisms

vacuoles fluid-filled structures within a cell, separated from the surrounding cytoplasm by a single membrane; more commonly found in plant cells

validity improved by the use of scientific journals when gathering information from secondary sources

variation physical or physiological, or behavioural difference between individuals in a population which may or may not make them more suited to prevailing environmental conditions

vascular tissue tissue which is organised into vessels (such as xylem and phloem in plants) to function in transport

vectors organisms that carry parasites and transmit them from one host to another

veins blood vessels that carry blood towards the heart, from other organs of the body

viviparous when offspring produced by sexual reproduction develop inside the maternal body and are released as live young or eggs; in plants seeds germinate while still attached to the plant (e.g. in mangroves)

volcanic aquifers an underground layer of water that interacts with volcanic activity

wall pressure an inward force exerted by the wall of a plant cell on the protoplasm (cell contents) to counteract the turgor pressure that it exerts

water balance	any mechanism regulating the concentration of water and dissolved substances within the cells or body fluids of an organism
WHO	World Health Organization
woodlands	a habitat with a sparse canopy of trees, usually with less rainfall than that of a rainforest
xylem	vascular tissue in plants that transports water and dissolved inorganic minerals upwards as ascending sap
xylem tracheids	non-living xylem elements formed from a single cell, with tapering ends and pitted walls thickened with lignin
xylem vessels	non-living xylem elements made up of a series of hollow cells placed end-to-end, with no cross walls separating them
zoologist	person who studies animals
zygote	a diploid cell resulting from the fusion of the male and female gametes

Index

Technetium-99, 180
technology, 73–80, 203–4, 215
teeth, 147–9, 152
telophase, *189, 192*
temperature, 6, 42, 259–60
terrestrial environments, 2, *3,* 4–9, *4–5,*
   286–7
Tersipes rostratus, 150, 296–7, *297*
Thylacinus cyanocephalus, 269, 276, 319
Tiktaalik, 40, *40*
timelines, 64–5, *79,* 207–8, 257
Tingamarra swamp crocodile, 271
tissues, defined, 81, 125
totipotency, 302
trachea, 163
tracheae, 167
tracheal tubes, 167
tracheids, 172
tracheoles, 167
transects, 11, 17, *56*
translocation, 175
transmission electron microscope, 75
transpiration, 41, 141, 172, 178–9
transport, 144, 155–6, 159–60, 169–71,
   174–5
trigger plants, 294–5, *295*

trophic interactions, *33–9,* 55, 58
turgidity, 84, 117

unicellular organisms, 67, 81, *82,* 125,
   125, 187
union, in fertilisation, 285
Urey, Harold, 199–203
urine, 160

vaccination, 64
vacuoles, 84, 172
variability in measurement, 61
variation, 252, 255–6
vascular tissue, 144, 171, 184
vegetation, 260, *313, see also* plants
vesicles, 91
Virchow, Rudolph, 64, 71
viruses, first observation of, 65
viviparity, 44, 306
volcanic aquifers, 203
volcano distribution, 244, *244, 246*
volume to surface area ratio, 120–3

Wallace, Alfred, 64–5, 234–5, 254–5,
   254, 272
waratahs, 294, *294*

water
   adapting to lack of, 42
   availability of, 7, 259–60
   in cells, 101
   movement in plants, 172–3
   nutrition and, 132, 139–41, *141*
   in osmosis, 116
   permeability to, 112
   in respiration, 163–4
   seed dispersal by, 298–9
   water balance, 160
Watson, James, 65
wattles (*Acacia* spp.), 41, *293, 299,* 307,
   307, 308
Wegener, Alfred, 235
Wilberforce, Samuel, 274–5, *274*
Wilkins, Maurice, 65
wind-dispersed seeds, 298–9
wind pollination, 292–3, *293,* 297
woodland, *5,* 260, *262*
World Health Organization, 65

xylem, 141, 171–5, *173, 174*

Zernike, Fritz, 65
zygotes, in fertilisation, 285